Machinists' and Metalworkers' Pocket Reference

Machinists' and Metalworkers' Pocket Reference

Ronald A. Walsh

McGraw-Hill

New York San Francisco Washington, D.C. Auckland Bogotá
Caracas Lisbon London Madrid Mexico City Milan
Montreal New Delhi San Juan Singapore
Sydney Tokyo Toronto

McGraw-Hill

A Division of The **McGraw-Hill** *Companies*

1 2 3 4 5 6 7 8 9 0 DOC/DOC 0 6 5 4 3 2 1 0

Library of Congress Cataloging-in-Publication Data

Walsh, Ronald A.

Machinists' and metalworkers' pocket reference / Ronald A. Walsh.

p. cm.

ISBN 0-07-136092-1

1. Machine-shop practice—Handbooks, manuals, etc. 2. Metalwork—Handbooks, manuals, etc. I. Title.

TJ1165.W16 2000

670.42'3—dc21

00-016292

The sponsoring editor for this book was Linda Ludewig, the editing supervisor was Frank Kotowski, Jr., and the production supervisor was Sherri Souffrance. It was set in Times Roman by North Market Street Graphics.

Printed and bound by R. R. Donnelley & Sons, Inc.

Contents

Chapter 6. Machining, Machine Tools, and Practices

Chapter 7. Tool Steels

Chapter 8. Sheet Metal Practices

Preface

This pocket reference for machinists and metalworkers will be of practical value to those who work in, or are associated with, the metalworking industries. Having been a product design engineer for many years, working with machinists, tool makers, and sheet metal workers, I could see the need for a small, handy pocket reference book containing the often-used data and procedures that are required on a daily basis for machinists, managers, and other support personnel. I hope that this pocket reference will make your work easier, save you time, and allow you to be more productive. For those who are involved with the mathematics used in machining and metalworking, the companion volume to this book, the *Machinists' and Metalworkers' Calculations Handbook,* is also available from McGraw-Hill.

Ronald A. Walsh

CHAPTER 1

MODERN METALWORKING MACHINERY, GAGES, AND MEASURING DEVICES

Metalworking machinery, tools, and measuring instruments have advanced considerably in the past 50 years. This chapter will show some of the new machines, tools, and instruments used throughout industry today that allow us to produce parts faster and more accurately than was possible in the past. The widespread use and implementation of microprocessors to control the actions of metalworking machinery will be evident in many of the photographs of modern equipment shown in this chapter.

Common measuring instruments such as the basic micrometer, vernier caliper, depth gage, and dial indicator also have advanced in design and features. Many of these instruments are now digital and provide a direct readout, without having to read a vernier scale, making their use more positive and less prone to reading errors by the machinist or machine operator. Although this is the case, in a later subsection the method of reading scales and vernier scales found on measuring instruments will be explained.

Some of the modern metalworking machines found today in manufacturing facilities and factories nationwide and worldwide include the following:

- Engine lathes
- Turning centers, automatic and computer-controlled
- Three-axis automatic milling machines, computer-controlled (machining centers)
- Bridgeport-type milling machines
- Surface grinders
- Automatic screw machines
- Threading machines
- Turret lathes
- Universal milling machines (column and knee type)
- EDMs (electric discharge machines), computer-controlled
- Roll-forming machines
- Slitting machines
- Automatic stock cutoff machines
- Cylindrical grinders, center and centerless
- Automatic punch presses, computer-controlled
- Automatic hydraulic brakes
- Hydraulic shears
- Hydropresses
- Yoder hammers
- Hydraulic broaches
- Flame cutters, automatic

- Laser cutters, automatic
- Chemical milling facilities
- Stretcher benders
- Offset or joggle machines
- Structural "ironworkers"
- High-pressure water cutting machines
- Pin routers
- Vertical and horizontal boring machines
- Shapers
- Planers
- Drill presses
- Sanders
- Manual grinders, motor-operated
- Leaf brakes

1.1 MODERN METALWORKING MACHINES AND METHODS OF METALWORKING

The two most basic machines in the modern machining industry are the lathe and milling machines. Figure 1.1 shows a modern gear-driven engine lathe, and Fig. 1.2(a) shows an automatic three-axis computer-controlled milling machine (machining center). The milling machine is equipped to automatically change its own cutting tools at the computer's command, in relation to the ongoing operation being performed, as dictated by the direct numerical control (DNC) system. The modern engine lathe is equipped with a quick-change variable-ratio gear box for threading operations, as shown in

FIGURE 1.1 Geared-head engine lathe with digital control panel.

Fig. 1.3. Three-jaw universal and four-jaw independent chucks are usually employed to hold the rotating stock for turning operations on the modern engine lathe. A digital readout display on the lathe shows the operator the carriage travel and cross-feed locations to three or four decimal places, thus eliminating the reading of the vernier scales on the lathe's controls. The readings are generally indicated for an *xy* axis system.

CAD/CAM (computer-aided design and computer-aided manufacturing) systems are employed throughout industry today to speed up production and balance profits. The following example is typical of operations of machinery in a modern manufacturing plant where computer-aided manufacturing is employed.

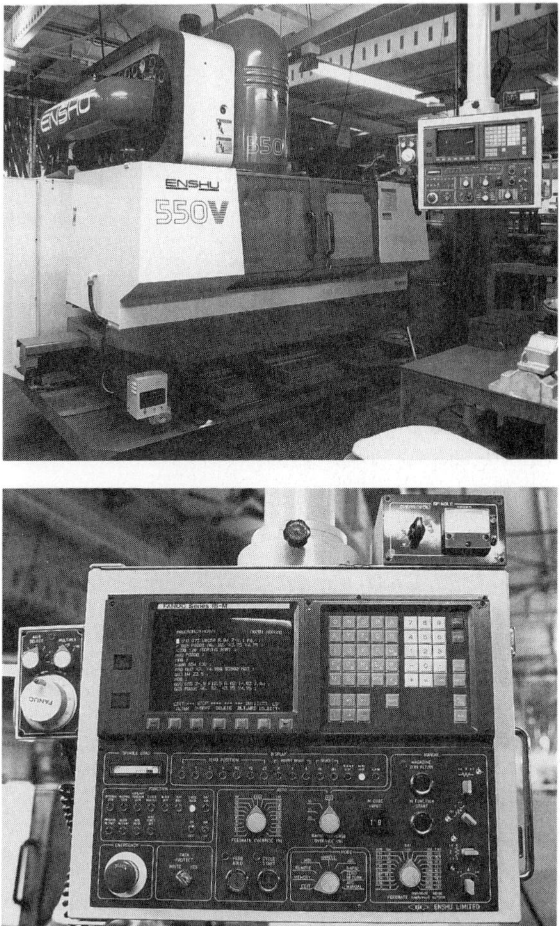

FIGURE 1.2(a) Vertical machining center with CNC controls.
FIGURE 1.2(b) Closeup view of CNC controller of the machine shown in Fig. 1.2(a).

FIGURE 1.3 Gear ratio, carriage and spindle speed controls, geared-head engine lathe.

Example of CAM through DNC. Direct numerical control (DNC) is the modern method used to control the actions of machine tools by way of computers. In this system, the tool engineering or other department writes the computer program for the "host" computer, which may control more than one machine tool. After the DNC program is written into the host computer, the machine operator in the shop can call the host computer and instruct it to load the memory of the controller on the machine for manufacturing a specific part. The machine operator then instructs the controller on his or her machine to begin the machining operations.

After the part is made, the controller on the machine sends the DNC program back to the host computer, where it is checked and verified. If the machine operator changed a sequence or operating instruction, the host computer will detect the change or changes and indicate this by way of the computer monitor or readout.

In previous times, such a machine was controlled by numerical (NC) tape which was fed into the machine's controller. The pattern of holes in the NC tape provided the digital code signals required to control the operations of the machine. Tapes were cumbersome, time-consuming, and could be damaged easily by handling in the shop. This method is now ancient history, although still in existence in some manufacturing facilities. Modern machines are controlled by fast electronic digital signals coming from the memory of the machines' controller, which, in turn, was loaded by the host computer. You can see a typical machine tool controller on the right of the Enshu 550-V machining center in Fig. 1.2(a) and a close-up view of the controller is shown in Fig. 1.2(b).

Some of the present methods of working metal include the following:

Machining-Type Operations

- *EDMs.* In electric discharge machines (EDMs), a moving vertical thin wire at a certain frequency voltage is used to do the actual cutting of the metal. A modern EDM is shown in Fig. 1.4(a). This machine is also DNC, as can be seen from its control panel. Dies of great accuracy are produced on these machines rapidly and economically. These modern machines can produce intricate dies easily that would have been difficult or impossible to produce using the old die-making methods and techniques. Dies produced on this machine can be seen in Fig. 1.4(b). Die clearance is very close, as can be seen by the paper part cut on

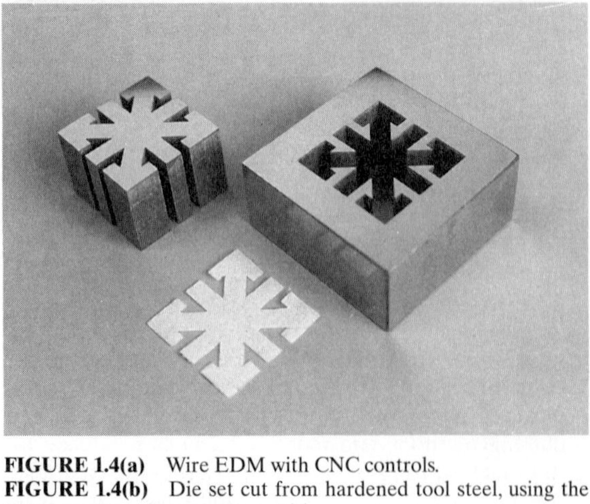

FIGURE 1.4(a) Wire EDM with CNC controls.
FIGURE 1.4(b) Die set cut from hardened tool steel, using the EDM machine shown in Fig. 1.4(a).

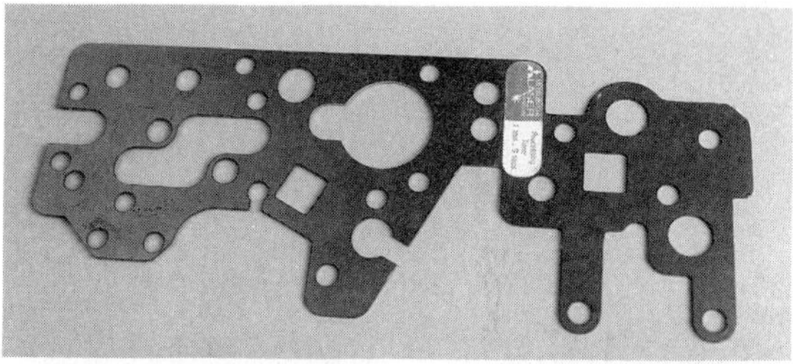

FIGURE 1.5(a) A laser-cut sheet metal part.
FIGURE 1.5(b) A modern laser cutting machine—CNC controlled.

these dies. The dies were cut by the EDM machine from hardened tool steel.

- *Laser.* A powerful laser light beam does the actual cutting of the metal, the light beam path being

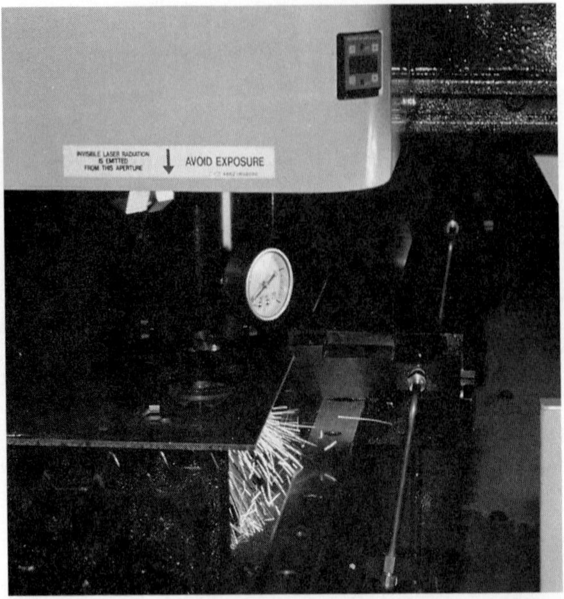

FIGURE 1.5(c) Close-up view of the laser cutting operation.

controlled by the axis controllers on the machine. Figure 1.5(a) shows a typical laser-cut metal part. This complex part was cut out in approximately 1.2 min. Complex thin parts whose quantity does not warrant a hard die are produced using this method. A modern laser cutting machine is shown in Fig. 1.5(b), and a close-up view of the cutting is shown in Fig. 1.5(c).

- *Chemical milling.* Large masses of metal may be removed effectively in producing a part by using the etching action of chemicals. Very thin and delicate parts also may be produced with chemical milling or etching. A tough photoresistive substance covers the parts of the metal that are not to be removed. Printed circuit board production is actually a chemical milling operation.
- *Automatic computer-controlled machining centers.* Described above.
- *High-pressure water cutting.* In this process, a very high pressure jet of water, loaded with microfine abrasives, is used to cut the material, metal, plastic, or other composition.
- *Flame cutting.* Parts may be cut automatically using the oxyacetylene flame to do the cutting on a machine with a tracer head or computer-controlled *xy* axes. Large, heavy metal parts are produced in this manner, although the outline accuracy is not great. This method has been superseded by laser cutting, where metal thickness allows.
- *Forging.*
- *Casting.*
- *Extruding.*
- *Impact extruding.*

Sheet Metal Parts Fabrication Methods

- *Hard dies.* A die set is used to stamp out the part in flat pattern. Progressive dies also bend the part into the required shape after it is stamped in flat pattern. This is the most common, economical method devised to mass produce large quantities of parts to great accuracy.

- *Punch press.* Large sheet metal parts may be made to accurate standards using modern computer-controlled automatic multistation punch presses. Figure 1.6 shows a large Amada "Coma" machine, which is computer-controlled, very fast, and accurate. A sheet metal panel weighing over 200 lb may be rapidly indexed and punched on this machine. Programmers write the DNC programs for a machine of this type, which are then loaded into the machine's computer or controller. The machine operator starts the program and stands back to watch the machine go through the sequence of operations required to produce the finished part in flat pattern.

FIGURE 1.6 High-speed CNC multistation turret punch press.

- *Roll forming.* Flat strips of sheet metal are fed into the roll-forming machine, where they progress through a set of sequenced rollers to produce a long sheet metal part of constant cross-sectional shape.
- *Hydropressing.* A sheet metal flat-pattern part is placed on a set of forming dies, being located correctly with locator pins, and is then pressed into shape by the action of the hydropress. Many aircraft sheet metal parts are produced in this manner. Lightening holes and shrink flutes are produced simultaneously with the part to control the metal along curved surfaces.

FIGURE 1.7 A high-tonnage press brake with digital back-gage controls.

- *Hydraulic brakes.* In this machine, the flat-pattern sheet metal part is given flanges or webs to produce the finished part. The modern brakes have automatic back gages and material-handling devices to assist the operator in making the various bends and flanges required on the part. See Fig. 1.7.
- *Hydraulic shears.* The standard hydraulic shear cuts sheet metal according to the back gage set by the machine operator and his or her accuracy in placing the sheet into the machine. In order to ensure a perfect 90° squaring of the flat sheet metal blank, modern shears such as the Weideman "Optishear" are employed. See Fig. 1.8.

FIGURE 1.8 A Weidemann squaring shear, CNC controlled.

- *Explosive forming.* A shaped charge of explosives is placed above the flat pattern of the part, usually under water, and when detonated, the explosive force rams the metal against a die or forming block to produce the final part or subsection of a compound part.
- *Spin forming.* A flat disk of sheet metal is held and spun in a lathe-type machine while forming tools are guided along the part as it is spinning. This is similar to a swaging action on the metal. Only certain shapes of parts are possible to produce with this method.

The preceding examples are the more common methods of producing metal parts today. Specialized metal handling such as Yoder hammering is used infrequently today because this requires a great deal of hand working and operator skill.

1.2 MEASURING INSTRUMENTS

The most common measuring devices used in the modern machine shop are, of course, the micrometer and the vernier caliper. These instruments have gone through a gradual change over the past 50 years to keep pace with modern machinery and higher productivity.

The basic micrometer has progressed from vernier scale to mechanical number dial scale to the present form of digital electronic display shown in Fig. 1.9. As can be seen, the accuracy of this instrument has increased gradually to the present state shown in Fig. 1.9. The vernier caliper also has gone through this gradual change. A modern vernier caliper is shown in Fig. 1.10,

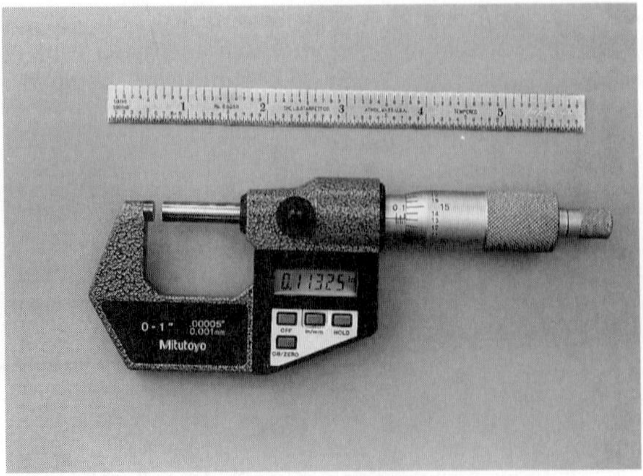

FIGURE 1.9 New class of LCD digital micrometer. Reading accuracy ±0.00005 in, or ±0.001 mm (5/100,000th in).

although it is no longer technically a "vernier" caliper because it has no working vernier scale. The application of digital electronics has improved these instruments a great deal, since operator reading errors due to misinterpretations in reading the vernier scale are eliminated. This translates into improved productivity.

Some of the measuring instruments used in industry today include the following:

- Digital micrometer, vernier micrometer (see Figs. 1.9 and 1.11)
- Digital caliper, vernier caliper (see Figs. 1.10 and 1.12)
- Dial indicator (see Fig. 1.13)
- Gage blocks (see Fig. 1.14)

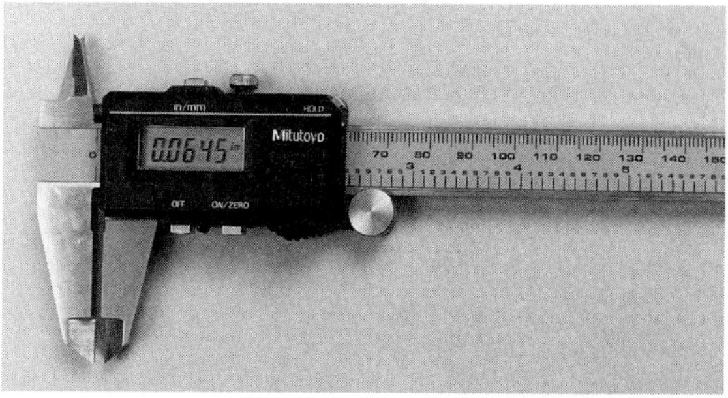

FIGURE 1.10 LCD digital caliper, 6-in size, reading to ±0.0005 in., (5/10,000th in.)

- Sine bar (5 and 10 in) and compound sine bar for double angles
- Sine plate and compound sine plate
- Vernier protractor (see Fig. 1.15 and Fig. 1.21)
- "Go, no-go" gages

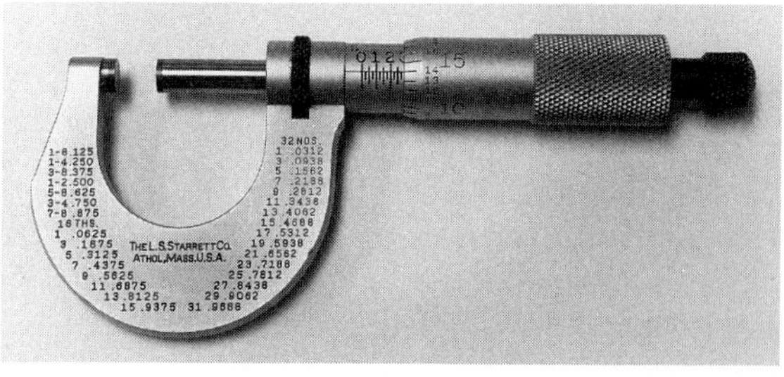

FIGURE 1.11 The standard "classic" 1-in micrometer, a high-quality instrument, reading to ±0.001 in.

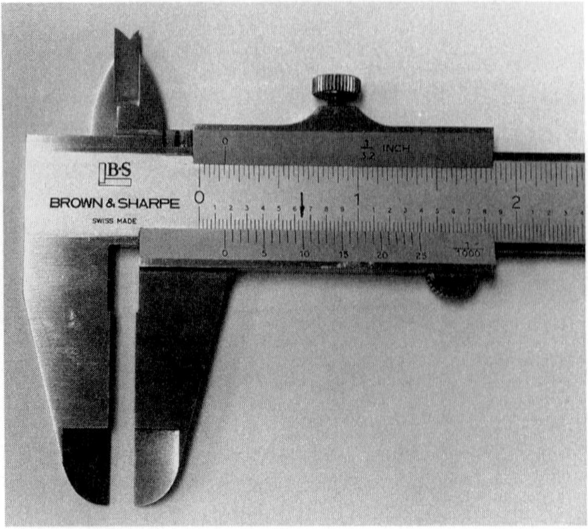

FIGURE 1.12 A standard "classic" vernier caliper (reading indicates 0.160 in). Arrow indicates vernier increment.

- Internal telescoping gages (see Fig. 1.16)
- Optical comparator
- Optical flat
- Vernier height gage
- Screw thread micrometers
- Universal surface gages
- Combination squares
- Diemakers' squares
- Steel and plastic scales (see Fig. 1.17)

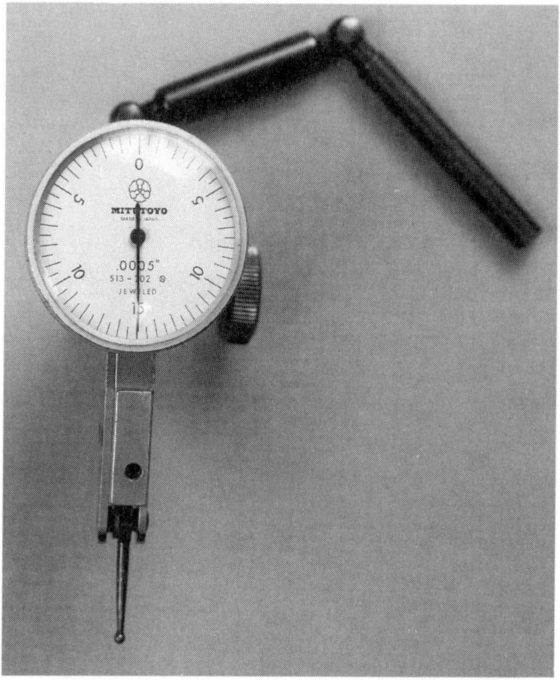

FIGURE 1.13 A dial indicator reading to ±0.0005-in accuracy.

- Vernier depth gage (see Fig. 1.18)
- Micrometer heads
- Hole gages
- Gear-tooth vernier calipers
- Digital and dial height gage (see Fig. 1.19)

FIGURE 1.14 A set of high-quality gage blocks ("Jo-blocks").

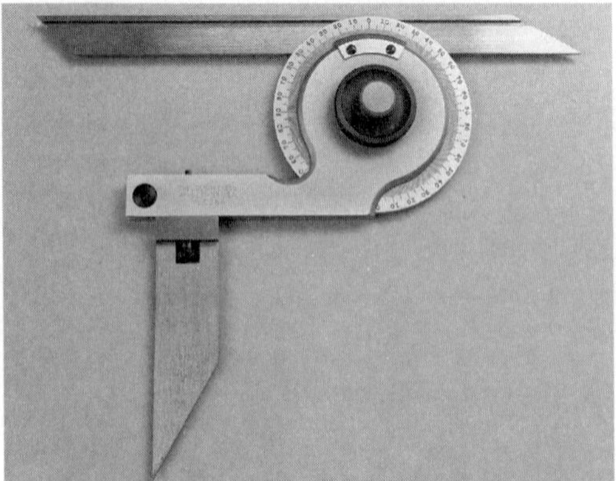

FIGURE 1.15 A typical vernier protractor for measuring angles to within 5 minutes.

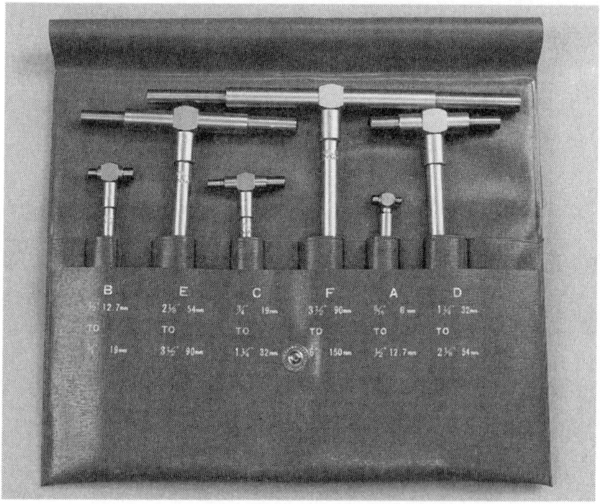

FIGURE 1.16 A set of inside, telescoping gages.

1.3 READING SCALES, DIALS, AND VERNIERS

Figure 1.17 shows a machinist's scale above and an engineer's scale below. The machinist's scale is graduated in thirty-seconds and sixty-fourths of an inch and is read by merely counting the divisions in either graduation. The back of the machinist's scale is often graduated in tenths and one-hundredths of an inch. The engineer's scale is graduated in fiftieths of an inch or 0.02 ins per division. This scale is marked (*) at 1.34 in with the arrow. The bottom of the engineer's scale is divided in millimeters, and 10 mm equals 1 cm. There are 2.54 cm, or 25.4 mm, to the inch, while 1 mm equals 0.03937 in.

Dial-type indicators (as in Fig. 1.13) are simple to read because the division values are given on the face of

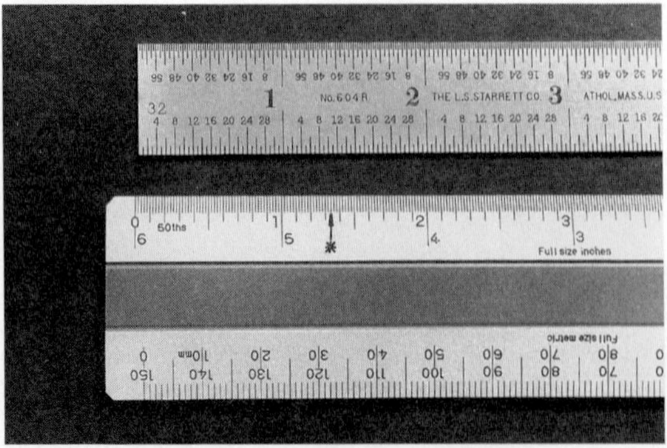

FIGURE 1.17 Machinist's scale (*top*), engineer's scale (*bottom*). The bottom scale is marked at 1.340 in.

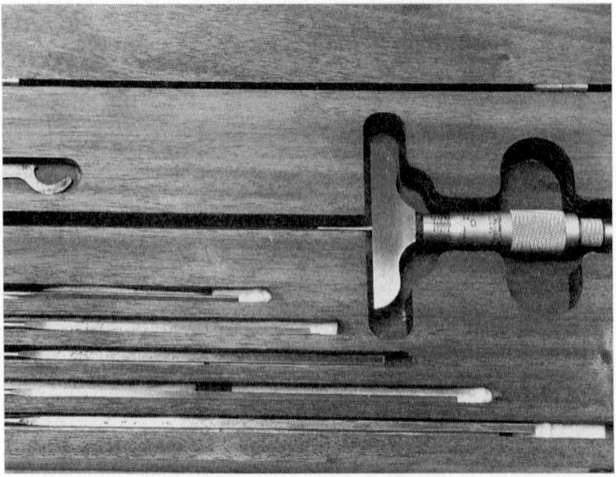

FIGURE 1.18 Standard micrometer depth-gage set.

the dial. Figure 1.19 illustrates a modern combination dial/digital height gage used in tooling and quality control departments. This instrument is set on a flat surface such as a "plane table" prior to taking height measurements. It is accurate to ±0.0005 in.

The vernier protractor is used to measure angles (see Fig. 1.15). This instrument is graduated to read in increments of 5 minutes of arc. A close-up photo of the protractor's vernier scale is shown in Fig. 1.20. The angle set on this scale is 62°45′ and is read to the left of the vernier zero position. The vernier protractor shown in Fig. 1.21 has a built-in magnifying glass and also reads to 5-minute increments of angle. Figure 1.22 explains how to properly read a vernier protractor.

The vernier caliper is read in a similar manner. The vernier caliper shown in Fig. 1.12 is set to 0.160 in. Each division on the caliper scale is equal to 0.025 in, so four divisions equals 0.10 in. The vernier scale is in alignment with the caliper scale at the tenth division on the vernier, so the final reading is $0.10 + (2 \times 0.025) + 0.010 = 0.160$ in. Figures 1.23 to 1.26 show the methods of reading the verniers on older-style micrometers. Figures 1.23 and 1.24 are for the inch system and Figures 1.26 and 1.27 are for the SL millimeter measure.

The vernier was invented by Pierre Vernier in 1631. The vernier consists of a small scale having a certain number of graduations which equal in combined length a different number of graduations, usually one more or one less, on a longer scale. Thus there is a small difference between a graduation on the vernier and a graduation on the scale. The readings depend on this difference between the graduations.

Utilizing the principles of the vernier, Joseph R. Brown (Brown & Sharpe) invented the vernier caliper

FIGURE 1.19 A modern dial/digital height gage used in tooling and quality control departments. Accuracy is ±0.0005 in.

FIGURE 1.20 Close-up view of vernier protractor scale.

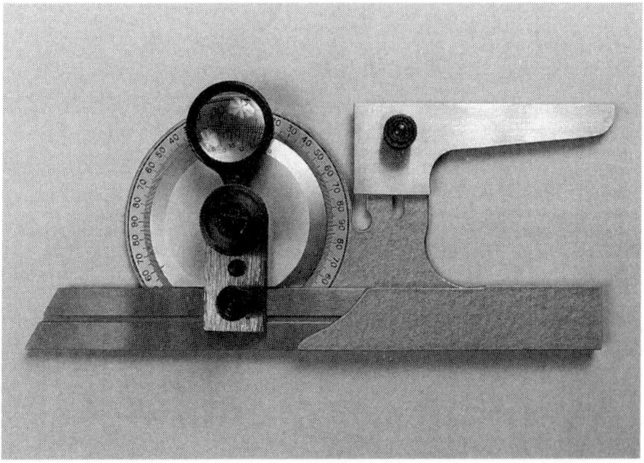

FIGURE 1.21 A vernier protractor with a scale magnifier.

How to Read Bevel Protractor with Vernier

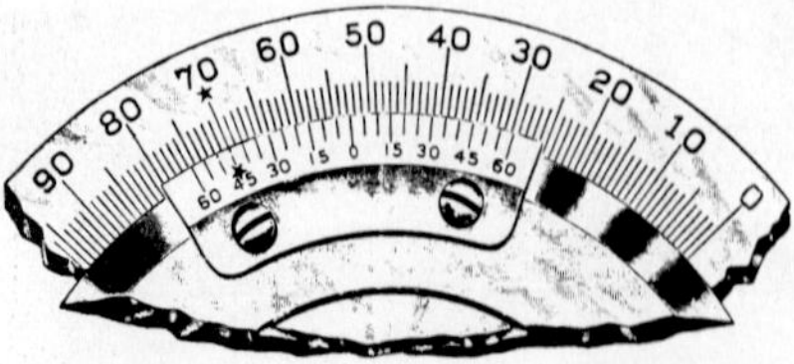

The disc of the Protractor, is graduated in degrees, and can be read to the left or to the right, from the 0 line to 90. It has clear figures each 10th space.

These figures represent degrees. The Vernier plate is graduated in such a way that 12 spaces on the plate occupy the same space as 23 spaces on the disc of the Protractor.

The difference between the width of one of the 12 spaces on the Vernier and two of the 23 spaces on the disc of the Protractor is therefore 1/12 of a degree, that is to say, 5 minutes (5′).

Each space on the Vernier is 1/12 of a degree, or say 5 minutes shorter than 2 spaces on the disc. The Vernier plate has clear figures each third space. These figures represent minutes.

When the 0 line of the Vernier plate coincides exactly with a graduated line on the disc of the Protractor, the reading is exact in whole degrees. When the 0 line of the Vernier plate does not coincide exactly with a graduated line of the disc of the Protractor, then find the line on the Vernier that does coincide exactly with a line on the disc of the Protractor.

This line on the Vernier indicates the number of twelfths of a degree or say 5 minutes (5′) that should be added to the reading in whole degrees.

In order to read the Protractor, note on the disc the number of whole degrees between 0 on the disc and 0 on the Vernier. Then count in the same direction the number of spaces from 0 on the Vernier to a graduated line that does coincide with a line on the disc. Multiply this number by 5 and the product will be the number of minutes that should be added to the number of whole degrees.

EXAMPLE: In the cut on this page the number of whole degrees between 0 on the disc and 0 on the Vernier is 52. The line marked 45 on the Vernier coincides with a line (70) on the disc, as is indicated by the asterisks, 9 being the number of spaces from 0 on the Vernier. By multiplying this number (9) by 5, 45 is obtained, the number of minutes that must be added to the number of whole degrees. The reading of the Protractor is, therefore, 52 degrees 45 minutes (52° 45′).

FIGURE 1.22 Instructions for reading vernier protractors. (*L.S. Starrett Co.*)

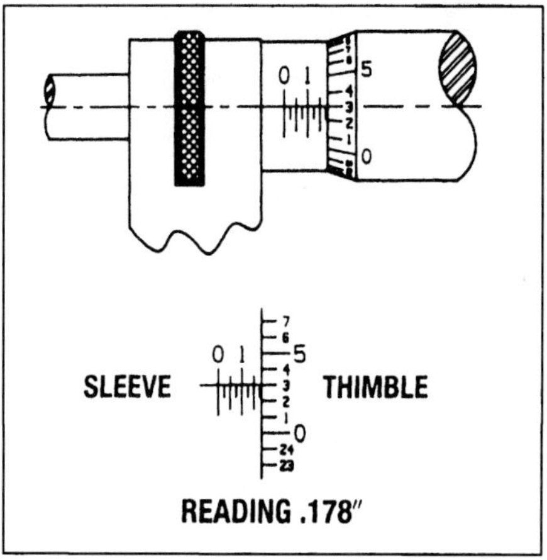

The "1" line on sleeve is visible,
representing .... .100″

There are 3 additional lines visible,
each representing .025″
3 x .025″ .... = .075″

Line "3" on the thimble coincides
with the reading line on the sleeve,
each line representing .001″
3 x .001″ .... = .003″

The micrometer reading is .... .178″

FIGURE 1.23

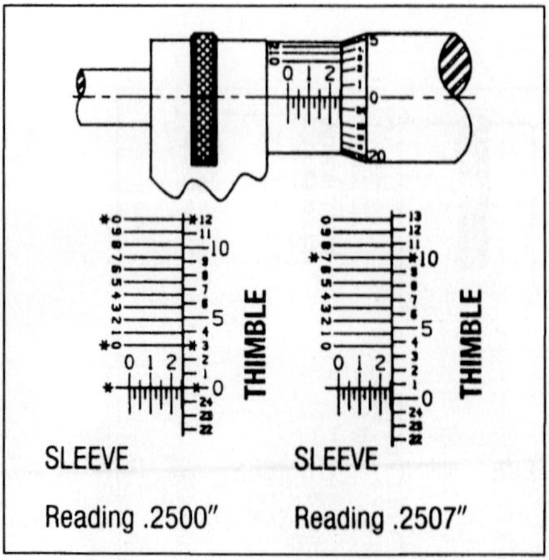

The "2" line on sleeve is visible, representing	.200″
There are two additional lines visible, each representing .025″	.050″
The reading line on the sleeve lies between the "0" and "1" on the thimble indicating that a vernier reading must be added	- - - -
The "7" line is the only line on the vernier that coincides with a line on the thimble, representing 7 x .0001″	= .0007″
The micrometer reading is	.2507″

FIGURE 1.24

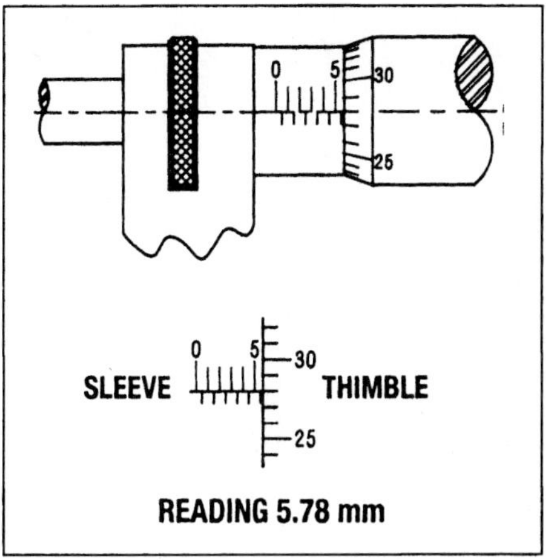

The 5 mm sleeve graduation is visible	5.00 mm
One additional 0.5 mm line is visible on the sleeve	0.50 mm
Line 28 on the thimble coincides with the reading line on the sleeve, so 28 x 0.01 mm	= 0.28 mm
The micrometer reading is	5.78 mm

FIGURE 1.25

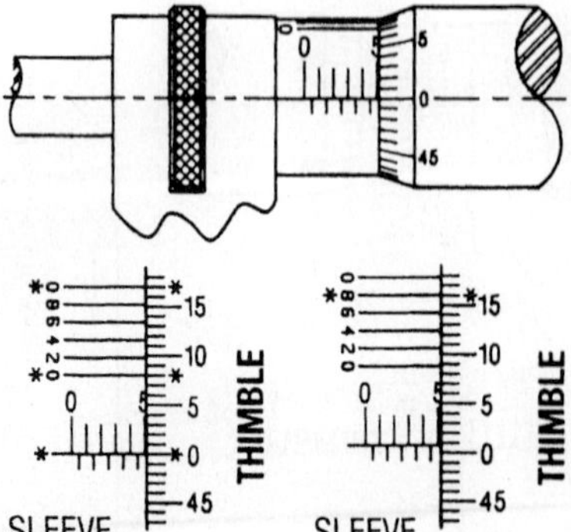

FIG. B
Reading 5.000 mm

FIG. C
Reading 5.008 mm

The 5 mm sleeve graduation is visible	5.000 mm
No additional lines on the sleeve are visible	0.000 mm
The reading line on the sleeve lies between zero and the first line on the thimble, indicating that a vernier reading must be added.	-----
Line 8 on the vernier is the only line that coincides with a line on the thimble	0.008 mm
The micrometer reading is	5.008 mm

FIGURE 1.26

Fraction	Decimals	Millimeters	Fraction	Decimals	Millimeters
1/64	0.015625	0.397	33/64	0.515625	13.097
1/32	.03125	0.794	17/32	.53125	13.494
3/64	.046875	1.191	35/64	.546875	13.891
1/16	.0625	1.588	9/16	.5625	14.288
5/64	.078125	1.984	37/64	.578125	14.684
3/32	.09375	2.381	19/32	.59375	15.081
7/64	.109375	2.778	39/64	.609375	15.478
1/8	.1250	3.175	5/8	.6250	15.875
9/64	.140625	3.572	41/64	.640625	16.272
5/32	.15625	3.969	21/32	.65625	16.669
11/64	.171875	4.366	43/64	.671875	17.066
3/16	.1875	4.763	11/16	.6875	17.463
13/64	.203125	5.159	45/64	.703125	17.859
7/32	.21875	5.556	23/32	.71875	18.256
15/64	.234375	5.953	47/64	.734375	18.653
1/4	.2500	6.350	3/4	.7500	19.050
17/64	.265625	6.747	49/64	.765625	19.447
9/32	.28125	7.144	25/32	.78125	19.844
19/64	.296875	7.541	51/64	.796875	20.241
5/16	.3125	7.938	13/16	.8125	20.638
21/64	.328125	8.334	53/64	.828125	21.034
11/32	.34375	8.731	27/32	.84375	21.431
23/64	.359375	9.128	55/64	.859375	21.828
3/8	.3750	9.525	7/8	.8750	22.225
25/64	.390625	9.922	57/64	.890625	22.622
13/32	.40625	10.319	29/32	.90625	23.019
27/64	.421875	10.716	59/64	.921875	23.416
7/16	.4375	11.113	15/16	.9375	23.813
29/64	.453125	11.509	61/64	.953125	24.209
15/32	.46875	11.906	31/32	.96875	24.606
31/64	.484375	12.303	63/64	.984375	25.003
1/2	.5000	12.700	1	1.000	25.400

MM	INCHES	MM	INCHES
.1	.0039	46	1.8110
.2	.0079	47	1.8504
.3	.0118	48	1.8898
.4	.0157	49	1.9291
.5	.0197	50	1.9685
.6	.0236	51	2.0079
.7	.0276	52	2.0472
.8	.0315	53	2.0866
.9	.0354	54	2.1260
1	.0394	55	2.1654
2	.0787	56	2.2047
3	.1181	57	2.2441
4	.1575	58	2.2835
5	.1969	59	2.3228
6	.2362	60	2.3622
7	.2756	61	2.4016
8	.3150	62	2.4409
9	.3543	63	2.4803
10	.3937	64	2.5197
11	.4331	65	2.5591
12	.4724	66	2.5984
13	.5118	67	2.6378
14	.5512	68	2.6772
15	.5906	69	2.7165
16	.6299	70	2.7559
17	.6693	71	2.7953
18	.7087	72	2.8346
19	.7480	73	2.8740
20	.7874	74	2.9134
21	.8268	75	2.9528
22	.8661	76	2.9921
23	.9055	77	3.0315
24	.9449	78	3.0709
25	.9843	79	3.1102
26	1.0236	80	3.1496
27	1.0630	81	3.1890
28	1.1024	82	3.2283
29	1.1417	83	3.2677
30	1.1811	84	3.3071
31	1.2205	85	3.3465
32	1.2598	86	3.3858
33	1.2992	87	3.4252
34	1.3386	88	3.4646
35	1.3780	89	3.5039
36	1.4173	90	3.5433
37	1.4567	91	3.5827
38	1.4961	92	3.6220
39	1.5354	93	3.6614
40	1.5748	94	3.7008
41	1.6142	95	3.7402
42	1.6535	96	3.7795
43	1.6929	97	3.8189
44	1.7323	98	3.8583
45	1.7717	99	3.8976
		100	3.9370

1 mm = .03937″ .001″ = .0254 mm

FIGURE 1.27

in 1851, the first functional tool available to the machinist for measuring 0.001-in increments accurately.

1.4 DECIMAL TO MILLIMETER CHART

Figure 1.27 is a standard decimal equivalent to millimeter chart which finds constant usage in the machine shop and other metalworking activities or procedures.

CHAPTER 2

U.S. CUSTOMARY AND METRIC (SI) MEASURES AND CONVERSIONS

2.1 CONVERSIONS FOR LENGTH, PRESSURE, VELOCITY, VOLUME, AND WEIGHT

To convert from:	to:	Multiply by:
	Length	
Centimeters	Inches	0.3937
Centimeters	Yards	0.01094
Feet	Inches	12.0
Feet	Meters	0.30481
Feet	Yards	0.333
Inches	Centimeters	2.540
Inches	Feet	0.08333
Inches	Meters	0.02540
Inches	Microns	25,400.
Inches	Millimeters	25.400
Inches	Yards	0.02778
Kilometers	Feet	3,281.
Kilometers	Miles (nautical)	0.5336
Kilometers	Miles (statute)	0.6214
Kilometers	Yards	1,094.

To convert from:	to:	Multiply by:
Meters	Feet	3.2809
Meters	Yards	1.0936
Microns	Inches	0.0000394
Microns	Meters	0.000001
Miles (statute)	Feet	5,280.
Miles (statute)	Kilometers	1.6093
Miles (statute)	Meters	1,609.34
Miles (statute)	Yards	1,760.
Miles (nautical)	Feet	6,080.2
Miles (nautical)	Kilometers	1.8520
Miles (nautical)	Meters	1,852.0
Millimeters	Inches	0.03937
Rods	Meters	5.0292
Yards	Centimeters	91.44
Yards	Feet	3.0
Yards	Inches	36.0
Yards	Meters	0.9144
	Pressure	
Dynes per cm^2	Pascals	0.1000
Grams per cm^3	Ounces per in^3	0.5780
Kilograms per cm^2	Pounds per in^2	14.223
Kilograms per cm^2	Pascals	98,066.5
Kilograms per m^2	Pascals	9.8066
Kilograms per m^2	Pounds per ft^2	0.2048
Kilograms per m^2	Pounds per yd^2	1.8433
Kilograms per m^3	Pounds per ft^3	0.06243
Ounces per in^3	Grams per cm^3	1.7300
Pounds per ft^3	Kilograms per cm^3	16.019
Pounds per ft^2	Kilograms per m^2	4.8824
Pounds per ft^2	Pascals	47.880
Pounds per in^2	Kilograms per cm^2	0.0703
Pounds per in^2	Pascals	6,894.76
Pounds per yd^2	Kilograms per m^2	0.5425
	Velocity	
Feet per minute	Meters per second	0.00508
Feet per second	Meters per second	0.3048
Inches per second	Meters per second	0.0254
Kilometers	Meters per second	0.2778
Knots	Meters per second	0.5144

Miles per hour	Meters per second	0.4470
Miles per minute	Meters per second	26.8224
	Volume	
Cubic centimeters	Cubic inches	0.06102
Cubic feet	Cubic inches	1,728.0
Cubic feet	Cubic meters	0.0283
Cubic feet	Cubic yards	0.0370
Cubic feet	Gallons	7.481
Cubic feet	Liters	28.32
Cubic feet	Quarts	29.9222
Cubic inches	Cubic centimeters	16.39
Cubic inches	Cubic feet	0.0005787
Cubic inches	Cubic meters	0.00001639
Cubic inches	Liters	0.0164
Cubic inches	Gallons	0.004329
Cubic inches	Quarts	0.01732
Cubic meters	Cubic feet	35.31
Cubic meters	Cubic inches	61,023.
Cubic meters	Cubic yards	1.3087
Cubic yards	Cubic feet	27.0
Cubic yards	Cubic meters	0.7641
Gallons	Cubic feet	0.1337
Gallons	Cubic inches	231.0
Gallons	Cubic meters	0.003785
Gallons	Liters	3.785
Gallons	Quarts	4.0
Liters	Cubic feet	0.03531
Liters	Cubic inches	61.017
Liters	Gallons	0.2642
Liters	Pints	2.1133
Liters	Quarts	1.057
Liters	Cubic meters	0.0010
Pints	Cubic meters	0.004732
Pints	Liters	0.4732
Pints	Quarts	0.50
Quarts	Cubic feet	0.03342
Quarts	Cubic inches	57.75
Quarts	Cubic meters	0.0009464
Quarts	Gallons	0.25
Quarts	Liters	0.9464
Quarts	Pints	2.0

To convert from:	to:	Multiply by:
	Weight	
Grams	Kilograms	0.001
Grams	Ounces	0.03527
Grams	Pounds	0.002205
Kilograms	Ounces	35.274
Kilograms	Pounds	2.2046
Ounces	Grams	28.35
Ounces	Kilograms	0.02835
Ounces	Pounds	0.0625
Pounds	Grams	453.6
Pounds	Kilograms	0.4536
Pounds	Ounces	16.0

Example: 3 pounds = 3 × 16 = 48 ounces
50 grams = 50 × 0.03527 = 1.7635 ounces

2.2 STANDARD CONVERSION TABLE: MEASURES ARE FOUND FROM THE TABLE

Multiply:	By:	To obtain:
Acres	43,560	Square feet
Acres	4047	Square meters
Acres	1.562×10^{-3}	Square miles
Acres	4840	Square yards
Acre—feet	43,560	Cubic feet
Acre—feet	325,851	Gallons
Acre—feet	1233.49	Cubic meters
Atmospheres	76.0	Centimeters of mercury
Atmospheres	29.92	Inches of mercury
Atmospheres	33.90	Feet of water
Atmospheres	10,333	Kilograms per square meter
Atmospheres	14.70	Pounds per square inch
Atmospheres	1.058	Tons per square foot
Barrels—oil	42	Gallons—oil
Barrels—cement	376	Pounds—cement
Bags or sacks—cement	94	Pounds—cement
Board feet	$144 \text{ in}^2 \times 1 \text{ in}$	Cubic inches
British thermal units (Btu)	0.2520	Kilogram-calories
British thermal units	777.5	Foot pounds
British thermal units	3.927×10^{-4}	Horsepower-hours
British thermal units	107.5	Kilogram-meters
British thermal units	2.928×10^{-4}	Kilowatt-hours
Btu/min	12.96	Foot pounds per second

Btu/min	0.02356	Horsepower
Btu/min	0.01757	Kilowatts
Btu/min	17.57	Watts
Centares (centiares)	1	Square meters
Centigrams	0.01	Grams
Centiliters	0.01	Liters
Centimeters	0.3937	Inches
Centimeters	0.01	Meters
Centimeters	10	Millimeters
Centimeters of mercury	0.01316	Atmospheres
Centimeters of mercury	0.4461	Feet of water
Centimeters of mercury	136.0	Kilograms per square meter
Centimeters of mercury	27.85	Pounds per square foot
Centimeters of mercury	0.1934	Pounds per square inch
Centimeters per second	1.969	Feet per minute
Centimeters per second	0.03281	Feet per second
Centimeters per second	0.036	Kilometers per hour
Centimeters per second	0.6	Meters per minute
Centimeters per second	0.02237	Miles per hour
Centimeters per second	3.728×10^{-4}	Miles per minute
Centimeters per second per second	0.03281	Feet per second per second
Cubic centimeters	3.531×10^{-5}	Cubic feet
Cubic centimeters	6.102×10^{-2}	Cubic inches
Cubic centimeters	10^{-6}	Cubic meters
Cubic centimeters	1.308×10^{-6}	Cubic yards
Cubic centimeters	2.642×10^{-4}	Gallons
Cubic centimeters	10^{-3}	Liters
Cubic centimeters	2.113×10^{-3}	Pints (liq.)
Cubic centimeters	1.057×10^{-3}	Quarts (liq.)
Cubic feet	2.832×10^{4}	Cubic centimeters
Cubic feet	1728	Cubic inches
Cubic feet	0.02832	Cubic meters
Cubic feet	0.03704	Cubic yards
Cubic feet	7.48052	Gallons
Cubic feet	28.32	Liters
Cubic feet	59.84	Pints (liq.)
Cubic feet	29.92	Quarts (liq.)
Cubic feet per minute	472.0	Cubic centimeters per second
Cubic feet per minute	0.1247	Gallons per second
Cubic feet per minute	0.4720	Liters per second
Cubic feet per minute	62.43	Pounds of water per minute
Cubic feet per second	0.646317	Millions gallons per day
Cubic feet per second	448.831	Gallons per minute
Cubic inches	16.39	Cubic centimeters
Cubic inches	5.787×10^{-4}	Cubic feet

Multiply:	By:	To obtain:
Cubic inches	1.639×10^{-5}	Cubic inches
Cubic inches	2.143×10^{-5}	Cubic yards
Cubic inches	4.329×10^{-3}	Gallons
Cubic inches	1.639×10^{-2}	Liters
Cubic inches	0.03463	Pints (liq.)
Cubic inches	0.01732	Quarts (liq.)
Cubic meters	10^{6}	Cubic centimeters
Cubic meters	35.31	Cubic feet
Cubic meters	61.023×10^{3}	Cubic inches
Cubic meters	1.308	Cubic yards
Cubic meters	264.2	Gallons
Cubic meters	10^{3}	Liters
Cubic meters	2113	Pints (liq.)
Cubic meters	1057	Quarts (liq.)
Cubic yards	7.646×10^{5}	Cubic centimeters
Cubic yards	27	Cubic feet
Cubic yards	46,656	Cubic inches
Cubic yards	0.7646	Cubic meters
Cubic yards	202.0	Gallons
Cubic yards	764.6	Liters
Cubic yards	1616	Pints (liq.)
Cubic yards	807.9	Quarts (liq.)
Cubic yards per minute	0.45	Cubic feet per second
Cubic yards per minute	3.367	Gallons per second
Cubic yards per minute	12.74	Liters per second
Decigrams	0.1	Grams
Deciliters	0.1	Liters
Decimeters	0.1	Meters
Degrees (angle)	60	Minutes
Degrees (angle)	0.01745	Radians
Degrees (angle)	3600	Seconds
Degrees per second	0.01745	Radians per second
Degrees per second	0.1667	Revolutions per minute
Degrees per second	0.002778	Revolutions per second
Dekagrams	10	Grams
Dekaliters	10	Liters
Dekameters	10	Meters
Drams	27.34375	Grains
Drams	0.0625	Ounces
Drams	1.771845	Grams
Fathoms	6	Feet
Feet	30.48	Centimeters
Feet	12	Inches
Feet	0.3048	Meters
Feet	⅓	Yards
Feet of water	0.02950	Atmospheres

Feet of water	0.8826	Inches of mercury
Feet of water	304.8	Kilograms per square meter
Feet of water	62.43	Pounds per square foot
Feet of water	0.4335	Pounds per square inch
Feet per minute	0.5080	Centimeters per second
Feet per minute	0.01667	Feet per second
Feet per minute	0.01829	Kilometers per hour
Feet per minute	0.3048	Meters per minute
Feet per minute	0.01136	Miles per hour
Feet per second	30.48	Centimeters per second
Feet per second	1.097	Kilometers per hour
Feet per second	0.5921	Knots
Feet per second	18.29	Meters per minute
Feet per second	0.6818	Miles per hour
Feet per second	0.01136	Miles per minute
Feet per second per second	30.48	Centimeters per second per second
Feet per second per second	0.3048	Meters per second per second
Foot pounds	1.286×10^{-3}	British thermal units
Foot pounds	5.050×10^{-7}	Horsepower-hours
Foot pounds	3.241×10^{-4}	Kilogram-calories
Foot pounds	0.1383	Kilogram-meters
Foot pounds	3.766×10^{-7}	Kilowatt-hours
Foot pounds per minute	1.286×10^{-3}	Btu per minute
Foot pounds per minute	0.01667	Foot pounds per second
Foot pounds per minute	3.030×10^{-5}	Horsepower
Foot pounds per minute	3.241×10^{-4}	Kilogram-calories per minute
Foot pounds per minute	2.260×10^{-5}	Kilowatts
Foot pounds per second	7.717×10^{-2}	Btu per minute
Foot pounds per second	1.818×10^{-3}	Horsepower
Foot pounds per second	1.945×10^{-2}	Kilogram-calories per minute
Foot pounds per second	1.356×10^{-3}	Kilowatts
Gallons	3785	Cubic centimeters
Gallons	0.1337	Cubic feet
Gallons	231	Cubic inches
Gallons	3.785×10^{-3}	Cubic meters
Gallons	4.95×10^{-3}	Cubic yards
Gallons	3.785	Liters
Gallons	8	Pints (liq.)
Gallons	4	Quarts (liq.)
Gallons—Imperial	1.20095	U.S. gallons
Gallons—U.S.	0.83267	Imperial gallons
Gallons water	8.3453	Pounds of water
Gallons per minute	2.228×10^{-3}	Cubic feet per second

Multiply:	By:	To obtain:
Gallons per minute	0.06308	Liters per second
Gallons per minute	8.0208	Cubic feet per hour
Gallons per minute	8.0208 area (sq. ft.)	Overflow rate (ft/h)
Gallons water per minute	6.0086	Tons water per 24 hours
Grains (troy)	L	Grains (avour.)
Grains (troy)	0.06480	Grams
Grains (troy)	0.04167	Pennyweights (troy)
Grains (troy)	2.0833×10^{-3}	Ounces (troy)
Grains per U.S. gallon	17.118	Parts per million
Grains per U.S. gallon	142.86	Pounds per million gallons
Grains per Imperial gallon	14.254	Parts per million
Grams	980.7	Dynes
Grams	15.43	Grains
Grams	10^{-3}	Kilograms
Grams	10^{3}	Milligrams
Grams	0.03527	Ounces
Grams	0.03215	Ounces (troy)
Grams	2.205×10^{-3}	Pounds
Grams per centimeter	5.600×10^{-3}	Pounds per inch
Grams per cubic centimeter	62.43	Pounds per cubic foot
Grams per cubic centimeter	0.03613	Pounds per cubic inch
Grams per liter	58.417	Grains per gallon
Grams per liter	8.345	Pounds per 1000 gallons
Grams per liter	0.062427	Pounds per cubic foot
Grams per liter	1000	Parts per million
Hectares	2.471	Acres
Hectares	1.076×10^{5}	Square feet
Hectograms	100	Grams
Hectoliters	100	Liters
Hectometers	100	Meters
Hectowatts	100	Watts
Horsepower	42.44	Btu per minute
Horsepower	33,000	Foot pounds per minute
Horsepower	550	Foot pounds per second
Horsepower	1.014	Horsepower (metric)
Horsepower	10.70	Kilogram-calories per minute
Horsepower	0.7457	Kilowatts
Horsepower	745.7	Watts

Horsepower (boiler)	33,479	Btu per hour
Horsepower (boiler)	9,803	Kilowatts
Horsepower-hours	2547	British thermal units
Horsepower-hours	1.98×10^6	Foot pounds
Horsepower-hours	641.7	Kilogram-calories
Horsepower-hours	2.737×10^5	Kilogram-meters
Horsepower-hours	0.7457	Kilowatthours
Inches	2.540	Centimeters
Inches of mercury	0.03342	Atmospheres
Inches of mercury	1.133	Feet of water
Inches of mercury	345.3	Kilograms per square meter
Inches of mercury	70.73	Pounds per square foot
Inches of mercury	0.4912	Pounds per square inch
Inches of water	0.002458	Atmospheres
Inches of water	0.07355	Inches of mercury
Inches of water	25.40	Kilograms per square meter
Inches of water	0.5781	Ounces per square inch
Inches of water	5.202	Pounds per square foot
Inches of water	0.03613	Pounds per square inch
Kilograms	980,665	Dynes
Kilograms	2.205	Pounds
Kilograms	1.102×10^{-3}	Tons (short)
Kilograms	10^3	Grams
Kilograms-calories	3.968	British thermal units
Kilograms-calories	3086	Foot pounds
Kilograms-calories	1.558×10^{-3}	Horsepower-hours
Kilograms-calories	1.162×10^{-3}	Kilowatthours
Kilogram-calories per minute	51.43	Foot pounds per second
Kilogram-calories per minute	0.09351	Horsepower
Kilogram-calories per minute	0.06972	Kilowatts
Kilograms per meter	0.6720	Pounds per foot
Kilograms per square meter	9.678×10^{-5}	Atmospheres
Kilograms per square meter	3.281×10^{-3}	Feet of water
Kilograms per square meter	2.896×10^{-3}	Inches of mercury
Kilograms per square meter	0.2048	Pounds per square foot
Kilograms per square meter	1.422×10^{-3}	Pounds per square inch
Kilograms per square millimeter	10^6	Kilograms per square meter

Multiply:	By:	To obtain:
Kiloliters	10^3	Liters
Kilometers	10^5	Centimeters
Kilometers	3281	Feet
Kilometers	10^3	Meters
Kilometers	0.6214	Miles
Kilometers	1094	Yards
Kilometers per hour	27.78	Centimeters per second
Kilometers per hour	54.68	Feet per minute
Kilometers per hour	0.9113	Feet per second
Kilometers per hour	0.5396	Knots
Kilometers per hour	16.67	Meters per minute
Kilometers per hour	0.6214	Miles per hour
Kilograms per hour per second	27.78	Centimeters per second per second
Kilograms per hour per second	0.9113	Feet per second per second
Kilograms per hour per second	0.2778	Meters per second per second
Kilowatts	56.92	Btu per minute
Kilowatts	4.425×10^4	Foot pounds per minute
Kilowatts	737.6	Foot pounds per second
Kilowatts	1.341	Horsepower
Kilowatts	14.34	Kilogram-calories per minute
Kilowatts	10^3	Watts
Kilowatthours	3415	British thermal units
Kilowatthours	2.655×10^6	Foot pounds
Kilowatthours	1.341	Horsepower-hours
Kilowatthours	860.5	Kilogram-calories
Kilowatthours	3.671×10^5	Kilogram-meters
Liters	10^3	Cubic centimeters
Liters	0.03531	Cubic feet
Liters	61.02	Cubic inches
Liters	10^3	Cubic meters
Liters	1.308×10^{-3}	Cubic yards
Liters	0.2642	Gallons
Liters	2.113	Pints (liq.)
Liters	1.057	Quarts (liq.)
Liters per minute	5.886×10^{-4}	Cubic feet per second
Liters per minute	4.403×10^{-3}	Gallons per second
Lumber width (in) $\times$ $\frac{\text{thickness (in)}}{12}$	Length (ft.)	Board feet
Meters	100	Centimeters
Meters	3.281	Feet
Meters	39.37	Inches

Meters	10^{-3}	Kilometers
Meters	10^{3}	Millimeters
Meters	1.094	Yards
Meters per minute	1.667	Centimeters per second
Meters per minute	3.281	Feet per minute
Meters per minute	0.05468	Feet per second
Meters per minute	0.06	Kilometers per hour
Meters per minute	0.03728	Miles per hour
Meters per second	196.8	Feet per minute
Meters per second	3.281	Feet per second
Meters per second	3.6	Kilometers per hour
Meters per second	0.06	Kilometers per minute
Meters per second	2.237	Miles per hour
Meters per second	0.03728	Miles per minute
Microns	10^{-6}	Meters
Miles	1.609×10^{5}	Centimeters
Miles	5280	Feet
Miles	1.609	Kilometers
Miles	1760	Yards
Miles per hour	44.70	Centimeters per second
Miles per hour	88	Feet per minute
Miles per hour	1.467	Feet per second
Miles per hour	1.609	Kilometers per hour
Miles per hour	0.8684	Knots
Miles per hour	26.82	Meters per minute
Miles per minute	2682	Centimeters per second
Miles per minute	88	Feet per second
Miles per minute	1.609	Kilometers per minute
Miles per minute	60	Miles per hour
Milliers	10^{3}	Kilograms
Milligrams	10^{-3}	Grams
Milliliters	10^{-3}	Liters
Millimeters	0.1	Centimeters
Millimeters	0.03937	Inches
Milligrams per liter	1	Parts per million
Million gallons per day	1.54723	Cubic feet per second
Miner's inches	1.5	Cubic feet per minute
Minutes (angle)	2.909×10^{-4}	Radians
Ounces	16	Drams
Ounces	437.5	Grains
Ounces	0.0625	Pounds
Ounces	28.349527	Grams
Ounces	0.9115	Ounces (troy)
Ounces	2.790×10^{-5}	Tons (long)
Ounces	2.835×10^{-5}	Tons (metric)
Ounces (troy)	480	Grains
Ounces (troy)	20	Pennyweights (troy)
Ounces (troy)	0.08333	Pounds (troy)

Multiply:	By:	To obtain:
Ounces (troy)	31.103481	Grams
Ounces (troy)	1.09714	Ounces (avoir.)
Ounces (fluid)	1.805	Cubic inches
Ounces (fluid)	0.02957	Liters
Ounces per square inch	0.0625	Pounds per square inch
Overflow rate (ft/h)	0.12468 × area (ft^2)	Gallons per minute
$\frac{1}{\text{Overflow rate (ft/h)}}$	8.0208	Square feet per gallon per minute
Parts per million	0.0584	Grains per U.S. gallon
Parts per million	0.07016	Grains per Imperial gallon
Parts per million	8.345	Pounds per million gallons
Pennyweights (troy)	24	Grains
Pennyweights (troy)	1.55517	Grams
Pennyweights (troy)	0.05	Ounces (troy)
Pennyweights (troy)	4.1667×10^{-3}	Pounds (troy)
Pounds	16	Ounces
Pounds	256	Drams
Pounds	7000	Grains
Pounds	0.0005	Tons (short)
Pounds	453.5924	Grams
Pounds	1.21528	Pounds (troy)
Pounds	14.5833	Ounces (troy)
Pounds (troy)	5760	Grains
Pounds (troy)	240	Pennyweights (troy)
Pounds (troy)	12	Ounces (troy)
Pounds (troy)	373.24177	Grams
Pounds (troy)	0.822857	Pounds (avoir.)
Pounds (troy)	13.1657	Ounces (avoir.)
Pounds (troy)	3.6735×10^{-4}	Tons (long)
Pounds (troy)	4.1143×10^{-4}	Tons (short)
Pounds (troy)	3.7324×10^{-4}	Tons (metric)
Pounds of water	0.01602	Cubic feet
Pounds of water	27.68	Cubic inches
Pounds of water	0.1198	Gallons
Pounds of water per minute	2.670×10^{-4}	Cubic feet per second
Pounds per cubic foot	0.01602	Grams per cubic centimeter
Pounds per cubic foot	16.02	Kilograms per cubic meters
Pounds per cubic foot	5.787×10^{-4}	Pounds per cubic inch
Pounds per cubic inch	27.68	Grams per cubic centimeter

Pounds per cubic inch	2.768×10^4	Kilograms per cubic meter
Pounds per cubic inch	1728	Pounds per cubic foot
Pounds per foot	1.488	Kilograms per meter
Pounds per inch	178.6	Grams per centimeter
Pounds per square foot	0.01602	Feet of water
Pounds per square foot	4.883	Kilograms per square meter
Pounds per square foot	6.945×10^{-3}	Pounds per square inch
Pounds per square inch	0.06804	Atmospheres
Pounds per square inch	2.307	Feet of water
Pounds per square inch	2.036	Inches of mercury
Pounds per square inch	703.1	Kilograms per square meter
Quadrants (angle)	90	Degrees
Quadrants (angle)	5400	Minutes
Quadrants (angle)	1.571	Radians
Quarts (dry)	67.20	Cubic inches
Quarts (liq.)	57.75	Cubic inches
Quintal, Argentine	101.28	Pounds
Quintal, Brazil	129.54	Pounds
Quintal, Castile, Peru	101.43	Pounds
Quintal, Chile	101.41	Pounds
Quintal, Mexico	101.47	Pounds
Quintal, Metric	220.46	Pounds
Quires	25	Sheets
Radians	57.30	Degrees
Radians	3438	Minutes
Radians	0.637	Quadrants
Radians per second	57.30	Degrees per second
Radians	0.1592	Revolutions per second
Radians per second	9.549	Revolutions per minute
Radians per second per second	573.0	Revolutions per minute per minute
Radians per second per second	0.1592	Revolutions per second per second
Revolutions	360	Degrees
Revolutions	4	Quadrants
Revolutions	6.283	Radians
Revolutions per minute	6	Degrees per second
Revolutions per minute	0.1047	Radians per second
Revolutions per minute	0.01667	Revolutions per second
Revolutions per minute per minute	1.745×10^{-3}	Radians per second per second
Revolutions per minute per minute	2.778×10^{-4}	Revolutions per second per second
Revolutions per second	360	Degrees per second
Revolutions per second	6.283	Radians per second

Multiply:	By:	To obtain:
Revolutions per second	60	Revolutions per minute
Revolutions per second per second	6.283	Radians per second per second
Revolutions per second per second	3600	Revolutions per minute per minute
Seconds (angle)	4.848×10^{-6}	Radians
Square centimeters	1.076×10^{-3}	Square feet
Square centimeters	0.1550	Square inches
Square centimeters	10^{-4}	Square meters
Square centimeters	100	Square millimeters
Square feet	2.296×10^{-5}	Acres
Square feet	929.0	Square centimeters
Square feet	144	Square inches
Square feet	0.09290	Square meters
Square feet	3.587×10^{-8}	Square miles
Square feet	$1/9$	Square yards
Square inches	6.452	Square centimeters
Square inches	6.944×10^{-3}	Square feet
Square inches	645.2	Square millimeters
Square kilometers	247.1	Acres
Square kilometers	10.76×10^{6}	Square feet
Square kilometers	10^{6}	Square meters
Square kilometers	0.3861	Square miles
Square kilometers	1.196×10^{6}	Square yards
Square meters	2.471×10^{-4}	Acres
Square meters	10.76	Square feet
Square meters	3.861×10^{-7}	Square miles
Square meters	1.196	Square yards
Square miles	640	Acres
Square miles	27.88×10^{6}	Square feet
Square miles	2.590	Square kilometers
Square miles	3.098×10^{6}	Square yards
Square millimeters	0.01	Square centimeters
Square millimeters	1.550×10^{-3}	Square inches
Square yards	2.066×10^{-4}	Acres
Square yards	9	Square feet
Square yards	0.8361	Square meters
Square yards	3.228×10^{-7}	Square miles
Temp. (°C.) + 273	1	Abs. temp. (°C.)
Temp. (°C.) + 17.78	1.8	Temp. (°F.)
Temp. (°F.) + 460	1	Abs. temp. (°F.)
Temp. (°F.) − 32	$5/9$	Temp. (°C.)
Tons (long)	1016	Kilograms
Tons (long)	2240	Pounds
Tons (long)	1.12000	Tons (short)
Tons (metric)	10^{3}	Kilograms

Tons (metric)	2205	Pounds
Tons (short)	2000	Pounds
Tons (short)	32,000	Ounces
Tons (short)	907.18486	Kilograms
Tons (short)	2430.56	Pounds (troy)
Tons (short)	0.89287	Tons (long)
Tons (short)	29166.66	Ounces (troy)
Tons (short)	0.90718	Tons (metric)
Tons of water per 24 hours	83.333	Pounds water per hour
Tons of water per 24 hours	0.16643	Gallons per minute
Tons of water per 24 hours	1.3349	Cubic feet per hour
Watts	0.05692	Btu per minute
Watts	44.26	Foot pounds per minute
Watts	0.7376	Foot pounds per second
Watts	1.341×10^{-3}	Horsepower
Watts	0.01434	Kilogram-calories per minute
Watts	10^{-3}	Kilowatts
Watthours	3.415	British thermal units
Watthours	2655	Foot pounds
Watthours	1.341×10^{-3}	Horsepower-hours
Watthours	0.8605	Kilogram-calories
Watthours	367.1	Kilogram-meters
Watthours	10^{-3}	Kilowatthours
Yards	91.44	Centimeters
Yards	3	Feet
Yards	36	Inches
Yards	0.9144	Meters

2.3 TEMPERATURE SYSTEMS AND CONVERSIONS

There are four common temperature systems used in engineering and design calculations: (°F) Fahrenheit, (°C) Celsius (formerly centigrade), (K) Kelvin, and (°R) Rankine.

The conversion equation for Celsius to Fahrenheit or Fahrenheit to Celsius is

$$\frac{5}{9} = \frac{°\mathrm{C}}{°\mathrm{F} - 32}$$

This exact relational equation is all that you need to convert from either system. Enter the known temperature, and solve the equation for the unknown value.

Example. You wish to convert 66°C to Fahrenheit.

$$\frac{5}{9} = \frac{66}{°F - 32}$$

$$5°F - 160 = 594$$

$$5°F = 754$$

$$°F = 150.8$$

This method is much easier than trying to remember the two equivalent equations, which are

$$°C = \frac{5}{9}\,(°F - 32)$$

and

$$°F = \frac{9}{5}\,°C + 32$$

The other two systems, Kelvin and Rankine, are converted as described here. The Kelvin and Celsius scales are related by

$$K = 273.18 + °C$$

Thus 0°C = 273.18 K. Absolute zero is equal to –273.18°C.

Example. A temperature of −75°C = 273.18 + (−75°C) = 198.18 K. The Rankine and Fahrenheit scales are related by

$$°R = 459.69 + °F$$

Thus 0°F = 459.69°R. Absolute zero is equal to −459.69°F.

Example. A temperature of 75°F = 459.69 + (+75°F) = 534.69°R.

2.4 SMALL WEIGHT EQUIVALENTS: U.S. CUSTOMARY (GRAINS AND OUNCES) VS. METRIC (GRAMS)

1 gram	= 15.43 grains
1 gram	= 15,430 milligrains
1 gram	= 15,430,000 micrograins
1 pound	= 7,000 grains
1 ounce	= 437.5 grains
1 ounce	= 28.35 grams
1 grain	= 0.0648 grams
1 grain	= 64.8 milligrams
0.1 grain	= 0.00648 grams
0.1 grain	= 6.48 milligrams
1 micrograin	= 0.0000000648 grams
1 micrograin	= 0.0000648 milligrams

1000 micrograins	= 0.0000648 grams
1000 micrograins	= 0.0648 milligrams
1 grain	= 0.002286 ounces
1 ounce	= 437.5 grains
10 grains	= 0.02286 ounces or 0.648 grams
100 grains	= 0.2286 ounces or 6.48 grams

Example. To obtain the weight in grams, multiply the weight in grains by 0.0648.

Example. To obtain the weight in grains, multiply the weight in grams by 15.43.

2.5 TORQUE CONVERSIONS

Torque Conversion Units

It is often necessary to convert torque from one system of units to another. The following charts will be useful and time-saving, and should also help prevent conversion errors. (See *Note* following these tables for explanation of asterisks.)

To convert pounds-feet to	Multiply by
Gram-inches	5443.1088
Ounce-inches	192.0
Pound-inches	12.0
Kilogram-centimeters*	13.8257
Kilogram-meters*	0.138257

To convert pound-inches to	Multiply by
Gram-inches	453.5924
Ounce-inches	16.0
Pounds-feet	0.08334
Kilogram-centimeters*	1.152
Kilogram-meters*	1.152×10^{-2}

To convert ounce-inches to	Multiply by
Gram-inches	28.3495
Pound-inches	0.0625
Pounds-feet	5.2087×10^{-3}
Kilogram-centimeters*	72.808×10^{-3}
Kilogram-meters*	728.08×10^{-6}

To convert gram-inches to	Multiply by
Ounce-inches	0.03527
Pound-inches	2.205×10^{-3}
Pounds-feet	1.8376×10^{-4}
Kilogram-centimeters*	2.54×10^{-3}
Kilogram-meters*	2.54×10^{-5}

To convert kilogram-centimeters to	Multiply by
Gram-inches	393.7
Ounce-inches	13.8858
Pound-inches	85.8108×10^{-2}
Pounds-feet	72.346×10^{-3}
Kilogram-meters*	0.01

To convert kilogram-meters to	Multiply by
Gram-inches	39370.0
Ounce-inches	1388.58
Pound-inches	85.8108
Pounds-feet	7.2346
Kilogram-centimeters*	100.0

[* *Note.* In the preceding conversions for the SI system, kilograms, when used as a weight or force, must be multiplied by 9.81 to arrive at the newton conversion. Many modern SI torque listings are given as newton-meters (N · m) instead of kilogram-meters (kg · m), or gram-centimeters (g · cm). This confusion was created when the kilogram was previously used as a force or weight. In modern SI systems, the kilogram is considered only as a unit of mass, while the newton is considered as a force. A 1-kg mass used as a force may be specified as $1 \times 9.81 = 9.81$ N. The kilogram is multiplied by the metric gravitational constant of 9.81 m/sec^2 to arrive at the newton measure of force. One newton is equal to 0.10197 kg-force.]

2.6 EQUIVALENT UNITS FOR ENERGY, WORK, POWER, AND FORCE

Units of energy or work and heat energy are:

U.S. customary system	SI (metric)
Foot-pound	Newton-meter
British thermal unit (Btu) [the amount of heat energy required to raise the temperature of 1 lb of water (H_2O) at 72°F by 1°F]	Joule (1 J = 1 N-m)
	Watthour, kilowatthour (Wh, kWh)
	Gram-calorie (the amount of heat energy required to raise the temperature of 1 g of water at 22°C by 1°C)
Horsepower-hour	

Units of power are:

U.S. customary system	SI (metric)
Horsepower	Watt, kilowatt (1 W = 1 J/sec = 1 N · m/sec)
Btu per second, minute, or hour	Joule per second, minute, or hour
Foot-pounds per second, minute, or hour	Newton-meters per second, minute, or hour

Units of force are:

U.S. customary system	SI (metric)
Pound-force, lbf (force)	Newton (force)
Pound-foot, pound-inch (torque or turning moment)	Newton-meter (torque or turning moment)
	Kilogram (when used as weight or force must be specified as mg · N; i.e., 5 kg weight expressed as a force is equal to $5 \times 9.81 = 49.05$ N)

Note: In units of force (SI) metric column above, "mg" represents mass × gravitational constant, not milligrams. The kilogram is used as mass in the SI system.

CHAPTER 3

MATERIALS: PHYSICAL AND CHEMICAL PROPERTIES, CHARACTERISTICS, AND USES

Materials, including metals, alloys, plastics, and other composite compounds are of prime importance to the mechanical designer, tool designer, machinist, and metalworker. The most important characteristics of materials to those who design and manufacture parts are the physical and chemical properties and the various uses to which the materials may be applied. This chapter discusses a great number of metals, alloys, plastics, and compounds, including elastomers. Included are composition, physical properties, and other characteristics useful for design and metalworking practices.

In design and metalworking practices, sometimes a material defect or failure occurs for unknown reasons. With the information provided in this chapter and in other appropriate American Society for Testing and Materials (ASTM), Society of Automotive Engineers (SAE), American Iron and Steel Institute (AISI), and American Gear Manufacturers Association (AGMA) standards, you may decide to have the material analyzed at a metallurgical and chemical laboratory to check the properties and compositions, or you may have the material checked mechanically at your company to determine if it meets the requirements of the material standards. Extensive lists of material standards, including types, chemical compositions, physical properties, and typical uses are contained in *The McGraw-Hill Machining and Metalworking Handbook*, second edition, 1999.

When producing the engineering drawings or specifications for a particular part, the appropriate American standard material designation or military/federal specification always must be indicated on the drawings or specifications. For example, if you are preparing a drawing for a mechanical spring and you wish to use music wire in its construction, you must indicate on the drawing that the material is to meet ASTM A-228 or appropriate SAE specifications. You also may request a certified material analysis data sheet from the supplier of the material, whether it comes from a mill, a foundry, a forge, or a materials processing plant. It is the responsibility of the material suppliers to provide the design engineer and purchasing departments with these analysis sheets when so directed, which is the usual procedure.

Chapter 6 of this book lists the machining characteristics and machine tool cutting and drilling speeds for many common and popular steels, alloys, other metals, plastics, and compositions. In this chapter, you will also find reference tables for current hardness scales or systems such as Rockwell and Brinell, Bhn. With these tables, you will be able to convert the hardness numbers for these common and currently used hardness measurement systems relative to each other.

An important characteristic of all stock mill products such as steel and alloy rod, bar, plate, and shapes is the tolerances or dimensional limits allowed during manufacturing. Tables of allowances or dimensional limits are therefore included in Section 3.3. A designer *cannot* design detail parts or assemblies effectively without these tolerance limits, making these tables particularly valuable in mechanical design, tool design work, and machining applications where dimensional limits of stock products *must* be known.

There are a tremendous number of different engineering materials for which the ASTM and SAE have listings: metallic, plastic, and composite. Steels alone account for hundreds of different alloys for wrought products, castings, and forgings. Although these alloys are listed in the standards and have specifications for composition and physical properties, they may not be readily available, except as special-order "mill run" quantities. A typical example would be austenitic stainless steel sheet in light gauges, cold-rolled to three-quarters hard, spring temper. This material is listed in the various 300 series stainless steels but may only be available from stock in 301 grade, three-quarters hard, spring temper. If you require 304 grade in your application, it may be available only in "mill run" quantities of a minimum of 500 to 2000 lb per single order.

As a general rule, during the early design stages of a particular part, alternate materials must be investigated. Due to limited availability of your chosen material, you may be forced to use another material which *is* stocked by the material vendors. When your anticipated material quantities are large, you will then have control of the material type as well as physical size and special characteristics such as finish, temper, and gauge. For example, some of the larger companies that order hundreds of tons of hot-rolled sheet steel per year may specify to the mill that they want to run light on the minus side of a particular gauge. The tolerance on hot-rolled sheet steel is such that when it is run on the minus side of the gauge limits, many thousands of pounds of steel can be saved. The same is true for wrought copper products such as bus bars and copper sheet, where the savings may be even more advantageous.

The list of steel alloys is so large that some authorities and regulating organizations have contemplated restricting the material standards to controlled listings, wherein redundant materials are eliminated. The large listings of alloys, plastics, and composites create a stocking problem for the distributors of mill products, plastics, and composite materials.

For designers, machinists, tool makers, and metalworkers employed at small to medium-sized companies, a good approach to the aforementioned materials problems is to obtain the stock materials catalogs available from the large distributors and manufacturers such as Ryerson, Vincent, Atlantic, Alcoa, Reynolds, Bethlehem Steel, U.S. Steel, Anaconda, General Electric, Dupont, Monsanto, and others.

3.1 STEELS

This section lists carbon and alloy steels, as well as the stainless steels, in their wrought form; that is to say, in the hot-rolled, cold-rolled, or cold-drawn forms. The usual shapes are sheets, plates, bars or strips, rounds, hexagons, tube, pipe, and structural configurations (beams, angles, channels, tees, square and rectangular tubes, and zees).

When carbon is added to iron in small quantities, carbon steel is produced. Besides carbon, a number of metallic elements can be added to iron to give the characteristics inherent in the various types of steels. The usual alloying elements are the following:

- Aluminum, which controls grain size in the steel
- Boron, which improves hardenability
- Chromium, which increases response to heat treatment as well as toughness (Chromium is used in stainless steels alone or with nickel.)
- Columbium, which is used in 18-8 stainless steels and welding electrodes
- Copper, which controls atmospheric corrosion and increases yield strengths
- Lead, which greatly improves machinability
- Manganese, which imparts strength and response to heat treatment
- Molybdenum, which increases depth of hardness and toughness
- Nickel, which increases strength and toughness but is not effective in improving hardenability
- Phosphorus, which is present in all steels and increases yield strength
- Silicon, which improves tensile strength and can improve hardenability
- Sulfur, which improves machinability but is detrimental to hot-forming properties

- Tellurium, which improves machinability in leaded steels
- Titanium, which is added to 18-8 stainless steels to prevent carbide precipitation
- Tungsten, which is used in good tool steels, making a fine grain structure when used in small amounts (When used in amounts from 17 to 20 percent, it produces a high-speed steel that retains hardness at high temperatures.)
- Vanadium, which is used to improve the shock strength of steels and retards grain growth even after hardening from high temperatures

Glossary of Steel Terms

Age hardening A process of aging that increases hardness and strength. Age hardening usually follows rapid cooling or cold working.

Annealing A process involving heat and cooling, usually applied to induce softening.

Austenite A term designating a metallurgical phase of steels, e.g., austenitic stainless steel.

Carburizing To introduce carbon while the steel is molten or while it is in the solid state by heating it in contact with carbonaceous material below its melting point.

Case hardening A process of hardening a ferrous alloy so that the "case," or surface layer, is much harder than the interior of the part. The typical case-hardening processes are carburizing and quenching, cyaniding, carbonitriding, nitriding, induction hardening, and flame hardening. Cases of Rockwell C 55 to 60 are readily obtained in medium- to high-carbon steels.

Ductility The property of a material that allows it to be drawn out of shape before fracturing by stress.

Elastic limit The maximum stress a material is capable of sustaining without a permanent set or deformation.

Fatigue The tendency for a metal to break after repeated or cyclic loadings which are below the ultimate tensile strength. Also known as *fatigue-endurance limit.*

Flame hardening A process of hardening a ferrous alloy by heating it above the transformation range by means of a flame and then by cooling as required.

Hardenability The property of a ferrous alloy that determines the depth and distribution of hardness induced by quenching.

Induction hardening A process of hardening a ferrous alloy by heating it above the transformation range by means of electrical induction and then cooling as required.

Killed steel Steel which is deoxidized with silicon or aluminum in order to reduce the oxygen content so that no reaction occurs between carbon and oxygen during solidification.

Martensite An unstable constituent in quenched steel formed without diffusion only during cooling below a certain temperature. Martensite is the hardest of the transformation products of austenite.

Nitriding A process of case hardening in which a ferrous alloy, usually of special composition, is heated in an atmosphere of ammonia or in direct contact with a nitrogenous material to produce surface hardening by absorbing nitrogen, without quenching.

Normalizing Heating a steel part of heavy section to a temperature 100°F above the critical range and then cooling in still air.

Pickeling Chemical or electrochemical removal of surface scale and oxides.

Quench hardening Heating a steel within or above the transformation range and cooling at a rate faster than the critical rate to increase the hardness substantially. Usually involves the formation of martensite.

Solution heat treatment A process in which an alloy is heated to a suitable temperature, held at this temperature long enough for certain constituents to enter into solid solution, and then cooled rapidly to hold the constituent in solution. The metal is left in a supersaturated state which is unstable and may subsequently exhibit age hardening.

Spheroidizing Any process of heating and cooling that produces a round or globular form of carbide in steels.

Strain hardening An increase in hardness and strength caused by plastic deformation at temperatures lower than the recrystallization range.

Stress relieving A process of reducing residual stresses in a metal part by heating the part to a suitable temperature and holding this temperature for a sufficient time. This process is applied to relieve stresses induced by casting, quenching, normalizing, machining, cold working (i.e., springs), or welding.

Temper A condition produced in a metal or alloy by mechanical or thermal treatment and having characteristic structure and mechanical properties. In addition to the annealed temper, conditions produced by thermal treatment are the solution heat-treated temper and the heat-treated and artificially aged temper.

Yield strength The stress at which a material exhibits a specified limited deviation from proportionality of stress to strain. In most steels, there is a proportionality between the amount of stress that produces a certain amount of strain. This phenomenon is known as *Hooke's law.* When a material passes its yield point, a lesser amount of stress produces a greater amount of strain, until the ultimate strength point is reached, where the material breaks.

Typical Uses for the Various Steels

Following is a listing of the popular and readily available steels that are usually stocked by suppliers and which have vast applications in industry. The ASTM/SAE tables have shown the chemical compositions, physical properties, and heat-treatment processes for a great number of American standard steels. The following list of applications will prove useful to many designers and mechanical engineers as well as to machinists, tool engineers, tool and die makers, and other metalworkers throughout industry. The following list is not all-inclusive, but rather is indicative of the much-used and readily available types of carbon, alloy, and stainless steels.

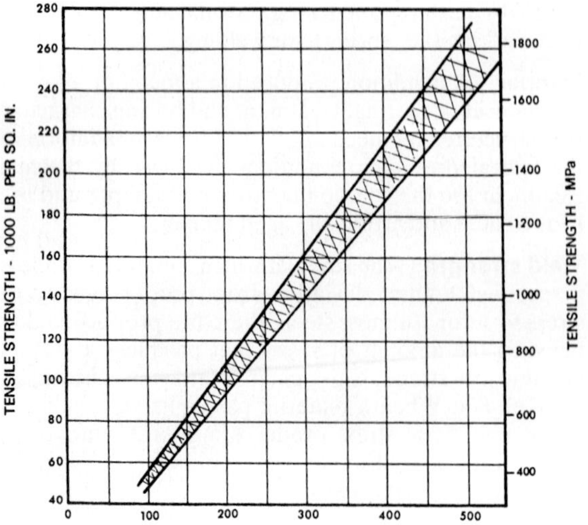

FIGURE 3.1 Brinell hardness numbers indicating approximate tensile strengths of steels

Carbon and Alloy Steels: Uses by SAE/AISI Number

SAE/AISI 1006 through 1015 Low-carbon, high-formability, weldable. Used for sheet metal, strip, wire, and rod. Excellent drawing qualities. Low-strength applications. Sheet metal structures, body and fender work, deep drawing parts of sheet steel.

SAE/AISI 1016 through 1030 Increased strength over the low-carbon group. These are known as the *case-hardening* or *carburizing grades*. The higher-manganese grades machine well. The higher-carbon types are used for thicker sections where a stronger core is desired. Type 1018 is used for a great many applications, may be easily case hardened, and is readily available. Grades 1020 and 1025 are used for low-strength bolts. All these steels are readily welded.

SAE/AISI 1030 through 1052 These are the medium-carbon types of steels used where higher strength than the lower-carbon grades is required. All these steels are used for forgings. Axles and shafts are made from the 1038 to 1045 group. Widely used for machined parts, both heat-treated and non-heat-treated. Welding is possible with precautions taken during the cooling process.

SAE/AISI 1055 through 1095 These are the high-carbon grades of carbon steel. Used for flat stampings, spring wire, cutting tools, flat springs, and many other high-strength applications. These steels are usually heat treated for their particular application and provide excellent wear resistance. Not recommended for welding applications.

Specific Applications

SAE/AISI 1018 Low-carbon, medium-manganese steel. Quality bar for carburized parts: gears, shafts, bolts, pins, etc.

SAE/AISI 1035 Medium-carbon, special-quality steel used for bolts, nuts, shafts, pins, etc. Can be heat treated.

SAE/AISI 1045 Medium-carbon, special-quality steel used for shafts, axles, gears, splines, etc. Can be heat treated.

SAE/AISI 12L14 Low-carbon, resulfurized, rephosphorized, and leaded steel used for screw-machine parts such as studs, nuts, and various fasteners. Can be case hardened.

SAE/AISI 1215 Similar to 12L14 and low-carbon steel used for studs, nuts, and fasteners. Can be case hardened.

SAE/AISI 12L15 Leaded version of 1215 used for screw-machine parts. Can be case hardened.

SAE/AISI 1117 Low-carbon, resulfurized, free-cutting steel used for shafts, gears, pins, nuts, etc. Can be carburized.

SAE/AISI 11L17 Leaded version of 1117 used for screw-machine parts, gears, shafts, pins, etc. Can be carburized.

SAE/AISI 1141 Medium-carbon, resulfurized, free-cutting steel used for shafts, nuts, bolts, etc. Can be hardened by heat treatment.

SAE/AISI 4140 Medium-carbon chromium-molybdenum alloy steel used for studs, nuts, bolts, gears, wrenches, shafts, etc. Can be heat treated.

SAE/AISI 41L40 Leaded version of 4140, free-machining type. Can be heat treated.

SAE/AISI 4145 Medium-carbon chromium-molybdenum steel used for studs, nuts, shafts, wrenches, gears, bolts, etc. Can be heat treated.

SAE/AISI 41L45 Leaded version of 4145, free-machining type. Can be heat treated.

SAE/AISI 4620 Low-carbon nickel-molybdenum steel used for gears, cams, pinions, and shafts. Excellent carburizing grade.

SAE/AISI 46L20 Leaded version of 4620, free-machining type. Can be carburized.

SAE/AISI 8620 Low-carbon nickel-chromium-molybdenum alloy steel used for gears, cams, shafts, and pinions. Excellent carburizing grade.

SAE/AISI 86L20 Leaded version of 8620, free-machining type. Can be carburized.

SAE/AISI 4340 Medium-carbon nickel-chromium-molybdenum alloy steel used for gears and shafting. Has high hardenability.

SAE/AISI 8642 Similar to 4340. Has high hardenability.

EF 4130 Aircraft-quality alloy steel.

EF 4140 Aircraft-quality alloy steel.

E 4340, EF 4620, EF 8740, and E 9310 All aircraft-quality alloy steels.

Stainless Steels: Uses by AISI Number

Note: The SAE numbers for stainless steels listed in SAE standards tables are typically known by their five-digit numbers, such as 30201, 30202, 30304, 51440A, 51502, etc. The last three digits of the listed SAE numbers are the standard industry identification numbers and are used as such in the following usage summary.

Chromium-Nickel Stainless Steels (Austenitic)

201 Low nickel, good corrosion resistance. High work-hardening rate. Excellent weldability.

202 General-purpose type equivalent to 302.

203 EZ Superior machinability. Good corrosion resistance.

216 Most corrosion resistant of all chromium-nickel-manganese stainless steels.

301 High work-hardening rate. Used in structural applications where high strength and resistance to atmospheric corrosion are required.

302 General-purpose stainless steel with good strength properties. Resistant to many corrosive conditions.

303 Free-machining type used in corrosive atmospheres, strong chemical solutions, many organic chemicals, most dyes, nitric acid, and foods.

303 Pb Leaded version of 303 used for high-volume automatic machining operations.

304 Low-carbon variation of 302. Weldable with caution. Excellent resistance to a high number of corrosive conditions and chemicals.

304 L Extra-low-carbon version of 304. Low carbon content prevents carbide precipitation during welding, which can produce cracks at the weld joints. Excellent weldability.

305 Good fabrication stainless for spinning, deep-drawing, and cold-heading operations.

309 Used in high-temperature applications. Resistant to most acids.

310 Higher alloy content improves the basic characteristics. Improved over 309 and 304 for corrosion resistance.

316 Best corrosion resistance of the standard stainless steels. Resists pitting and most chemicals. Used for paper-mill machinery parts and photographic industry parts and containers. High-temperature strength.

316 L Low-carbon version of 316 which is more easily welded without carbide precipitation.

317 Higher alloy content than 316, providing more corrosion resistance.

321 This alloy is stabilized with titanium for weldments subject to severe corrosion. Excellent corrosion resistance to a wide variety of organic and inorganic substances.

347 Stabilized with Cb and Ta for use in the carbide precipitation range of 800 to 1500°F, with no impairment to corrosion resistance.

Chromium Stainless Steels (Ferritic)

409 Developed for automotive muffler service and used in non-critical exterior parts. Economical and easily fabricated.

430 Most widely used of the nonhardenable types of chromium stainless steel. Good mechanical properties and heat resistance. Resistant to nitric acid, sulfur gases, and many organic chemicals, including foods.

430 F Free-machining version of 430. Similar in properties to 430.

446 High resistance to corrosion and scaling at high temperatures. Excellent in sulfuric atmospheres.

Chromium Stainless Steels (Martensitic)

403 Excellent for highly stressed parts such as turbine blades. Good resistance to water and atmospheric corrosion.

405 Variation of 410 for improved weldability. Same corrosion resistance as 403.

410 Low-cost general-purpose stainless steel that is heat treatable. Used where corrosion is not severe.

410 L Same as 410 except lower carbon range for improved weldability.

414 Modification of 410 with more nickel to improve corrosion resistance. Can be heat treated to Rockwell C25 to C43.

416 Free-machining version of 410 with corrosion resistance to food acids, basic salts, water, and most atmospheric corrosion products.

440 A, B, C, and F Series of high-carbon stainless steels. All are the same basic composition except carbon content. Can be heat treated for high strength and high hardness. These steels are corrosion resistant only in the hardened conditions. 440 F is a free-machining type used in many applications.

Low-Chromium Stainless Steels

PH13-8Mo

15-5PH

PH15-7Mo

17-4PH

17-7PH

All these grades provide excellent corrosion resistance similar to the austenitic stainless steels and are capable of being heat treated to various high-strength conditions with minimum distortion. They are generally furnished in the annealed condition for ease of machining. They develop their high-strength properties by aging at selected temperatures. They may be provided as vacuum arc–

remelted steels for more demanding applications. They are used in many high-strength, anticorrosion applications. Among other uses, 17-7PH (ASTM A313) round wire is used for helical spring applications and 17-7PH (ASTM A693) strip is used for flat spring applications.

Identification of Stainless Steels

A method for determining whether a steel is stainless steel or carbon steel is as follows:

1. Make up a 7 to 8% aqueous solution of copper sulfate ($CuSO_4$). Use the proportion of 86 g copper sulfate per liter of water. Specific gravity = 1.08 g/ml, or 11^0 Be′.
2. Clean a small area of the metal to be tested. All grease and residue must be removed. Lightly abrade the area if necessary.
3. Apply a few drops of the copper sulfate solution to the cleaned area on the test sample.
4. If the material is regular carbon steel or iron, a metallic copper film will form where the test solution contacts the sample. Stainless steel will show *no* copper deposit.
5. Applying a small magnet to the test sample will show the difference between austenitic and ferritic stainless steels, with ferritic materials being attracted to the magnet. Austenitic stainless steels are normally nonmagnetic, although cold-worked austenitic stainless steels will show a very weak magnetic attraction. Also, hardened chromium stainless steels are magnetic.

3.2 TABLES OF COMMON STEELS, ALLOY STEELS, METALS, AND OTHER METAL ALLOYS

Table 3.1 lists the common carbon, wrought alloy, and stainless steels.

Table 3.2 lists the common wrought aluminum alloys.

Table 3.3 lists the common wrought copper alloys.

Table 3.4 lists the common magnesium alloys, nickel alloys, and special alloys.

Table 3.5 lists the wrought titanium alloys.

Machining Characteristics of Metals and Alloys (see also Chapter 6)

TABLE 3.1 Mechanical Properties of Common Carbon, Wrought Alloy, and Stainless Steels

AISI no.	Form	Yield strength (psi)	Tensile strength (psi)	% Elongation (in 2 inches)	Brinell hardness (BH)
1010	HR	30000	47000	30	91
1015	HR	45000	61000	39	126
1018	HR	40000	69000	38	143
1018	CD	70000	82000	20	163
1020	HR	40000	69000	38	143
1025	HR	45000	67000	36	143
1030	HR	50000	80000	32	179
1042	CD	89000	102000	16	207
1045	HR	59000	98000	24	212
1045	WQ	90000	120000	18	240
1060	HR	70000	118000	17	241
1095	HR	83000	140000	8	293
1095	WQ	138000	201000	12	388
1117	HR	42000	71000	33	143
1137	CD	90000	105000	15	207
1141	CD	95000	112000	16	223
1144	CD	97000	112000	14	235
4130	N	63000	97000	25	197
4140	N	95000	148000	17	302
4150	N	106000	167000	11	321
4340	T	182000	162000	15	363
4620	N	53000	83000	29	174
1141	T	176000	237000	6	461
1215	CD	68000	76000	15	239
1340	T	231000	262000	11	505
Chromium-nickel stainless steels, (austenitic). Non-heat-treatable					
201	A	55000	115000	55	210
301	A	40000	110000	50	180
303	A	35000	90000	50	161
304	A	35000	85000	55	180
309	A	40000	95000	45	201
310	A	45000	95000	50	181
316	A	35000	85000	60	199
321	A	35000	85000	55	199
Chromium stainless steels, (martensitic). Heat treatable. (The values shown reflect heat treatment for maximum strength)					
410	HT	180000	200000	—	400
416	HT	115000	140000	—	280
440C	HT	275000	285000	—	600

FORM KEY:

HR = hot rolled; CD = cold drawn; N = normalized; WQ = water quenched; T = tempered; HT = heat treated; A = annealed. % Elongation is for 0.50 inch diameter specimens.

NOTE: The 300 series is non-heat-treatable. Maximum strength may only be obtained through cold working.

These steels may be ordered cold-worked in 1/4, 1/2, 3/4 and full hard conditions. In this case, max. strength for cold drawn wires is 340000/350000 psi, and for strips 248000/250000 psi. Austenitic stainless steels are nonmagnetic; martensitic and ferritic stainless steels are magnetic.

TABLE 3.2 Mechanical Properties of Common Wrought Aluminum Alloys

Alloy & temper	Yield strength	Tensile strength	% Elongation	Brinell hardness
1100-O	5000	13000	45	23
1100-H14	16000	18000	20	32
2014-O	14000	27000	18	45
2014-T6, T651	60000	70000	13	135
2024-O	11000	27000	22	47
2024-T4, T351	47000	68000	19	120
3003-O	6000	16000	40	28
3003-H14	21000	22000	16	40
5005-H14	22000	23000	6	42
5052-O	13000	28000	30	47
5052-H32	28000	33000	18	60
5083-H112	28000	44000	16	80
5054-O	17000	36000	25	62
5054-H34	35000	44000	16	81
6061-O	8000	18000	30	30
6061-T6, T651	40000	45000	17	95
6063-O	7000	13000	18	25
6063-T6	31000	35000	18	73
7075-O	15000	33000	16	60
7075-T6, T651	73000	83000	11	150
7178-T6, T651	78000	88000	11	160

NOTE: 5 Elongation is for 0.50-inch-thick specimens.

TABLE 3.3 Mechanical Properties of Common Wrought Copper Alloys

Copper or copper alloy	Yield strength (psi)	Tensile strength (psi)	% Elongation (in. in 2 inches)	Rockwell hardness (B scale)
102 ●	45000	50000	12	50
110 ●	44000	48000	16	47
210 ▲	40000	48000	12	51
220 ▲	45000	52000	11	57
230 ▲	49000	57000	12	65
260 ▲	52000	62000	23	70
268, 270 ▲	50000	61000	23	70
342, 343 ▲	50000	61000	20	70
360 ▲	45000	58000	25	78
464, 467 ▲	53000	75000	20	82
510 ▲	55000	68000	28	78
614 ●	45000	78000	40	84
655 ▲	45000	78000	17	87
715 ▲	70000	75000	15	81
◆ 175, Temper AT	90000	110000	15	96
◆ 170, Temper AT	145000	165000	7	RC36
◗ UNS-N033, (HT)	232000	272000	7	86 (15N)

NOTE: Above are typical values for 0.50-inch-diameter specimens, ½ hard (▲), and full hard (●). ◆ = Beryllium alloy; ◗ = Beryllium nickel alloy; HT = heat treated.

TABLE 3.4 Mechanical Properties of Common Magnesium Alloys, Nickel Alloys, and Special Alloys

Alloy	Form	Yield strength (psi)	Tensile strength (psi)	% Elongation (in. in 2 in.)	Brinell hardness (BH)
Wrought Magnesium					
SAE 51	Hard	—	32000	4	—
SAE 510	Hard	25000	36000	4	—
SAE 511	Hard	28000	39000	3	—
SAE 52	Extrud	25000	37000	11	—
SAE 520	Extrud	26000	40000	11	—
Cast Magnesium					
SAE 50	As cast	10000	24000	4	—
SAE 500	As cast	10000	20000	1.5	—
Nickel Alloys					
Monel	CD	80000	99000	25	190
R-Monel	CD	75000	90000	25	180
K-Monel	CD	85000	112000	25	209
S-Monel	As cast	100000	129000	1.5	320
Inconel	CD	90000	115000	19	200
Inconel "X"	HR	120000	180000	25	360
Special Copper Alloys					
Tellurium copper 165	H	39000	47000	11	RB50*
Hitenso 1622	H	49000	57000	14	RB65*
Chromium copper	HT	60000	69000	20	RB85*
HS Bronze 316	H	60000	70000	11	RB78*

NOTE: H = Hard; HT = Heat treated; HS = High strength; * = Rockwell B scale.

TABLE 3.5 Mechanical Properties of Wrought Titanium Alloys

Nominal composition (%)	Condition	Tensile strength (ksi)	Room temperature Yield strength (ksi)	Elongation (%)	Reduction in area (%)
		Commercially Pure			
99.5 Ti	Annealed	48	35	30	55
99.2 Ti	Annealed	63	50	28	50
99.1 Ti	Annealed	75	65	25	45
99.0 Ti	Annealed	96	85	20	40
99.2 Ti[a]	Annealed	63	50	28	50
98.9[b]	Annealed	75	65	25	42
		Alpha Alloys			
5 Al, 2.5 Sn	Annealed	125	117	16	40
5 Al, 2.5 Sn (low O_2)	Annealed	117	108	16	—
		Near Alpha Alloys			
8 Al, 1 Mo, 1 V	Duplex annealed	145	138	15	28
11 Sn, 1 Mo, 2.25 Al, 5.0 Zr, 1 Mo, 0.2 Si	Duplex annealed	160	144	15	35
6 Al, 2 Sn, 4 Zr, 2 Mo	Duplex annealed	142	130	15	35
5 Al, 5 Sn, 2 Zr, 2 Mo, 0.25 Si	975°C (1785°F) (½ h), AC + 595°C (1100°F) (2 h), AC	152	140	13	—
6 Al, 2 Nb, 1 Ta, 1 Mo	As-rolled 2.5-cm (1-in) plate	124	110	13	34
6 Al, 2 Sn, 1.5 Zr, 1 Mo, 0.35 Bi, 0.1 Si	Beta forge + duplex anneal	147	137	11	—
		Alpha-Beta Alloys			
8 Mn	Annealed	137	125	15	32
3 Al, 2.5 V	Annealed	100	85	20	—
6 Al, 4 V	Annealed	144	134	14	30
	Solution + age	170	160	10	25
6 Al, 4 V (low O_2)	Annealed	130	120	15	35
6 Al, 6 V, 2 Sn	Annealed	155	145	14	30
	Solution + age	185	170	10	20
7 Al, 4 Mo	Solution + age	160	150	16	22
6 Al, 2 Sn, 4 Zr, 6 Mo	Solution + age	184	170	10	23
6 Al, 2 Sn, 2 Zr, 2 Mo, 2 Cr, 0.25 Si	Solution + age	185	165	11	33
10 V, 2 Fe, 3 Al	Solution + age	185	174	10	19
		Beta Alloys			
13 V, 11 Cr, 3 Al	Solution + age	177	170	8	—
	Solution + age	185	175	8	—
8 Mo, 8 V, 2 Fe, 3 Al	Solution + age	190	180	8	—
3 Al, 8 V, 6 Cr, 4 Mo, 4 Zr	Solution + age	210	200	7	—
	Annealed	128	121	15	—
11.5 Mo, 6 Zr, 4.5 Sn	Solution + age	201	191	11	—

[a] Also contains 0.2 Pd.

[b] Also contains 0.8 Ni and 0.3 Mo.

SOURCE: Titanium Metals Corp. of America and RMI Co.

3.3 MILL TOLERANCES OF WROUGHT METALS AND METAL ALLOY STOCK PRODUCTS

See Tables 3.6 through 3.33.

Tables 3.6 to 3.10 Hot-rolled and cold-rolled carbon steel sheets, plates, rounds, hexagons, squares and flats.

Table 3.11 Cold-finished alloy steel bars, rounds, hexagons, squares and flats.

Table 3.12 Tool steels (drill rods) bars, rounds, flats, hexagons, squares and octagons.

Tables 3.13 and 3.14 Carbon and alloy steel tubing.

Tables 3.15 to 3.22 Stainless steel sheets, plates, squares, hexagons, octagons, rounds and flats.

Table 3.23 Stainless steel tubing.

Tables 3.24 to 3.28 Aluminum alloy sheets, plates, bars, hexagons, octagons, and rounds.

Tables 3.29 and 3.30 Aluminum alloy drawn and extruded tubing.

Tables 3.31 and 3.32 Copper strip, sheet, plate, bar, rounds, hexagons and octagons.

Table 3.33 Copper round seamless tubing.

TABLE 3.6 Hot-Rolled Carbon Steel Sheet Thickness Tolerances for Coils and Cut Lengths, Including Pickled Sheets

Specified width in inches	Specified thickness in inches					
	0.2299 to 0.1800	0.1799 to 0.0972	0.0971 to 0.0710	0.0709 to 0.0568	0.0567 to 0.0509	0.0508 to 0.0449
Over 12 to 20 incl.	0.007	0.007	0.006	0.006	0.005	0.005
Over 20 to 40 incl.	0.008	0.007	0.007	0.006	0.005	0.005
Over 40 to 48 incl.	0.009	0.008	0.007	0.006	0.006	0.005
Over 48 to 60 incl.	—	0.008	0.007	0.007	0.006	—
Over 60 to 72 incl.	—	0.008	0.008	0.007	0.007	—
Over 72		0.008	0.008			

NOTE: Thickness tolerances are over and under in inches.

TABLE 3.7 Hot-Rolled Carbon Steel Sheet Thickness Tolerances for Coils over 0.180 in Thick

Specified width in inches	Specified thickness in inches			
	0.5000 to 0.3751	0.3750 to 0.3125	0.3124 to 0.2300	0.2299 to 0.1800
Over 12 to 15 incl.	0.014	0.012	0.010	—
Over 15 to 20 incl.	0.015	0.014	0.011	—
Over 20 to 32 incl.	0.016	0.015	0.012	—
Over 32 to 40 incl.	0.017	0.016	0.013	—
Over 40 to 48 incl.	0.018	0.017	0.014	—
Over 48 to 60 incl.	0.019	0.019	0.015	0.011
Over 60 to 72 incl.	0.021	0.021	0.016	0.012

NOTE: Thickness tolerances are over and under in inches.

TABLE 3.8 Cold Rolled Carbon Steel Sheet Thickness Tolerances for Coils and Cut Lengths—Over 12 in Wide

Specified width in inches	Specified thickness in inches					
	0.1419 to 0.0972	0.0971 to 0.0710	0.0709 to 0.0568	0.0567 to 0.0389	0.0388 to 0.0195	0.0194 to 0.0142
Over 12 to 15 incl.	0.005	0.005	0.005	0.004	0.003	0.002
Over 15 to 72 incl.	0.006	0.005	0.005	0.004	0.003	0.002
Over 72	0.007	0.006	0.005	0.004	0.003	—

NOTE: Thickness tolerances are over and under in inches.

TABLE 3.9 Size Tolerances for Cold-Finished Carbon Steel Bars, Cold Drawn or Turned and Polished*

Size, in†	Maximum of carbon range 0.28% or less	Maximum of carbon range over 0.28% to 0.55%, incl	Maximum of carbon range to 0.55%, incl, stress relieved or annealed after cold finishing	Maximum of carbon range over 0.55% or all grades quenched and tempered or normalized and tempered before cold finishing
	All tolerances are in inches† and are minus			
	Rounds—Cold Drawn (to 4 in) or Turned and Polished			
To 1½, incl	0.002	0.003	0.004	0.005
Over 1½ to 2½, incl	0.003	0.004	0.005	0.006
Over 2½ to 4, incl	0.004	0.005	0.006	0.007
Over 4 to 6, incl	0.005	0.006	0.007	0.008
Over 6 to 8, incl	0.006	0.007	0.008	0.009
Over 8 to 9, incl	0.007	0.008	0.009	0.010
	Hexagons			
To ¾, incl	0.002	0.003	0.004	0.006
Over ¾ to 1½, incl	0.003	0.004	0.005	0.007
Over 1½ to 2½, incl	0.004	0.005	0.006	0.008
Over 2½ to 3⅛, incl	0.005	0.006	0.007	0.009
Over 3⅛ to 4, incl	0.005	0.006	—	—
	Squares‡			
To ¾, incl	0.002	0.004	0.005	0.007
Over ¾ to 1½, incl	0.003	0.005	0.006	0.008
Over 1½ to 2½, incl	0.004	0.006	0.007	0.009
Over 2½ to 4, incl	0.006	0.008	0.009	0.011
Over 4 to 5, incl	0.010	—	—	—
Over 5 to 6, incl	0.014	—	—	—
	Flats‡			
Width,† in				
To ¾, incl	0.003	0.004	0.006	0.008
Over ¾ to 1½, incl	0.004	0.005	0.008	0.010
Over 1½ to 3, incl	0.005	0.006	0.010	0.012
Over 3 to 4, incl	0.006	0.008	0.011	0.016
Over 4 to 6, incl	0.008	0.010	0.012	0.020
Over 6	0.013	0.015	—	—

* This table includes tolerances for bars that have been annealed, spheroidize annealed, normalized, normalized and tempered, or quenched and tempered before cold finishing. This table does not include tolerances for bars that are annealed, spheroidize annealed, normalized, normalized and tempered, or quenched and tempered after cold finishing; the producer should be consulted for tolerances for such bars.

† Width governs the tolerances for both width and thickness of flats. For example, when the maximum of carbon range is 0.28% or less, for a flat 2 in wide and 1 in thick, the width tolerance is 0.005 in and the thickness tolerance is the same, namely, 0.005 in.

‡ Tolerances may be ordered all plus, or distributed plus and minus with the sum equivalent to the tolerances listed.

SOURCE: Reprinted with permission from the *Annual Book of ASTM Standards,* copyright 1992, American Society for Testing and Materials.

TABLE 3.10 Size Tolerances for Cold-Finished Carbon Steel Round Bars Cold Drawn, Ground, and Polished or Turned, Ground, and Polished

Size, in, cold drawn, ground, and polished	Turned, ground, and polished	Tolerances from specified size, minus only, in
To 1½, incl	To 1½, incl	0.001
Over 1½ to 2½, excl	Over 1½ to 2½, excl	0.0015
2½ to 3, incl	2½ to 3, incl	0.002
Over 3 to 4, incl	Over 3 to 4, incl	0.003
—	Over 4 to 6, incl	0.004*
—	Over 6	0.005*

* For nonresulfurized steels (steels specified to maximum sulphur limits under 0.08%) or for steels thermally treated, the tolerance is increased by 0.001 in.

SOURCE: Reprinted with permission from the *Annual Book of ASTM Standards,* copyright 1992, American Society for Testing and Materials.

TABLE 3.11 Size Tolerance for Cold-Finished Alloy Steel Bars, Cold Drawn or Turned and Polished

Size, in*	Maximum of carbon range 0.28% or less	Maximum of carbon range over 0.28% to 0.55%, incl	Maximum of carbon range to 0.55%, incl, stress relieved or annealed after cold finishing	Maximum of carbon range over 0.55% with or without stress relieving or annealing after cold finishing; also, all carbons, quenched and tempered (heat treated), or normalized and tempered, before cold finishing
	All tolerances are in inches and are minus			
	Rounds—Cold Drawn (to 4 in) or Turned and Polished			
To 1, incl, in coils	0.002	0.003	0.004	0.005
Cut lengths:				
To 1½, incl	0.003	0.004	0.005	0.006
Over 1½ to 2½, incl	0.004	0.005	0.006	0.007
Over 2½ to 4, incl	0.005	0.006	0.007	0.008
Over 4 to 6, incl	0.006	0.007	0.008	0.009
Over 6 to 8, incl	0.007	0.008	0.009	0.010
Over 8 to 9, incl	0.008	0.009	0.010	0.011
	Hexagons			
To ¾, incl	0.003	0.004	0.005	0.007
Over ¾ to 1½, incl	0.004	0.005	0.006	0.008
Over 1½ to 2½, incl	0.005	0.006	0.007	0.009
Over 2½ to 3⅛, incl	0.006	0.007	0.008	0.010
Over 3⅛ to 4, incl	0.006	—	—	—
	Squares			
To ¾, incl	0.003	0.005	0.006	0.008
Over ¾ to 1½, incl	0.004	0.006	0.007	0.009
Over 1½ to 2½, incl	0.005	0.007	0.008	0.010
Over 2½ to 4, incl	0.007	0.009	0.010	0.012
Over 4 to 5, incl	0.011	—	—	—
	Flats*			
To ¾, incl	0.004	0.005	0.007	0.009
Over ¾ to 1½, incl	0.005	0.006	0.009	0.011
Over 1½ to 3, incl	0.006	0.007	0.011	0.013
Over 3 to 4, incl	0.007	0.009	0.012	0.017
Over 4 to 6, incl	0.009	0.011	0.013	0.021
Over 6	0.014	—	—	—

* Width governs the tolerances for both width and thickness of flats. For example, when the maximum of carbon range is 0.28% or less, for a flat 2 in wide and 1 in thick, the width tolerance is 0.006 in and the thickness tolerance is the same, namely, 0.006 in.

SOURCE: Reprinted with permission from the *Annual Book of ASTM Standards,* copyright 1992, American Society for Testing and Materials.

TABLE 3.12 Tool Steel Bars: Size Tolerances

	Standard manufacturing tolerance, in, plus and minus	Precision tolerance, in, plus and minus
	Drill Rod Rounds	
0.500 to 1.500, incl.	0.001	0.0005
0.125 to 0.499, incl.	0.0005	0.00025
Up to 0.124, incl.	0.0003	0.0002
	Drill Rod Flats, Squares, Hexagons, Octagons	
1 to ¾ in incl.	Plus or minus 0.0015	—
Under ¾ to ¼ in	Plus or minus 0.001	—
Under ¼ in	Plus or minus 0.0005	—

SOURCE: Ryerson, Inc.

TABLE 3.13 Carbon Steel and Alloy Steel Seamless Mechanical Tubing: Diameter and Wall-Thickness Tolerances—Cold Worked

Specified size of OD in inches	As drawn or finish annealed				Quenched and tempered		Wall thickness
	OD, in		ID, in		OD, in	ID, in	in %
	Over	Under	Over	Under	Over/under	Over/under	Over/under
3/16 to 1/2 excl (2, 3)	0.004	0.000	—	—	0.010	—	15
1/2 to 1 1/2 (1–4)	0.005	0.000	0.000	0.005	0.015	0.015	10
1 1/2 to 3 1/2 (1–4)	0.010	0.000	0.000	0.010	0.030	0.030	10
3 1/2 to 5 1/2 (1, 4)	0.015	0.000	0.005	0.015	0.045	0.045	10
5 1/2 to 8 excl							
When wall less than 5% of OD (4)	0.030	0.030	0.035	0.035	—	—	10
When wall is from 5 to 7.5% of OD	0.020	0.020	0.025	0.025	—	—	10
When wall is over 7.5% of OD (1)	0.030	0.000	0.015	0.030	—	—	10
8 to 10 3/4 incl							
When wall less than 5% of OD (4)	0.045	0.045	0.050	0.050	—	—	10
When wall is from 5 to 7.5% of OD	0.035	0.035	0.040	0.040	—	—	10
When wall is over 7.5% of OD (1)	0.045	0.000	0.015	0.040	—	—	10
Over 10 3/4 to 12 3/4 incl							
When wall less than 5% of OD (4)	0.060	0.060	0.060	0.060	—	—	10
When wall is from 5 to 7.5% of OD	0.050	0.050	0.050	0.050	—	—	10
When wall is over 7.5% of OD (1)	0.045	0.045	0.045	0.045	—	—	10

(1) Many tubes with ID less than 50% of OD, or with wall thickness over 25% of OD, or weighing more than 90 lb/ft are difficult to draw over mandrel. Therefore, ID can vary over or under by an amount equal to 10% of wall and wall may vary 12½% over or under that specified.

(2) Footnote a doesn't apply for tubes less than ½ in (or less than ⅝ in when wall thickness is more than 20% of OD) that are not commonly drawn over mandrel. Therefore, the wall thickness can vary 15% over and under that specified and the ID is governed by the OD and wall thickness tolerances in table.

(3) For tubes less than ½ in (or less than ⅝ in when wall thickness is more than 20% of OD), which can be drawn over mandrel, the tolerances shown in the table apply, except that the wall thickness tolerances are 10% over and under the specified wall thickness.

(4) Tubing with wall thickness less than 3% of OD cannot be straightened properly without a certain amount of distortion. Consequently, an ovality tolerance of ½% over and under nominal OD applies in addition to the tolerances in above table.

SOURCE: Ryerson, Inc.

TABLE 3.14 Round, Drawn-over-Mandrel Carbon Steel Welded Mechanical Tubing

	Wall Thickness Tolerances									
	Outside diameter of tubing in inches									
	⅜ to ⅞		Over ⅞ to 1⅞		Over 1⅞ to 3¾		Over 3¾ to 5½		Over 5½ to 7¼	
Wall	Plus	Minus	Plus	Minus	Plus	Minus	Plus	Minus	Plus	Minus
22 (0.028)	0.002	0.002	0.002	0.002	0.002	0.002	—	—	—	—
20 (0.035)	0.002	0.002	0.002	0.002	0.002	0.002	—	—	—	—
18 (0.049)	0.002	0.003	0.002	0.003	0.002	0.003	—	—	—	—
16 (0.065)	0.002	0.002	0.002	0.003	0.002	0.003	0.004	0.004	—	—
14 (0.083)	0.002	0.002	0.002	0.003	0.003	0.003	0.004	0.005	—	—
13 (0.095)	0.002	0.002	0.002	0.003	0.003	0.003	0.004	0.005	0.004	0.005
12 (0.109)	0.002	0.003	0.002	0.004	0.003	0.003	0.005	0.005	0.005	0.005
11 (0.120)	0.003	0.003	0.002	0.004	0.003	0.003	0.005	0.005	0.005	0.005
10 (0.134)	—	—	0.002	0.004	0.003	0.003	0.005	0.005	0.005	0.005
9 (0.148)	—	—	0.002	0.004	0.003	0.003	0.005	0.005	0.005	0.005
8 (0.165)	—	—	0.003	0.004	0.003	0.004	0.005	0.006	0.005	0.006
7 (0.180)	—	—	0.004	0.004	0.003	0.005	0.006	0.006	0.006	0.006
6 (0.203)	—	—	0.004	0.005	0.004	0.005	0.006	0.007	0.006	0.007
5 (0.220)	—	—	—	—	0.004	0.006	0.007	0.007	0.007	0.007
4 (0.238)	—	—	—	—	0.005	0.006	0.007	0.007	0.007	0.007
3 (0.259)	—	—	—	—	0.005	0.006	0.007	0.007	0.007	0.007
2 (0.284)	—	—	—	—	0.005	0.006	0.007	0.007	0.007	0.007
1 (0.300)	—	—	—	—	0.006	0.006	0.008	0.008	0.008	0.008
0.320	—	—	—	—	0.007	0.007	0.008	0.008	0.008	0.008
0.344	—	—	—	—	0.008	0.008	0.009	0.009	0.009	0.009

TABLE 3.14 (*Continued*)

OD size range in inches	OD and ID Tolerances			
	OD inches*		ID inches*	
	Over	Under	Over	Under
Up to 0.499	0.004	0.000	—	—
0.500–1.699 incl.	0.005	0.000	0.000	0.005
1.700–2.099 incl.	0.006	0.000	0.000	0.006
2.100–2.499 incl.	0.007	0.000	0.000	0.007
2.500–2.899 incl.	0.008	0.000	0.000	0.008
2.900–3.299 incl.	0.009	0.000	0.000	0.009
3.300–3.699 incl.	0.010	0.000	0.000	0.010
3.700–4.099 incl.	0.011	0.000	0.000	0.011
4.100–4.499 incl.	0.012	0.000	0.000	0.012
4.500–4.899 incl.	0.013	0.000	0.000	0.013
4.900–5.299 incl.	0.014	0.000	0.000	0.014
5.300–5.549 incl.	0.015	0.000	0.000	0.015
5.550–5.999 incl.	0.010	0.010	0.010	0.010
6.000–6.499 incl.	0.013	0.013	0.013	0.013
6.500–6.999 incl.	0.015	0.015	0.015	0.015
7.000 and over	0.018	0.018	0.018	0.018

* When the wall thickness is less than 3% of the OD, an ovality tolerance of plus and minus ½% of the mean outside diameter is required. This tolerance is in addition to the OD and ID tolerances indicated above.

SOURCE: Ryerson, Inc.

TABLE 3.15 Stainless Steel Sheets: Standard Gauge Tolerances—All Finishes

Thickness in inches Under Incl.	Gauge number	Tolerance in inches (plus or minus) Continuous mill 48 in max. width	Hand mill Over 48 to 60 in width incl.	Hand mill Over 60 in width
0.1875 to 0.146	8 & 9	0.007	0.0105	0.014
0.146 to 0.131	10	0.006	0.009	0.012
0.131 to 0.115	11	0.005	0.0075	0.010
0.115 to 0.099	12	0.005	0.007	0.009
0.099 to 0.084	13	0.004	0.006	0.008
0.084 to 0.073	14	0.004	0.0055	0.007
0.073 to 0.059	15 & 16	0.003	0.0045	0.006
0.059 to 0.041	17 & 19	0.003	—	—
0.041 to 0.030	20 to 22	0.002	—	—
0.030 to 0.017	23 to 27	0.0015	—	—
0.017 to 0.008	28 to 34	0.0015	—	—
0.008 to 0.006	35 to 38	0.0015	—	—
0.006	39	0.001	—	—

SOURCE: Ryerson, Inc.

TABLE 3.16 Stainless Steel Plates: Thickness Tolerances

Thickness in inches	Thickness tolerance over variation* in inches: Widths to 84 in incl.	Widths over 84 to 120 in incl.	Widths over 120 to 144 in incl.	Widths over 144 in incl.
3/16 to 3/8 excl.	0.046	0.050	—	—
3/8 to 3/4 excl.	0.054	0.058	0.075	0.090
3/4 to 1 excl.	0.060	0.064	0.083	0.100
1 to 2 incl.	0.070	0.074	0.095	0.115
Over 2	OA	OA	OA	OA

* No plate shall vary more than 0.01 in under the thickness ordered. OA = On application. Spot grinding not to exceed 0.01 in under the specified thickness is permitted to remove surface imperfections.

SOURCE: Ryerson, Inc.

TABLE 3.17 Stainless Steel Bars: Hot-Rolled Rounds and Squares: Tolerances

Specified sizes, in	Variation from size, in		Out of round or square,* in
Over 5/16 to 7/16 incl.	+0.006	−0.006	0.009
Over 7/16 to 5/8 incl.	+0.007	−0.007	0.010
Over 5/8 to 7/8 incl.	+0.008	−0.008	0.012
Over 7/8 to 1 incl.	+0.009	−0.009	0.013
Over 1 to 1 1/8 incl.	+0.010	−0.010	0.015
Over 1 1/8 to 1 1/4 incl.	+0.011	−0.011	0.016
Over 1 1/4 to 1 3/8 incl.	+0.012	−0.012	0.018
Over 1 3/8 to 1 1/2 incl.	+0.014	−0.014	0.021
Over 1 1/2 to 2 incl.	+1/64	−1/64	0.023
Over 2 to 2 1/2 incl.	+1/32	−0	0.023
Over 2 1/2 to 3 1/2 incl.	+3/64	−0	0.035
Over 3 1/2 to 4 1/2 incl.	+1/16	−0	0.046
Over 4 1/2 to 5 1/2 incl.	+5/64	−0	0.058
Over 5 1/2 to 6 1/2 incl.	+1/8	−0	0.070
Over 6 1/2 to 8 incl.	+5/32	−0	0.085

* Out-of-round is the difference between the maximum and minimum diameters of the bar, measured at the same cross section. Out-of-square is the difference in the two dimensions at the same cross section of a square bar, each dimension being the distance between opposite faces.

SOURCE: Ryerson, Inc.

TABLE 3.18 Stainless Steel Hot-Rolled Hexagons and Octagons: Tolerances

Specified sizes measured between opposite sides, in	Variation from size, in		Maximum difference, three measurements for hexagons only, in
1/4 to 1/2 incl.	+0.007	−0.007	0.011
Over 1/2 to 1 incl.	+0.010	−0.010	0.015
Over 1 to 1 1/2 incl.	+0.021	−0.021	0.025
Over 1 1/2 to 2 incl.	+1/32	−1/32	1/32
Over 2 to 2 1/2 incl.	+3/64	−3/64	3/64
Over 2 1/2 to 3 1/2 incl.	+1/16	−1/16	1/16

SOURCE: Ryerson, Inc.

TABLE 3.19 Stainless Steel Hot-Rolled Flats: Tolerances

Specified widths, in	Variations from thickness for thickness given, in			Variations from width, in	
	⅛ to ½, incl.	Over ½ to 1, incl.	Over 1 to 2, incl.	Over	Under
⅜ to 1 incl.	±0.008	±0.010	—	1/64	1/64
Over 1 to 2 incl.	±0.012	±0.015	±1/32	1/32	1/32
Over 2 to 4 incl.	±0.015	±0.020	±1/32	1/16	1/32
Over 4 to 6 incl.	±0.015	±0.020	±1/32	3/32	1/16
Over 6 to 8 incl.	±0.016	±0.025	±1/32	⅛	5/32
Over 8 to 10 incl.	±0.021	±0.031	±1/32	5/32	3/16

SOURCE: Ryerson, Inc.

TABLE 3.20 Stainless Steel Centerless Ground Rounds and Cold-Drawn Rounds: Tolerances

Sizes, in	Variation, in	
	Over	Under
3/32 to 5/16 excl.	0.001	0.001
5/16 to 1/2 excl.	0.0015	0.0015
1/2 to 1 excl.	0.002	0.002
1 to 1 1/2 excl.	0.0025	0.0025
1 1/2 to 4 incl.	0.003	0.003

SOURCE: Ryerson, Inc.

TABLE 3.21 Stainless Steel Cold-Drawn Squares, Hexagons, and Octagons: Tolerances

Specified size, in	Variations from size, in	
	Over	Under
3/32 to 5/16 excl.	0	0.002
5/16 to 1/2 excl.	0	0.003
1/2	0	0.004
Over 1/2 to 1 incl.	0	0.004
Over 1 to 2 incl.	0	0.006
Over 2 to 3 incl.	0	0.008
Over 3	0	0.010

SOURCE: Ryerson, Inc.

TABLE 3.22 Stainless Steel Cold-Drawn* Flats: Tolerances

Size width or thickness, in	Variations from width over or under, in		Variations from thickness over or under, in
	For thicknesses 1/4 and under	For thicknesses over 1/4	
Over 3/8 to 1 incl.	0.004	0.002	0.002
Over 1 to 2 incl.	0.006	0.003	0.003
Over 2 to 3 incl.	0.008	0.004	0.004
Over 3 to 4 1/2 incl.	0.010	0.005	0.005

* When it is necessary to heat treat or heat treat and pickle after cold finishing, because of special hardness or mechanical property requirements, tolerances are double those shown in the table.

SOURCE: Ryerson, Inc.

TABLE 3.23 Cold-Drawn Annealed Seamless Mechanical Stainless Steel Tubing: Outside Diameter, Ovality, and Wall Thickness Variations

OD size, in	Variations from OD, in		Ovality, double OD tolerance when wall is:	Wall thickness, percent*,†	
	Over	Under		Over	Under
Under ½	0.005	0.005	Lighter than 0.015 in	15	15
½ to 1½ excl.	0.005	0.005	Lighter than 0.065 in	10	10
1½ to 3½ excl.	0.010	0.010	Lighter than 0.095 in	10	10
3½ to 5½ excl.	0.015	0.015	Lighter than 0.150 in	10	10
5½ to 8 excl.	0.030	0.030	Lighter than 0.240 in	10	10

* Tubes with wall thicknesses more than 25% of their outside diameter or with wall thicknesses greater than 1¼ in or weighing more than 90 lb/ft may vary in wall thickness plus and minus 12½%.

† For tubes with inside diameter less than ½ in (or less than ⅝ in when wall thickness is more than 20% of the outside diameter), which cannot be successfully drawn over a mandrel, the wall thickness may vary 15% over or under that specified and the inside diameter will be governed by the outside diameter and wall thickness variations.

SOURCE: Ryerson, Inc.

TABLE 3.24 Thickness Tolerances: Aluminum Alloy Sheet and Plate

Thickness in inches*	Width in inches: Up to 18	Over 18 to 36	Over 36 to 48	Over 48 to 54	Over 54 to 60	Over 60 to 66	Over 66 to 72
0.006–0.010	0.001	0.0015	0.0025	0.0025	—	—	—
0.011–0.017	0.0015	0.0015	0.0025	0.0035	—	—	—
0.018–0.028	0.0015	0.002	0.0025	0.0035	0.004	0.004	0.004
0.029–0.036	0.002	0.002	0.0025	0.004	0.005	0.005	0.005
0.037–0.045	0.002	0.0025	0.003	0.004	0.005	0.005	0.005
0.046–0.068	0.0025	0.003	0.004	0.005	0.006	0.006	0.006
0.069–0.076	0.003	0.003	0.004	0.005	0.006	0.006	0.006
0.077–0.096	0.0035	0.0035	0.004	0.005	0.006	0.006	0.006
0.097–0.108	0.004	0.004	0.005	0.005	0.007	0.007	0.007
0.109–0.125	0.0045	0.0045	0.005	0.005	0.007	0.007	0.007
0.126–0.140	0.0045	0.0045	0.005	0.005	0.007	0.010	0.012
0.141–0.172	0.006	0.006	0.008	0.008	0.009	0.012	0.014
0.173–0.203	0.007	0.007	0.010	0.010	0.011	0.014	0.016
0.204–0.249	0.009	0.009	0.011	0.011	0.013	0.016	0.018
0.250–0.320	0.013	0.013	0.013	0.013	0.015	0.018	0.020
0.321–0.438	0.019	0.019	0.019	0.019	0.020	0.020	0.023
0.439–0.625	0.025	0.025	0.025	0.025	0.025	0.025	0.025
0.626–0.875	0.030	0.030	0.030	0.030	0.030	0.030	0.030
0.876–1.125	0.035	0.035	0.035	0.035	0.035	0.035	0.035
1.126–1.375	0.040	0.040	0.040	0.040	0.040	0.040	0.040
1.376–1.625	0.045	0.045	0.045	0.045	0.045	0.045	0.045
1.626–1.875	0.052	0.052	0.052	0.052	0.052	0.052	0.052
1.876–2.250	0.060	0.060	0.060	0.060	0.060	0.060	0.060
2.251–2.750	0.075	0.075	0.075	0.075	0.075	0.075	0.075
2.751–3.000	0.090	0.090	0.090	0.090	0.090	0.090	0.090
3.001–4.000	0.110	0.110	0.110	0.110	0.110	0.110	0.110
4.001–5.000	0.125	0.125	0.125	0.125	0.125	0.125	0.125
5.001–6.000	0.135	0.135	0.135	0.135	0.135	0.135	0.135

Thickness in inches*	Width in inches: Over 72 to 78	Over 78 to 84	Over 84 to 90	Over 90 to 96	Over 96 to 132	Over 132 to 144	Over 144 to 156	Over 156 to 168
0.029–0.036	0.006	0.006	0.007	0.009	—	—	—	—
0.037–0.045	0.006	0.006	0.007	0.011	—	—	—	—
0.046–0.068	0.007	0.007	0.008	0.012	0.013	—	—	—
0.069–0.076	0.007	0.007	0.012	0.012	0.016	—	—	—
0.077–0.096	0.007	0.007	0.012	0.012	0.016	—	—	—
0.097–0.108	0.008	0.008	0.016	0.018	0.020	—	—	—
0.109–0.125	0.008	0.008	0.016	0.018	0.020	—	—	—
0.126–0.140	0.013	0.014	0.016	0.018	0.020	—	—	—
0.141–0.172	0.015	0.016	0.017	0.019	0.023	—	—	—
0.173–0.203	0.017	0.017	0.017	0.022	0.026	—	—	—
0.204–0.249	0.018	0.018	0.018	0.024	0.028	—	—	—
0.250–0.320	0.020	0.020	0.020	0.025	0.030	0.035	0.042	0.053
0.321–0.438	0.023	0.025	0.025	0.026	0.033	0.038	0.045	0.057
0.439–0.625	0.030	0.030	0.030	0.035	0.035	0.043	0.049	0.067

TABLE 3.24 Thickness Tolerances: Aluminum Alloy Sheet and Plate (*Continued*)

	Width in inches (*Continued*)							
Thickness in inches*	Over 72 to 78	Over 78 to 84	Over 84 to 90	Over 90 to 96	Over 96 to 132	Over 132 to 144	Over 144 to 156	Over 156 to 168
0.626–0.875	0.037	0.037	0.037	0.045	0.045	0.054	0.059	0.077
0.876–1.125	0.045	0.045	0.045	0.055	0.055	0.065	0.070	0.088
1.126–1.375	0.052	0.052	0.052	0.065	0.065	0.075	0.080	0.098
1.376–1.625	0.060	0.060	0.060	0.075	0.075	0.085	0.090	0.108
1.626–1.875	0.070	0.070	0.070	0.088	0.088	—	—	—
1.876–2.250	0.080	0.080	0.080	0.100	0.100	—	—	—
2.251–2.750	0.100	0.100	0.100	0.125	0.125	—	—	—
2.751–3.000	0.120	0.120	0.120	0.150	0.150	—	—	—
3.001–4.000	0.140	0.140	0.140	0.160	0.160	—	—	—
4.001–5.000	0.150	0.150	0.150	0.160	—	—	—	—
5.001–6.000	0.160	0.160	0.160	0.170	—	—	—	—

* Sheets have a thickness less than 0.250 in, plates 0.250 in and larger.

NOTE: All tolerances are plus and minus. When a dimension tolerance is specified other than as an equal bilateral tolerance, the value of the standard tolerance is that which would apply to the mean of the maximum and minimum dimensions permissible under the tolerance. Standard tolerances conform to or are within those specified in "Recommended Standards for Wrought Aluminum Mill Products," as issued by The Aluminum Association.

SOURCE: Ryerson, Inc.

TABLE 3.25 Diameter Tolerances: Aluminum Alloy Round Wire and Rods

	Allowable deviation from specified diameter, in			
			Rolled rod*	
Specified diameter in inches	Drawn wire*	Cold-finished rod*	Plus	Minus
Up to 0.035	0.0005	—	—	—
0.036–0.064	0.001	—	—	—
0.065–0.374	0.0015	—	—	—
0.375–0.500	—	0.0015	—	—
0.501–1.000	—	0.002	—	—
1.001–1.500	—	0.0025	—	—
1.501–2.000	—	0.004	0.006	0.006
2.001–3.000	—	0.006	0.008	0.008
3.001–3.499	—	0.008	0.012	0.012
3.500–5.000	—	0.012	0.031	0.016
5.001–8.000	—	—	0.062	0.031

* Tolerances are plus and minus.

SOURCE: Ryerson, Inc.

TABLE 3.26 Distance Across Flats Tolerances: Aluminum Alloy Square, Hexagonal, and Octagonal Wire and Bars

Specified distance across flats in inches	Allowable deviation from specified distance across flats, in		
	Drawn wire*	Cold-finished bar*	Rolled bar*
Up to 0.035	0.001	—	—
0.036–0.064	0.0015	—	—
0.065–0.374	0.002	—	—
0.375–0.500	—	0.002	—
0.501–1.000	—	0.0025	—
1.001–1.500	—	0.003	—
1.501–2.000	—	0.005	0.016
2.001–3.000	—	0.008	0.020
3.001–4.000	—	—	0.020

* All tolerances are plus or minus.
SOURCE: Ryerson, Inc.

TABLE 3.27 Thickness and Width Tolerances: Aluminum Alloy Rectangular Wire and Bars

Specified thickness and width in inches	Allowable deviation from specified thickness and width, in			
	Drawn wire and cold-finished bar*		Rolled bar*	
	Thickness	Width	Thickness	Width
Up to 0.035	0.001	—	—	—
0.036–0.064	0.0015	—	—	—
0.065–0.500	0.002	0.002	0.006	—
0.501–0.750	0.0025	0.0025	0.008	0.016
0.751–1.000	0.0025	0.0025	0.012	0.016
1.001–1.500	0.003	0.003	0.016	0.016
1.501–2.000	0.005	0.005	0.016	0.031
2.001–3.000	0.008	0.008	0.020	0.031
3.001–4.000	—	0.010	—	—
4.001–6.000	—	—	—	0.047
6.001–10.000	—	—	—	0.062

NOTE: Standard tolerances conform to or are within those specified in "Recommended Standards for Wrought Aluminum Mill Products," as issued by The Aluminum Association.
* All tolerances are plus or minus.
SOURCE: Ryerson, Inc.

TABLE 3.28 Diameter Tolerances and Distance Across Flats Tolerances: Aluminum Alloy Round and Hexagonal Screw-Machine Stock

Round stock		Hexagonal stock	
Specified diameter in inches	Allowable deviation from specified diameter, in	Specified distance across flats in inches	Allowable deviation from specified distance across flats, in
0.125–0.500	0.0015	0.125–0.500	0.002
0.501–1.000	0.002	0.501–1.000	0.0025
1.001–1.500	0.0025	1.001–1.500	0.003
1.501–2.000	0.004	1.501–2.000	0.005
2.001–2.500	0.006	2.001–3.000	0.008
2.501–3.000	0.008		
3.001–3.375	0.012		

NOTE: All tolerances are plus or minus. Standard tolerances conform to those specified in "Recommended Standards for Wrought Aluminum Products," as issued by The Aluminum Association.

SOURCE: Ryerson, Inc.

TABLE 3.29 Aluminum Alloy Round Drawn Tubing: Tolerances

(a) Diameter Tolerances

Specified OD or ID, in	Allowable deviation of mean diameter from specified diameter (size), in	Allowable deviation of diameter at any point from specified diameter (ovalness), in*	
		Not heat treated†	Heat treated†
Under 0.501	0.003	0.003	0.006
0.501–1.000	0.004	0.004	0.008
1.001–2.000	0.005	0.005	0.010
2.001–3.000	0.006	0.006	0.012
3.001–5.000	0.008	0.008	0.016
5.001–6.000	0.010	0.010	0.020
6.001–8.000	0.015	0.015	0.030
8.001–10.000	0.020	0.020	0.040
10.001–12.000	0.025	0.025	0.050

(b) Wall Thickness Tolerances

Specified wall thickness, in	Allowable deviation of mean wall thickness from specified wall thickness, in	Allowable deviation of wall thickness at any point from specified wall thickness (eccentricity)	
		Not heat treated‡	Heat treated
0.010–0.035	0.002	0.002	Plus or minus 10% of specified wall thickness min. ±0.003
0.036–0.049	0.003	0.003	
0.050–0.083	0.004	0.004	
0.084–0.120	0.005	0.006	
0.121–0.203	0.006	0.008	
0.204–0.300	0.008	0.012	
0.301–0.375	0.015	0.020	
0.376–0.500	0.020	0.030	

* Not applicable to annealed (–0 temper) tubing, coiled tubing, or tubing having a wall thickness less than 0.020 in or 2½% of the OD—or equivalent round diameter (dia. of a circle having a circumference equal to the perimeter of a tube).

† For the –T8 tempers of 6063 alloy, the tolerances for non-heat-treated apply.

‡ For coiled tubing, use tolerances in heat treated column.

NOTE: All tolerances are plus or minus. Standard tolerances conform to or are within those specified in "Recommended Standards for Wrought Aluminum Mill Products," as issued by The Aluminum Association. When outside diameter, inside diameter, and wall thickness are all specified, standard tolerances are applicable to any two of these dimensions, but not to all three. When a dimension tolerance is specified other than as an equal bilateral tolerance, the value of the standard tolerance is that which would apply to the mean of the maximum and minimum dimensions permissible under the tolerance. Mean diameter is the average of two diameter measurements taken at right angles to each other at any point along the length. Mean wall thickness is the average of two measurements taken opposite each other. When dimensions specified are outside and inside, rather than wall thickness, the allowable deviation at any point (eccentricity) is plus or minus 10% of the mean wall thickness, but not less than ±0.003 in.

SOURCE: Ryerson, Inc.

TABLE 3.30 Aluminum Alloy Round Extruded Tubing: Tolerances

(a) Diameter Tolerances

Specified OD or ID, in	Allowable deviation of mean diameter from specified diameter (size), in	Allowable deviation of diameter at any point from specified diameter (ovalness), in
0.500–0.999	0.010	0.020
1.000–1.999	0.012	0.025
2.000–3.999	0.015	0.030
4.000–5.999	0.025	0.050
6.000–7.999	0.035	0.075
8.000–9.999	0.045	0.100
10.000–11.999	0.055	0.125
12.000–13.999	0.065	0.150
14.000–15.999	0.075	0.175
16.000–17.999	0.085	0.200

(b) Wall Thickness Tolerances

Specified wall thickness, in	Allowable deviation of mean wall thickness from specified wall thickness in inches				Allowable deviation of wall thickness at any point from mean wall thickness (eccentricity)
	Outside diameter in inches				
	Under 1.250	1.250–2.999	3.000–4.999	5.000 and over	
Under 0.047	0.006	—	—	—	Plus or minus 10% of mean wall thickness max. ±0.060 min. ±0.010
0.047–0.061	0.007	0.008	0.008	0.010	
0.062–0.077	0.008	0.008	0.009	0.012	
0.078–0.124	0.009	0.009	0.010	0.015	
0.125–0.249	0.009	0.009	0.013	0.020	
0.250–0.374	0.011	0.011	0.016	0.025	
0.375–0.499	—	0.015	0.021	0.035	
0.500–0.749	—	0.020	0.028	0.045	
0.750–0.999	—	—	0.035	0.055	
1.000–1.499	—	—	0.045	0.065	
1.500–2.000	—	—	—	0.075	
2.001–2.499	—	—	—	0.085	±0.120
2.500–2.999	—	—	—	0.095	
3.000–3.499	—	—	—	0.105	
3.500–4.000	—	—	—	0.115	

NOTE: See page 3-43.

NOTE: All tolerances are plus or minus. Tolerances are applicable to the average shape; wider tolerances may be required for some shapes and closer tolerances may be possible for others. When outside diameter, inside diameter, and wall thickness (or their equivalent dimensions in other-than-round tube) are all specified, standard tolerances are applicable to any two of these dimensions, but not to all three. When a dimension tolerance is specified other than as an equal bilateral tolerance, the value of the standard tolerance is that which would apply to the mean of the maximum and minimum dimensions permissible under the tolerance. Mean diameter is the average of two diameter measurements taken at right angles to each other at any point along the length. Not applicable in the annealed (–0) temper or if wall thickness is less than 2½ percent of the outside diameter or equivalent round diameter. The equivalent round diameter is the diameter of a circle having a circumference equal to the perimeter of the tube. The mean wall thickness of round tube is the average of two measurements taken opposite each other. The mean wall thickness of other-than-round tube is the average of two measurements taken opposite each other at approximate center line of tube and perpendicular to the longitudinal axis of the cross section. When dimensions specified are outside and inside, rather than wall thickness itself, allowable deviation of any point (eccentricity) applies to mean wall thickness. Standard tolerances conform to or are within those specified in "Recommended Standards for Wrought Aluminum Mill Products," as issued by The Aluminum Association.

SOURCE: Ryerson, Inc.

TABLE 3.31 Copper Mill Products: Thickness Tolerances (in inches)—Cold Rolled

Thickness, in	Width, in: Up to 8 incl.	Over 8 to 12 incl.	Over 12 to 14 incl.	Over 14 to 20 incl.	Over 20 to 28 incl.	Over 28 to 36 incl.	Over 36 to 48 incl.	Over 48 to 60 incl.
Nonrefractory Alloys								
	Strip				*Sheet*			
Up to 0.004 incl.	0.0003	0.0006	0.0006	—	—	—	—	—
Over 0.004 to 0.006 incl.	0.0004	0.0008	0.0008	0.0013	—	—	—	—
Over 0.006 to 0.009 incl.	0.0006	0.0010	0.0010	0.0015	—	—	—	—
Over 0.009 to 0.013 incl.	0.0008	0.0013	0.0013	0.0018	0.0025	0.003	0.0035	0.004
Over 0.013 to 0.017 incl.	0.0010	0.0015	0.0015	0.002	0.0025	0.003	0.0035	0.0045
Over 0.017 to 0.021 incl.	0.0013	0.0018	0.0018	0.002	0.003	0.0035	0.004	0.005
Over 0.021 to 0.026 incl.	0.0015	0.002	0.002	0.0025	0.003	0.0035	0.004	0.005
Over 0.026 to 0.037 incl.	0.002	0.002	0.002	0.0025	0.0035	0.004	0.005	0.006
Over 0.037 to 0.050 incl.	0.002	0.0025	0.0025	0.003	0.004	0.005	0.006	0.007
Over 0.050 to 0.073 incl.	0.0025	0.003	0.003	0.0035	0.005	0.006	0.007	0.008
Over 0.073 to 0.130 incl.	0.003	0.0035	0.0035	0.004	0.006	0.007	0.008	0.010
Over 0.130 to 0.188 incl.	0.0035	0.004	0.004	0.0045	0.007	0.008	0.010	0.012
	Bar		*Plate*					
Over 0.188 to 0.205 incl.	0.0035	0.004	0.004	0.0045	0.007	0.008	0.010	0.012
Over 0.205 to 0.300 incl.	0.004	0.0045	0.0045	0.005	0.009	0.010	0.012	0.014
Over 0.300 to 0.500 incl.	0.0045	0.005	0.005	0.006	0.012	0.013	0.015	0.018
Over 0.500 to 0.750 incl.	0.0055	0.007	0.007	0.009	0.015	0.017	0.019	0.023
Over 0.750 to 1.00 incl.	0.007	0.009	0.009	0.011	0.018	0.021	0.024	0.029
Over 1.00 to 1.50 incl.	0.022	0.022	0.022	0.022	0.022	0.025	0.029	0.036
Over 1.50 to 2.00 incl.	0.026	0.026	0.026	0.026	0.026	0.030	0.036	0.044

	Refractory Alloys							
	Strip				*Sheet*			
Up to 0.004 incl.	0.0004	0.0008	0.0008	—	—	—	—	—
Over 0.004 to 0.006 incl.	0.0006	0.0010	0.0010	0.0015	—	—	—	—
Over 0.006 to 0.009 incl.	0.0008	0.0013	0.0013	0.002	—	—	—	—
Over 0.009 to 0.013 incl.	0.0010	0.0015	0.0015	0.0025	—	—	—	—
Over 0.013 to 0.017 incl.	0.0013	0.002	0.002	0.0025	—	—	—	—
Over 0.017 to 0.021 incl.	0.0015	0.0025	0.0025	0.003	—	—	—	—
Over 0.021 to 0.026 incl.	0.002	0.0025	0.0025	0.003	0.004	0.005	0.006	0.007
Over 0.026 to 0.037 incl.	0.0025	0.003	0.003	0.0035	0.005	0.006	0.007	0.008
Over 0.037 to 0.050 incl.	0.003	0.0035	0.0035	0.004	0.006	0.007	0.008	0.010
Over 0.050 to 0.073 incl.	0.0035	0.004	0.004	0.0045	0.007	0.008	0.010	0.012
Over 0.073 to 0.130 incl.	0.004	0.0045	0.0045	0.005	0.008	0.010	0.012	0.014
Over 0.130 to 0.188 incl.	0.0045	0.005	0.005	0.006	0.010	0.012	0.014	0.016
	Bar		*Plate*					
Over 0.188 to 0.205 incl.	0.0045	0.005	0.005	0.006	0.010	0.012	0.014	0.016
Over 0.205 to 0.300 incl.	0.005	0.006	0.006	0.007	0.012	0.014	0.016	0.018
Over 0.300 to 0.500 incl.	0.006	0.007	0.007	0.008	0.015	0.017	0.019	0.023
Over 0.500 to 0.750 incl.	0.008	0.010	0.010	0.012	0.019	0.021	0.024	0.029
Over 0.750 to 1.00 incl.	0.010	0.012	0.012	0.015	0.023	0.026	0.030	0.037
Over 1.00 to 1.50 incl.	0.028	0.028	0.028	0.028	0.028	0.032	0.037	0.045
Over 1.50 to 2.00 incl.	0.033	0.033	0.033	0.033	0.033	0.038	0.045	0.055

SOURCE: Vincent, Inc.

TABLE 3.32 Copper Round Rod, Hexagonal and Octagonal Stock: Tolerances

Diameter or distance between parallel surfaces, in	Nonrefractory alloys		Refractory alloys	
	Round	Hexagonal, octagonal	Round	Hexagonal, octagonal
Up to 0.150 incl.	0.0013	0.0025	0.002	—
Over 0.150 to 0.500 incl.	0.0015	0.003	0.002	0.004
Over 0.500 to 1.00 incl.	0.002	0.004	0.003	0.005
Over 1.00 to 2.00 incl.	0.0025	0.005	0.004	0.006
Over 2.00	0.15%*	0.30%*	0.20%*	0.40%*

* Expressed to the nearest 0.001 in.

SOURCE: Vincent, Inc.

TABLE 3.33 Copper Round Seamless Tube: Wall-Thickness Tolerances

Wall thickness, in	Outside diameter, in						
	1/32 to 1/8 incl.	Over 1/8 to 5/8 incl.	Over 5/8 to 1 incl.	Over 1 to 2 incl.	Over 2 to 4 incl.	Over 4 to 7 incl.	Over 7 to 10 incl.
Nonrefractory Alloys							
Up to 0.018	0.002	0.001	0.0015	0.002	—	—	—
Incl. 0.018 to 0.025	0.003	0.002	0.002	0.0025	—	—	—
Incl. 0.025 to 0.035	0.003	0.0025	0.0025	0.003	0.004	—	—
Incl. 0.035 to 0.058	0.003	0.003	0.0035	0.0035	0.005	0.007	—
Incl. 0.058 to 0.083	—	0.0035	0.004	0.004	0.006	0.008	0.010
Incl. 0.083 to 0.120	—	0.004	0.005	0.005	0.007	0.009	0.011
Incl. 0.120 to 0.165	—	0.005	0.006	0.006	0.008	0.010	0.012
Incl. 0.165 to 0.220	—	0.007	0.0075	0.008	0.010	0.012	0.014
Incl. 0.220 to 0.284	—	—	0.009	0.010	0.012	0.014	0.016
Incl. 0.284 to 0.380	—	—	0.011	0.012	0.014	0.016	0.018
Incl. 0.380 and over	—	—	—	5%*	5%*	6%*	6%*
Refractory Alloys							
Up to 0.018	0.0025	0.0015	0.002	0.0025	—	—	—
Incl. 0.018 to 0.025	0.004	0.0025	0.0025	0.003	—	—	—
Incl. 0.025 to 0.035	0.004	0.003	0.003	0.004	0.005	—	—
Incl. 0.035 to 0.058	0.004	0.004	0.0045	0.0045	0.0065	0.009	—
Incl. 0.058 to 0.083	—	0.0045	0.005	0.005	0.0075	0.010	0.013
Incl. 0.083 to 0.120	—	0.005	0.0065	0.0065	0.009	0.011	0.014
Incl. 0.120 to 0.165	—	0.007	0.007	0.0075	0.010	0.013	0.015
Incl. 0.165 to 0.220	—	—	0.009	0.010	0.013	0.015	0.018
Incl. 0.220 to 0.284	—	—	0.012	0.013	0.015	0.018	0.020
Incl. 0.284 to 0.380	—	—	—	0.015	0.018	0.020	0.023
Incl. 0.380 and over	—	—	—	6%*	6%*	8%*	8%*

* Expressed to the nearest 0.001 in.
SOURCE: Vincent, Inc.

3.4 AISI-SAE STEEL DESIGNATIONS

The AISI-SAE designation numbers for carbon and alloy steels are shown in Table 3.34.

3.5 HARDNESS TESTS AND HARDNESS NUMBER CONVERSIONS

Brinell Hardness Numbers:

The Brinell hardness system is one of the most widely used systems for indicating the hardness of metals and alloys. The Brinell hardness number of a material may be calculated from the following equation:

$$\mathrm{HB} = \frac{P}{\frac{\pi D}{2}\left(D - \sqrt{D^2 - d^2}\right)}$$

where HB = Brinell hardness number
P = load applied to the test ball, kg
D = diameter of ball, mm
d = measured diameter at the rim of the impression in the material, mm

The standard ball is 10 mm in diameter, and standard loads are 500, 1500, and 3000 kg. The test is not valid when the hardness of the test material is above the anticipated Brinell hardness number of 630.

The Brinell system and the Rockwell hardness systems are the most widely used hardness systems in the metalworking industries. Table 3.35 shows the comparison of hardness between the Brinell and Rockwell sys-

TABLE 3.34 AISI-SAE Steel Designation System

Numbers and digits	Numbers and digits
Carbon Steels:	88xx
10xx 11xx 12xx 15xx	93xx 94xx 97xx 98xx
Manganese Steels:	Nickel-Molybdenum Steels:
13xx	46xx 48xx
Nickel Steels:	
23xx	**Chromium steels:**
25xx	50xx 51xx
Nickel-Chromium Steels:	5xxx
31xx 32xx 33xx 34xx	51xxx 52xxx **Chromium-vanadium steels:** 61xx
Molybdenum Steels:	
40xx 44xx	**Tungsten-chromium steels:** 72xx
Chromium-Molybdenum Steels:	Silicon-Manganese Steels:
41xx	92xx
Nickel-Chromium-Molybdenum Steels:	High-Strength Low-Alloy Steels:
43xx 47xx 81xx 86xx 87xx (Continued, next column)	9xx **Boron steels:** xxBxx **Leaded steels:** xxLxx

NOTE: "xx" in the last two or three digits of these designations indicates the carbon content in hundredths of a percent; i.e., 8620 steel indicates 0.20% carbon; 1045 steel indicates 0.45% carbon, etc.

AISI = American Iron and Steel Institute; SAE = Society of Automotive Engineers.

TABLE 3.35 Hardened Steel and Hard Alloys

Scale				
C	A	15-N	30-N	Brinell
Load (kg)				
150	60	15	30	3000
80	92.0	96.5	92.0	—
78	91.0	96.0	91.0	—
76	90.0	95.5	90.0	—
74	89.0	95.0	88.5	—
72	88.0	94.5	87.0	—
70	86.5	94.0	86.0	—
68	85.5	—	84.5	—
66	84.5	92.5	83.0	—
64	83.5	—	81.0	—
62	82.5	91.0	79.0	—
60	81.0	90.0	77.5	614
58	80.0	—	75.5	587
56	79.0	88.5	74.0	560
54	78.0	87.5	72.0	534
52	77.0	86.5	70.5	509
50	76.0	85.5	68.5	484
49	75.5	85.0	67.5	472
48	74.5	84.5	66.5	460
47	74.0	84.0	66.0	448
46	73.5	83.5	65.0	437
45	73.0	83.0	64.0	426
44	72.5	82.5	63.0	415
43	72.0	82.0	62.0	404
42	71.5	81.5	61.5	393
41	71.0	81.0	60.5	382
40	70.5	80.5	59.5	372
39	70.0	80.0	58.5	362
38	69.5	79.5	57.5	352
37	69.0	79.0	56.5	342
36	68.5	78.5	56.0	332
35	68.0	78.0	55.0	322
34	67.5	77.0	54.0	313
33	67.0	76.5	53.0	305
32	66.5	76.0	52.0	297
31	66.0	75.5	51.5	290
30	65.5	75.0	50.5	283
29	65.0	74.5	49.5	276
28	64.5	74.0	48.5	270
27	64.0	73.5	47.5	265
26	63.5	72.5	47.0	260
25	63.0	72.0	46.0	255
24	62.5	71.5	45.0	250
23	62.0	71.0	44.0	245
22	61.5	70.5	43.0	240
21	61.0	70.0	42.5	235
20	60.5	69.5	41.5	230

Unhardened & Soft Tempered Steel Cast Irons & Non-Ferrous Metals

Scale		
B	Brinell	
Load (kg)		
100	500	3000
100	201	240
98	189	228
96	179	216
94	171	205
92	163	195
90	157	185
88	151	176
86	145	169
84	140	162
82	135	156
80	130	150
78	126	144
76	122	139
74	118	135
72	114	130
70	110	125
68	107	121
66	104	117
64	101	114
62	98	110
60	95	107
58	92	104
56	90	101
54	87	—
52	85	—
50	83	—
48	81	—
44	78	—
42	76	—
38	73	—
34	70	—
30	67	—
28	66	—
26	65	—
25	64	—
23	63	—
21	62	—
19	61	—
17	60	—
15	59	—
13	58	—
10	57	—
7	56	—
5	55	—
2	54	—
0	53	—

tems. The Brinell system contains only one scale of hardness, while the Rockwell system contains numerous scales, according to the material being tested. The Rockwell systems consist of scales, A, B, C and other designations as shown in Table 3.35. Use Table 3.35 to convert from one system of hardness to the other.

Figure 3.2 shows a typical testing machine for measuring the hardness numbers for various materials in Brinell or Rockwell hardness.

3.6 PLASTICS: THERMOPLASTICS AND THERMOSET PLASTICS

Common Plastics and Typical Uses

Acetal (Delrin, Celcon) *Properties:* High modulus of elasticity, low coefficient of friction, excellent abrasion and impact resistance, low moisture absorption, excellent machinability, ablative. *Typical uses:* Bearings, gears, antifriction parts, electrical components, washers, seals, insulators, cams.

Acetate (Cellulose) *Properties:* Odorless, tasteless, nontoxic, grease resistant, high impact strength. *Typical uses:* Badges, blister packaging, displays, optical covers, book covers.

Acrylic (Plexiglas, Lucite) *Properties:* Unusual optical clarity, high tensile strength, weatherability, good electrical properties, ablative. *Typical uses:* Displays, signs, models, lenses, electrical and electronic parts.

Benelex (Laminate) *Properties:* High compressive strength, machinable, resists corrosion (alkalis or

FIGURE 3.2 A Standard type hardness testing instrument for the Rockwell and Brinell scales.

acids), good electrical insulation, high flexural, shear, and tensile strength. *Typical uses:* Work surfaces, electrical panels and switch gear, bus braces (low voltage only), neutron shielding.

Diallyl phthalate, Melamine *Properties:* High strength, chemical resistant, low water absorption, medium-high temperature use. *Typical uses:* Terminal blocks and strips, dishware, automotive applications, aerospace applications.

Epoxy glass *Properties:* High strength, high-temperature applications, flame retardant, low coefficient of thermal expansion, low water absorption. *Typical uses:* High-quality printed-circuit boards, microwave stripline applications, VHF and UHF applications, electrical insulation, services in temperature range –400 to 500°F.

Mylar (polyester film, polyethylene terephthalate) *Properties:* High dielectric strength, chemical resistance, high mechanical strength, moisture resistant, temperature range—70° to 105°C, does not embrittle with age. *Typical uses:* Electrical and industrial applications, graphic arts applications.

Nylon *Properties:* Wear resistant, low friction, high tensile strength, excellent impact resistance, high fatigue resistance, easy machining, corrosion resistant, lightweight. *Typical uses:* Bearings, bushings, valve seats, washers, seals, cams, gears, guides, wheels, insulators, wear parts.

Phenol formaldehyde (Bakelite) *Properties:* Wear resistant, rigid, moldable to precise dimensions, strong, excellent electrical properties, economical, will not support combustion. *Typical uses:* Electrical and electronic parts, handles, housings, insulator parts, mechanism parts, parts that are to resist temperatures to 250°C.

Phenolic laminates *Properties:* Immune to common solvents, lightweight, strong, easily machined. *Typical uses:* Bearings, machined parts, insulation, gears, cams, sleeves, electrical and electronic parts.

Polycarbonate (Lexan) *Properties:* Virtually unbreakable, weather resistant, optically clear, lightweight, self-extinguishing, thermoformable, machinable, solvent cementable. *Typical uses:* High voltage insulation, impact resistant injection moldings, glazing, bulletproof glazing, plumbing fittings. The strongest thermoplastic.

Polyester glass *Properties:* Extremely tough, high dielectric strength, heat resistant, low water absorption, antitracking electrically, self-extinguishing, machinable. *Typical uses:* Insulators and bus braces, switch phase barriers, general electrical insulation, mechanical insulated push rods for switches and breakers, contact blocks, terminal blocks.

Polyethylene *Properties:* Transparent in thin sheets, water resistant. *Typical uses:* Bags for food storage, vapor barriers in construction, trays, rollers, gaskets, seals, radiation shielding.

Polypropylene *Properties:* Good tensile strength, low water absorption, excellent chemical resistance, stress-crack resistant, electrical properties. *Typical uses:* Tanks, ducts, exhaust systems, gaskets, laboratory and hospital ware, wire coating, sporting goods.

Polystyrene *Properties:* Outstanding electrical properties, excellent machinability, ease of fabrication, excellent chemical resistance, oil resistant, clarity, rigidity, hardness, dimensional stability. *Typical uses:* Lighting panels, tote boxes, electronic components, door panels (refrigerators), drip pans, displays, furniture components.

Polysulfone *Properties:* Tough, rigid, high-strength, high-temperature thermoplastic, temperature range –150° to +300°F, excellent electrical characteristics, good chemical resistance, low creep and cold-flow properties, capable of being repeatedly autoclaved. *Typical uses:* Food-processing and medical industries, electrical and electronics, appliance, automotive, aircraft, and aerospace uses.

Polyurethane *Properties:* Elastomeric to rock-hard forms available, high physical characteristics, toughness, durability, broad hardness range, withstands severe use, abrasion resistant, weather resistant, radiation resistant, temperature range—80° to 250°F, resistant to common solvents, available also in foam types. *Typical uses:* Replaces a host of materials that are not performing well; extremely broad range of usage; replaces rubber parts, plastic parts, and some metallic parts.

Polyvinyl chloride (PVC) *Properties:* Corrosion resistant, formable, lightweight, excellent electrical properties, impact resistant, low water absorption, cementable, machinable, weldable. *Typical uses:* Machined parts, nuts, bolts, PVC pipe and fittings, valves, strainers.

RTV silicone rubber *Properties:* Resistant to temperature extremes (–75° to 400°F), excellent electrical characteristics, weather resistant, good chemical resistance, FDA, USDA, and UL approved. *Typical uses:* General-purpose high-quality sealant, gasket cement, food contact surfaces, electrical insulation, bonding agent, glass-tank construction, and countless other applications.

Styrofoam *Properties:* Low water absorption, floats, thermal insulator, extremely lightweight. *Typical uses:* Insulation board for homes and buildings, cups, containers, thermos containers, shock absorbing packaging, plates (food), flotation logs.

Teflon (PTFE) *Properties:* Unexcelled chemical resistance, cryogenic service, electrical insulation, very low friction, high dielectric strength, very low dissipation factor, very high resistivity, machinability. *Typical uses:* Valve components, gaskets (with caution, due to cold flow), pump parts, seal rings, insulators (electrical), terminals, bearings, rollers, bushings, electrical tapes, plumbing tapes, machined parts, bondable with special etchant preparations.

Urea-formaldehyde *Properties:* Hard, strong, molds accurately, low water absorption, excellent electrical properties, ablative, economical, will not support combustion. *Typical uses:* Electrical and electronic parts, insulators, small parts, housings.

3.6.1 General Mechanical Properties of Plastics

Table 3.36 shows the mechanical strength "ranges" of the common plastics. The figures in the table reflect both "filled" and "unfilled" types. The fillers are commonly fiberglass, carbon fibers, mica, etc.

TABLE 3.36 Mechanical Properties of Plastics

Plastic	Tensile strength, lb/in^2	Compressive strength, lb/in^2	Flexural yield, lb/in^2
ABS	3000 9000	6000 22,000	5000 28,000
Acetal	6000 12,000	10,000 19,000	12,000 29,000
Acetate (cellulose)	4400 8000	—	5000 11,000
Epoxy glass	9000 21,000	24,000 41,000	9000 61,000
Epoxy resin	3000 14,000	14,000 26,000	13,000 22,000
Nylon	7000 26,000	12,000 30,000	6000 42,000
Phenolic resins	4000 10,000	10,000 35,000	4000 18,000
Polycarbonate	7000 26,000	8000 22,000	13,000 33,000
Polyester glass	3000 52,000	11,000 52,000	5000 82,000
Polyethylene	500 5600	1500 5600	1500 7200
Polypropylene	2700 15,000	3600 8200	4000 12,000
Polystyrene	1500 22,000	3000 23,000	2500 27,000
Polysulfone	10,000 12,000	13,000 15,000	14,000 16,000
Polyurethane	4000 12,000	19,000 21,000	700 20,000
Polyvinyl chloride (PVC)	5000 8000	7000 14,000	9000 17,000
Rigid Teflon (TFE)	4000 7000	4500 7500	7000 10,000

3.7 ELASTOMERS AND THEIR PROPERTIES

The elastomers consist of elastic, resilient materials such as natural rubber, butyl, neoprene, and polyurethane and silicone compounds. Elastomers are distinguished from other polymer compounds by their ability to return to their original shape after being deformed or stretched beyond their original length. Classification and physical properties of the elastomers are shown in Table 3.37 and 3.38.

Elastomers are rated for degree of hardness on the Shore durometer scales. There are three Shore hardness scales: Shore OO, Shore A, and Shore D. Figure 3.3 shows the relationship between the different scales and gives examples of typical elastomer ratings. This gives the specifier of an elastomer a feel for the hardness relationships between the different elastomers and the three Shore durometer scales. In many applications it is extremely important that the elastomer selected possess the correct hardness for the use specified, and that the correct specifications (Durometer scale and number) be given to the supplier when the elastomer is ordered. This need for accuracy also holds true for the type of elastomer and its chemical composition or physical properties.

Elastomers are used for damping and shock-absorbing purposes, as gaskets and seals, in tool-and-die work, and for numerous other commercial and consumer product applications throughout industry.

TABLE 3.37 Elastomer Properties

Base	Weight, lb/in³	Tensile strength, lb/in²	Percent elongation in 2 in	Durometer	Temperature range, °F
Acrylic	0.039 0.041	1900 2100	400 500	40 90	−40 +350
Butyl	0.032 0.034	2400 2600	450 550	30 100	−65 +250
Fluorocarbon	0.045 0.075	2400 2600	250 350	60 95	−40 +500
Natural rubber	0.032 0.034	3400 3600	550 650	30 100	−65 +212
Synthetic rubber	0.032 0.034	3400 3600	550 650	40 80	−65 +212
Neoprene	0.043 0.045	2900 3100	550 650	40 100	−65 +212
Nitrile	0.035 0.037	2900 3100	550 650	20 90	−65 +300
Silicone	0.035 0.037	1400 1600	650 750	20 90	−120 +600
Urethane	0.038 0.040	4400 4600	600 700	50 100	−65 +212

TABLE 3.38 Hardness Values for Selected Elastomers

Material	Hardness, Shore A durometer
Thermoplastic elastomers	35–100
Urethane, polyether, and polyester	80–100
Styrene butadiene	40–100
Natural rubber	30–100
Chlorinated polyethylene	40–95
Chloroprene	40–95
Chlorosulfonated polyethylene	50–95
Silicone	25–90
Polyacrylate	40–90
Nitrile	20–90
Fluorocarbon	65–90
Ethylene propylene, ethylene propylene diene	30–90
Epichlorohydrin	40–80
Isobutylene-isoprene	40–80
Polybutadiene	45–80
Polysulfide	20–80
Propylene oxide	40–80
Synthetic isoprene	40–80

SOURCE: *1995 Materials Selector Issue, Machine Design,* 12 Dec. 1994, Part 2, p. 14.

3.8 PROPERTIES OF MATERIALS: GENERAL AND SPECIFIC

Table 3.39 lists the properties of various metals and alloys. Table 3.40 lists the properties of other metals and alloys. Table 3.41 lists the coefficients of linear expansion for common materials. Table 3.42 lists the properties of common woods. Table 3.43 lists the chemical symbols for metals. Table 3.44 lists the standard pipe dimensions. Figure 3.3 shows the relationships of the three Shore durometer scales, OO, A, and D.

TABLE 3.39 Properties of Various Metals and Alloys

Material	Relative resistivity* at 20°C	Density g/cm³	Thermal conductivity at 20°C W/cm·°C	Thermal expansion × 10^{-6}/°C, in	Melting, °C
Aluminum	1.54	2.70	2.22	23.6	660
Beryllium	2.3	1.85	1.46	11.6	1277
Bismuth	67.0	9.80	0.084	13.3	271
Brass, yellow	3.7	8.47	1.17	20.3	930
Cadmium	4.3	8.65	0.92	29.8	321
Carbon, graphite	790	2.25	0.24	0.6–4.3	Sublimes
Chromium	7.4	7.19	0.67	6.2	1875
Cobalt	3.6	8.85	0.69	13.8	1495
Columbium (see Niobium)					
Constantan	29.0	8.9	0.21	14.9	1290
Copper, hard drawn	1.03	8.94	3.91	16.8	1083
Gallium	4.7	5.91	0.29	18.0	30
Germanium	2.7×10^{6}	5.33	0.59	5.75	937
Gold	1.36	19.32	2.96	14.2	1063
Inconel, 17–16–8	56.9	8.51	0.15	11.5	1425
Indium	4.9	7.31	0.24	33.0	156
Invar, 64–36	46.0	8.00	0.11	0–2	1425
Iron	5.6	7.87	0.75	11.8	1536
Lead	12.0	11.34	0.35	29.3	327
Magnesium	2.58	1.74	1.53	27.1	650
Mercury	55.6	13.55	0.082	—	–38.9
Molybdenum	3.3	10.22	1.42	4.9	2610
Monel, 67–30	27.9	8.84	0.26	14.0	1325
Nichrome, 80–20	62.5	8.4	0.134	13.0	1400
Nickel	5.5	8.89	0.61	13.3	1440
Niobium	7.2	8.57	0.52	7.31	2468
Palladium	6.3	12.02	0.70	11.8	1552
Phosphor bronze 95–5	6.4	8.86	0.71	17.8	1000
Platinum, 99.9%	6.16	21.45	0.69	8.9	1769
Silicon	10^{11}	2.33	1.25	2.5	1420
Silver	0.922	10.49	4.18	19.7	961
Steel, 0.4–0.5 C	7–12	7.8	0.5	11.0	1480
Steel, stainless 304	42	7.9	0.16	17.0	1430
Steel, stainless 410	33	7.7	0.24	11	1500
Tantalum	7.4	16.6	0.54	6.6	3000
Thorium	8.1	11.6	0.37	12.5	1750
Tin	7.0	7.30	0.63	23	232
Titanium	24.2	4.51	0.41	8.4	1670
Tungsten	3.2	19.3	1.67	4.6	3410
Uranium	17.5	18.7	0.3	6.8–14.1	1132
Zinc	3.5	7.14	1.10	27	420
Zirconium	23	6.5	0.21	5.8	1852

* Standard resistivity of 100% IACS copper at 20°C = 1.7241×10^{-6} Ω·cm.

TABLE 3.40 Other Common Metals and Alloys

Metal	Density g/cm³	Weight, lb/in³	Young's modulus (E), tension, lb/in²	Torsional modulus (*G*), lb/in²	Poisson's ratio	Electrical resistivity, Ω·cm
Aluminum (pure)	2.70	0.098	9×10^6	—	0.33	2.6×10^{-6}
Aluminum alloy (high strength)	2.78	0.101	$10–11 \times 10^6$	—	0.33	2.8×10^{-6}
Antimony	6.69	0.242	11.3×10^6	2.9×10^6	—	3.1×10^{-6}
Beryllium copper (C170) alloy	8.4	0.303	19×10^6	—	0.29	3.1×10^{-6}
Bismuth	9.75	0.352	—	—	—	119×10^{-6}
Brass (80Cu/20Zn)	8.6	0.311	16×10^6	6×10^6	0.34	7×10^{-6}
Cadmium	8.65	0.312	10.1×10^6	3.5×10^6	—	7.5×-6
Cast iron (gray)	7.2	0.260	14×10^6	5.6×10^6	0.21	—
Chromium	7.19	0.260	—	—	—	2.6×10^{-6}
Copper	8.96	0.324	17×10^6	6.4×10^6	0.34	1.72×10^{-6}
Gold	19.32	0.698	11.4×10^6	—	—	2.44×10^{-6}
Iron	7.87	0.284	28×10^6	11.2×10^6	0.30	10×10^{-6}
Iron (malleable)	7.85	0.284	25×10^6	11.5×10^6	0.17	—
Lead	11.35	0.410	2.4×10^6	—	0.43	22×10^{-6}
Lithium	0.53	0.019	—	—	—	9×10^{-6}
Magnesium (cast)	1.74	0.063	6.5×10^6	2.4×10^6	0.35	44×10^{-6}
Mercury	13.55	0.490	—	—	—	96×10^{-6}
Molybdenum	10.22	0.369	—	—	—	5.7×10^{-6}
Nickel	8.90	0.322	30×10^6	10.6×10^6	0.32	7.8×10^{-6}
Palladium	12.02	0.434	—	—	—	11×10^{-6}
Phosphor bronze	8.90	0.322	16×10^6	6×10^6	0.35	10×10^{-6}
Platinum	21.45	0.775	24.2×10^6	9.3×10^6	—	10×10^{-6}
Silver	10.50	0.379	11.2×10^6	3.8×10^6	—	1.6×10^{-6}
Steel (medium) carbon)	7.80	0.282	30×10^6	11.4×10^6	0.30	15×10^{-6}
Steel (stainless 300 type)	8.03	0.290	28×10^6	11.4×10^6	0.28	30×10^{-6}
Steel (stainless 400 type)	7.75	0.280	29×10^6	12.6×10^6	0.28	30×10^{-6}
Tin	7.31	0.264	7×10^6	2.4×10^6	—	11.5×10^{-6}
Titanium	4.54	0.164	16×10^6	—	—	—
Tungsten	19.30	0.697	51.5×10^6	21.5×10^6	—	5.51×10^{-6}
Zinc	7.13	0.258	14.5×10^6	5×10^6	0.11	6×10^{-6}

TABLE 3.41 Coefficients of Linear Expansion for Common Materials

Metal, alloy, or other material	Linear expansion	
	in per 1°F	in per 1°C
Aluminum, wrought	0.0000128	0.0000231
Brass	0.0000104	0.0000188
Bronze	0.0000101	0.0000181
Copper	0.0000093	0.0000168
Cast iron, gray	0.0000059	0.0000106
Wrought iron	0.0000067	0.0000120
Lead	0.0000159	0.0000286
Magnesium alloy	0.0000160	0.0000290
Nickel	0.0000070	0.0000126
Cast steel	0.0000061	0.0000110
Hard steel	0.0000073	0.0000132
Medium steel	0.0000067	0.0000120
Soft steel	0.0000061	0.0000110
Stainless steel	0.0000099	0.0000178
Zinc, rolled	0.0000173	0.0000263
Concrete	0.0000079	0.0000143
Granite	0.0000047	0.0000084
Marble	0.0000056	0.0000100
Plaster	0.0000092	0.0000166
Slate	0.0000058	0.0000104
Fir	0.0000021	0.0000037
Maple	0.0000036	0.0000064
Oak	0.0000027	0.0000049
Pine	0.0000030	0.0000054
Plate glass	0.0000050	0.0000089
Hard rubber	0.0000044	0.0000080
Porcelain	0.0000009	0.0000016
Silver	0.0000104	0.0000188
Tin	0.0000148	0.0000269
Tungsten	0.0000024	0.0000043

TABLE 3.42 Properties of Common Woods

Name	Density, g/cm^3	Density, lb/ft^3	Modulus of elasticity, $lb/in^2 \times 10^6$
Ash, white	0.64	39.8	1.77
Birch, paper	0.60	37.5	1.59
Cedar, eastern	0.49	30.7	0.87
Cherry, black	0.53	33.3	1.49
Cyprus	0.48	30.1	1.44
Ebony, African	0.77	48.0	1.43
Eucalyptus, Australian	0.83	51.8	2.64
Fir, Douglas	0.51	32.0	1.93
Gum, black	0.55	34.5	1.19
Hemlock, eastern	0.43	27.0	1.20
Hickory, shagbark	0.84	52.2	2.17
Ironwood, black	1.01	67.3	2.99
Locust, black	0.71	44.2	2.06
Mahogany, African	0.67	41.7	1.53
Maple, sugar	0.68	42.2	1.83
Oak, black	0.67	41.8	1.64
Oak, red	0.66	41.1	1.81
Oak, white	0.71	44.3	1.78
Pine, longleaf	0.64	39.8	2.06
Poplar, yellow	0.43	26.7	1.50
Redwood	0.44	27.2	1.36
Spruce, white	0.43	27.0	1.42
Teak, Indian	0.58	36.3	1.70
Walnut, black	0.56	35.1	1.69

TABLE 3.43 Chemical Symbols for Metals

Metal	Symbol	Metal	Symbol
Aluminum	Al	Manganese	Mn
Antimony	Sb	Mercury	Hg
Beryllium	Be	Molybdenum	Mo
Bismuth	Bi	Nickel	Ni
Boron	B	Platinum	Pt
Cadmium	Cd	Selenium	Se
Carbon	C	Silicon	Si
Chromium	Cr	Silver	Ag
Cobalt	Co	Tellurium	Te
Copper	Cu	Tin	Sn
Gold	Au	Titanium	Ti
Iridium	Ir	Tungsten	W
Iron	Fe	Vanadium	V
Lead	Pb	Zinc	Zn
Lithium	Li	Zirconium	Zr
Magnesium	Mg		

TABLE 3.44 Standard Pipe Sizes

To determine the weight per linear foot of a pipe, use the relationship

$$W = \pi\left(\frac{OD}{2}\right)^2 - \pi\left(\frac{ID}{2}\right)^2 \times 12 \times \text{density in } \frac{\text{lbs}}{\text{in}^3}$$

Nominal pipe size	OD, in	ASA pipe schedules (wall thickness, in) 40	60	80	100	120	160
⅛	0.405	0.068		0.095			
¼	0.540	0.088		0.119			
⅜	0.675	0.091		0.126			
½	0.840	0.109		0.147			
¾	1.050	0.113		0.154			
1	1.315	0.133		0.179			
1¼	1.660	0.140		0.191			
1½	1.900	0.145		0.200			
2	2.375	0.154		0.218			
2½	2.875	0.203		0.276			
3	3.500	0.216		0.300			
3½	4.000	0.226		0.318			
4	4.500	0.237		0.337		0.438	0.531
5	5.563	0.258		0.375		0.500	0.625
6	6.625	0.280		0.432		0.562	0.719
8	8.625	0.322	0.406	0.500	0.594	0.719	0.906
10	10.750	0.365	0.500	0.594	0.719	0.844	1.125
12	12.750	0.406	0.562	0.688	0.844	1.000	1.312

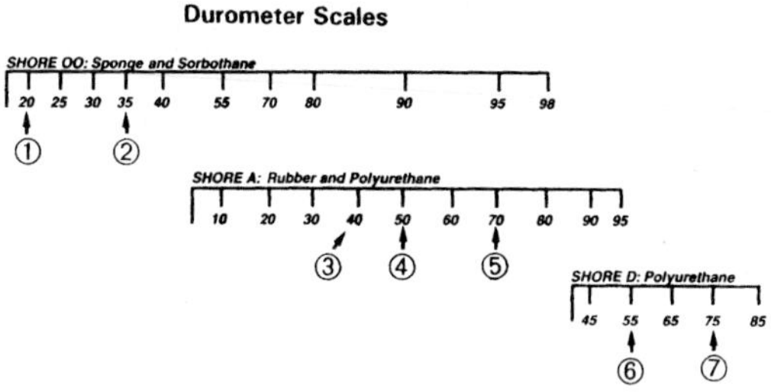

Durometer is the International standard for measuring the hardness of rubber, sponge rubber, plastic, and other nonmetallic materials. Very soft materials, such as gels, foam and sponge are rated on the "Shore OO" scale. Rubber, soft plastic, polyurethane, leather and felt are rated on the "Shore A" scale. Hard materials such as hard rubber, rigid PVC, nylon, acrylic, polyurethane, and ABS are rated on the "Shore D" scale. There is some overlap on the Shore scales. Example: A rubber rated at 30 on the Shore A is also rated at 80 on the Shore OO; a rubber rated at 45 on the Shore D is also rated at 90 on the Shore A.

Hardness examples from the Shore scales: ① = chewing gum or soft gum; ② = racquet ball; ③ = rubber band; ④ = tire tread; ⑤ = shoe heel; ⑥ = bowling ball; ⑦ = table top.

FIGURE 3.3 Relationship between the Shore durometer scales OO, A, and D for elastomers

CHAPTER 4
THREAD SYSTEMS

4.1 THREAD SYSTEMS: AMERICAN STANDARD AND METRIC (60° V)

The international standard screw threads consist of the unified inch series and the metric series. The metric series is standardized into the M and MJ profiles. The unified series is designated as UN (unified national). Another unified profile is designated as UNR, which has a rounded root on the external thread. The metric profile MJ also has a rounded root on the external thread and a larger minor diameter of both the internal and external threads. Both the UNR and MJ profiles are used for applications requiring high fatigue strength; they are also employed in aerospace applications.

A constant-pitch unified series is also standardized and consists of 4, 6, 8, 12, 16, 20, 28, and 32 threads per inch. These are used for sizes over 1 in diameter, and the 8UN, 12UN, and 16UN are the *preferred* pitches. Figure 4.1 shows that both the unified and metric M series use the same profile geometry. The other common screw thread systems are shown in Fig. 4.5(a) through (k) of this chapter and include the Acme, stub Acme, Whitworth, buttress, and others.

The unified threads are further classified as UN A for external threads and UN B for internal threads. The UN series contains three fit classes, 1A, 2A, and 3A for external threads and 1B, 2B, and 3B for internal threads. A typical engineering drawing call-out for a coarse external class 2 thread in ¼–20 size would be

0.250–20UNC–2A

A ¼–20 size class 2 tapped hole would be shown as

0.250–20UNC–2B

Thread Fit Classes

Unified thread series and interference-fit threads.

Unified Thread Series

1A External, loose fit for easy assembly and noncritical uses

2A External, general applications where plating may be applied

3A External, tight fit used for great accuracy; no plating allowance provided

1B Internal, loose fit for easy assembly and noncritical uses

2B Internal, general applications where plating may be applied

3B Internal, tight fit used for great accuracy; no plating allowance provided

Class 5 Interference-Fit Threads

Class 5 External

NC5 HF For driving in hard ferrous materials over 160 BHN

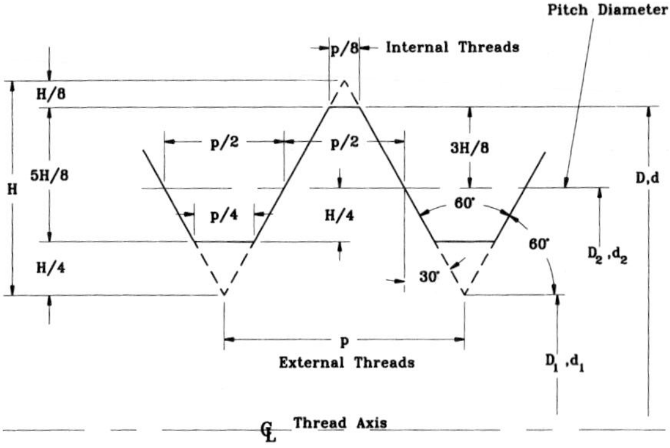

Basic Thread Profile for Unified (UN) and Metric (M) Threads (ISO 68)

D, (d) = basic major diameter of internal (external) thread
D_1, (d_1) = basic minor diameter of internal (external) thread
D_2, (d_2) = basic pitch diameter of internal (external) thread
p = pitch
$H = 0.5\sqrt{3}\ p$

FIGURE 4.1 Basic thread profiles (unified and metric).

NC5 CSF For driving in copper alloys and soft ferrous materials under 160 BHN

NC5 ONF For other nonferrous materials, any hardness

Class 5 Internal

NC5 IF Entire ferrous material range

NC5 INF Entire nonferrous material range

Interference-fit threads are commonly used on threaded studs to ensure a tight, vibration-resistant fit. The internal thread of a class 5 interference-fit application should be lubricated for best results when torquing the stud or externally threaded part into the class 5 fit tapped hole.

Unified and Metric Thread Data

Unified National Tap Drill Sizes. You may calculate the tap drill diameter for 75 and 100 percent thread for the unified and metric M profile series by using the following equations. For unified national threads,

$$D_{\mathrm{td}} = D_{\mathrm{m}} - \frac{0.947}{n} \quad \text{(Tap drill diameter for 75 percent thread)}$$

$$D_{\mathrm{td}} = D_{\mathrm{m}} - \frac{1.299}{n} \quad \text{(Tap drill diameter for 100 percent thread)}$$

For metric threads, M profile,

$$D_{\mathrm{d}} = D_{\mathrm{M}} - \frac{(\%T)(p)}{76.98}$$

$$\%T = \frac{76.98}{p}(D_{\mathrm{M}} - D_{\mathrm{d}})$$

In the preceding equations, symbolism is as follows. For unified threads,

D_{td} = diameter of tap drill, in

D_{m} = major diameter of external thread or tap diameter, in

n = number of threads per inch

For metric threads,

D_d = drilled hole diameter, mm

D_M = basic major diameter, mm

$\%T$ = percent of thread (i.e., 70 percent = 0.70, etc.)

p = pitch of thread, mm

Note: Recommended percent of thread is usually taken as 70 to 75 percent for a general class 2 type fit, which allows for electroplatings such as zinc, cadmium, chrome, nickel, etc.

Unified national coarse (UNC) screw thread data are presented in Fig. 4.2. Unified national fine (UNF) screw thread data are presented in Fig. 4.3. Metric M profile thread data are presented in Fig. 4.4. American national drill sizes, and metric sizes are shown in Chapter 6, under Drilling. Both number and letter size drills are indicated with their decimal equivalents.

The standard limits of size for both American and unified national series screw threads are fully covered in Table 4.1. The sections of the table are extracted from National Bureau of Standards Handbook H-28 and are in general agreement with current ANSI standards. The table covers or includes sizes 0–80 through 1½–18.

Calculating the Pitch Diameter of Unified and Metric Threads

It is often necessary to find the pitch diameter of various unified (UN) and metric (M) thread sizes. This would be necessary for threads that are not listed in the preceding tables of thread sizes and where the thread is

Screw Thread Data, Unified National Coarse (UNC)

Thread	Tap Drill	Decimal, in.	Stress Area Sq. inches	Basic Pitch Diameter, inches
#1-64	#53	0.0595	0.0026	0.0629
#2-56	#50	0.0700	0.0037	0.0744
#3-48	#47	0.0785	0.0048	0.0855
#4-40	#43	0.0890	0.0060	0.0958
#5-40	#38	0.1015	0.0080	0.1088
#6-32	#36	0.1065	0.0090	0.1177
#8-32	#29	0.1360	0.0140	0.1437
#10-24	#25	0.1495	0.0175	0.1629
¼-20	#7	0.2010	0.0318	0.2175
5/16-18	F	0.2570	0.0524	0.2764
3/8-16	5/16	0.3125	0.0775	0.3344
7/16-14	T	0.3580	0.1063	0.3911
½-13	27/64	0.4219	0.1419	0.4500
9/16-12	31/64	0.4844	0.1820	0.5084
5/8-11	17/32	0.5312	0.2260	0.5660
¾-10	41/64	0.6406	0.3340	0.6850
7/8-9	49/64	0.7656	0.4620	0.8028
1-8	7/8	0.8750	0.6060	0.9188

FIGURE 4.2 Screw thread data (UNC).

Screw Thread Data, Unified National Fine (UNF)

Thread	Tap Drill	Decimal, in.	Stress Area, Sq. inches	Basic Pitch Diameter, inches
#0-80	3/64	0.0469	0.0018	0.0519
#1-72	#53	0.0595	0.0027	0.0640
#2-64	#50	0.0700	0.0039	0.0759
#3-56	#45	0.0820	0.0052	0.0874
#4-48	#42	0.0935	0.0066	0.0985
#5-44	#37	0.1040	0.0083	0.1102
#6-40	#33	0.1130	0.0102	0.1218
#8-36	#29	0.1360	0.0147	0.1460
#10-32	#21	0.1590	0.0200	0.1697
¼-28	#3	0.2130	0.0364	0.22268
5/16-24	I	0.2720	0.0580	0.2854
3/8-24	Q	0.3320	0.0878	0.3479
7/16-20	25/64	0.3906	0.1187	0.4050
½-20	29/64	0.4531	0.1599	0.4675
9/16-18	33/64	0.5156	0.2030	0.5264
5/8-18	9/16	0.5625	0.2560	0.5889
¾-16	11/16	0.6875	0.3730	0.7094
7/8-14	13/16	0.8125	0.5090	0.8286
1-12	29/32	0.9063	0.6630	0.9459

FIGURE 4.3 Screw thread data (UNF).

Metric Thread Data, M Profile, Internal and External

Thread Designation Dia. X Pitch (Millimeters)	Tap Drill, mm	Pitch Dia. 6H, Internal, mm	Pitch Dia. 6G, External, mm
M1.6 x 0.35	1.25	1.373	1.291
M2 x 0.4	1.60	1.740	1.654
M2.5 x 0.45	2.05	2.208	2.117
M3 x 0.5	2.50	2.675	2.580
M3.5 x 0.6	2.90	3.110	3.004
M4 x 0.7	3.30	3.545	3.433
M5 x 0.8	4.20	4.480	4.361
M6 x 1	5.00	5.350	5.212
M8 x 1.25	6.70	7.188	7.042
M8 x 1	7.00	7.350	7.212
M10 x 1.5	8.50	9.026	8.862
M10 x 1.25	8.70	9.188	9.042
M10 x 0.75	------	9.513	9.391
M12 x 1.75	10.20	10.863	10.679
M12 x 1.5	------	11.026	10.854
M12 x 1.25	10.80	11.188	11.028
M12 x 1	------	11.350	11.206
M14 x 2	12.00	12.701	12.503
M14 x 1.5	12.50	13.026	12.854
M15 x 1	------	14.350	14.206
M16 x 2	14.00	14.701	14.503
M16 x 1.5	14.50	15.026	14.854
M17 x 1	------	16.350	16.206
M18 x 1.5	16.50	17.026	16.854
M20 x 2.5	17.50	18.376	18.164
M20 x 1.5	18.50	19.026	18.854
M20 x 1	------	19.350	19.206
M22 x 2.5	19.50	20.376	20.164
M22 x 1.5	20.50	21.026	20.854
M24 x 3	21.00	22.051	21.803
M24 x 2	22.00	22.701	22.493
M25 x 1.5	------	24.026	23.854

FIGURE 4.4 Metric thread data.

larger than that normally listed in handbooks. These would include threads on large bolts and threads on jack screws and lead screws used on various machinery or machine tools. To calculate the pitch diameters, refer to Fig. 4.1.

$$H = 0.5\sqrt{3} \cdot p = 0.866025p$$

where p = pitch of the thread. In the unified system this is equal to the reciprocal of the number of threads per in (i.e., for a ⅜-16 thread the pitch would be $1/16 = 0.0625$ in). For the metric system the pitch is given in millimeters on the thread listing (i.e., on an M12 × 1.5 metric thread, the pitch would be 1.5 mm or 1.5×0.03937 in $= 0.059055$ in).

d = basic diameter of the external thread (i.e., ⅜-16 would be 0.375 in; a #8-32 would be 0.164 in, etc.).

Example: Find the pitch diameter of a 0.375-16 UNC-3A thread:
Using Fig. 4.1,

where d = basic outside diameter of the thread $= 0.375$ in

$$H = 0.866025 \times \text{p} = 0.866025 \times 0.0625 = 0.054127 \text{ in (for this case only)}$$

We would next perform the following:

$$\text{Pitch dia} = [d/2 - (5H/8) + (H/4)] \times 2$$

$$= [.375/2 - (5 \times 0.054127/8) + (0.054127/4)] \times 2$$

$$= (.1875 - 0.033829 + 0.013532) \times 2$$

$$= 0.3344 \text{ in pitch dia for a } 3/8\text{-16 UNC-3A thread.}$$

If you check the basic pitch diameter for this thread in a table of pitch diameters, you will find that this is the correct answer when the thread is a class 3A and the pitch diameter is maximum. This may be verified by checking Table 4.1. Thus, you may calculate any pitch diameter for the different classes of fits on any UN or metric M profile thread, since the thread geometry is shown in Fig. 4.1. Pitch diameters for other classes or types of thread systems may be calculated when you know the basic thread geometry, as in this case for the (UN) and metric (M) thread systems.

4.2 INTERNATIONAL THREAD SYSTEMS

The various thread systems used worldwide are shown in Fig. 4.5(a) through (k), and include: ISO-M and UN; UNJ (controlled root radii); Whitworth (BSW); American Buttress (7° face); NPT (American National Pipe Thread); BSPT (British Standard Pipe Thread); Acme (29°); Acme (Stub 29°); API (Taper 1:6); TR DIN 103; and RD DIN 405 (Round).

The Fastener Quality Act

The Fastener Quality Act Defined. In 1988 bill HR3000 was introduced in Congress to deter the sale of substandard or counterfeit fasteners by requiring that the appropriate items be inspected and tested by an accredited laboratory prior to sale. Signed into law by President Bush in 1990, it is now known as Public Law 101-592, or the Fastener Quality Act (FQA).

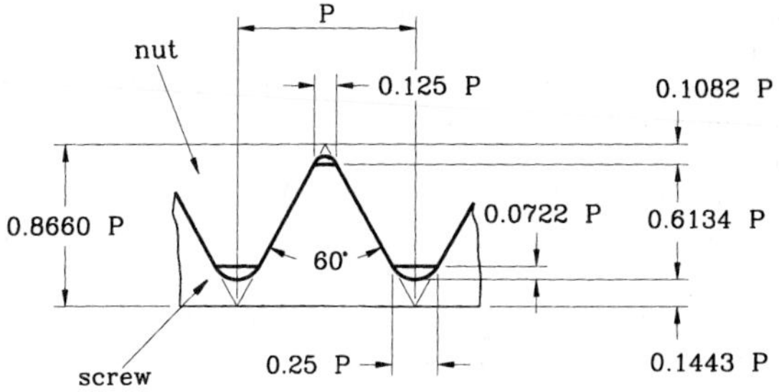

(a) ISO - M (Metric)
(UN) (Unified National)

FIGURE 4.5(a)

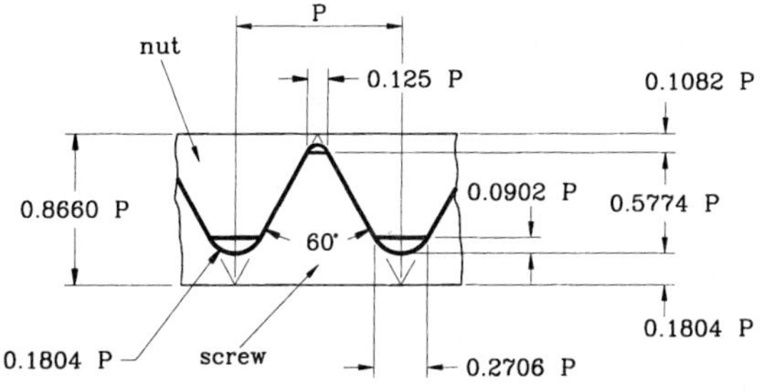

(b) UNJ
(Controlled root radii)

FIGURE 4.5(b)

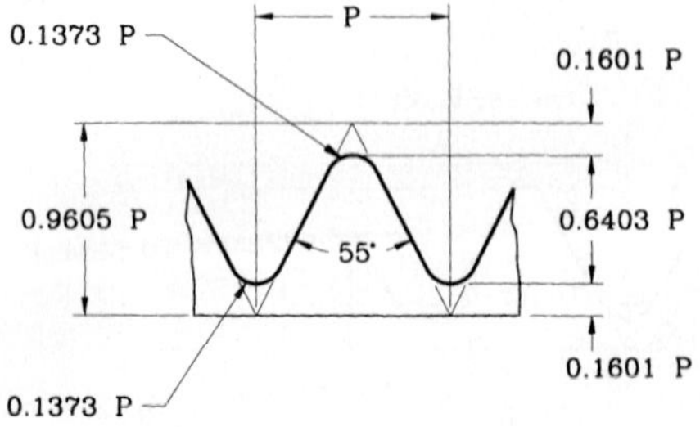

(c) Whitworth (BSW)

FIGURE 4.5(c)

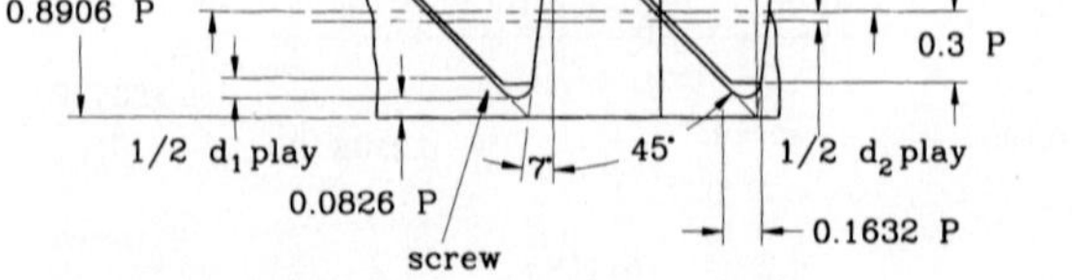

(d) American Buttress (7° face)

FIGURE 4.5(d)

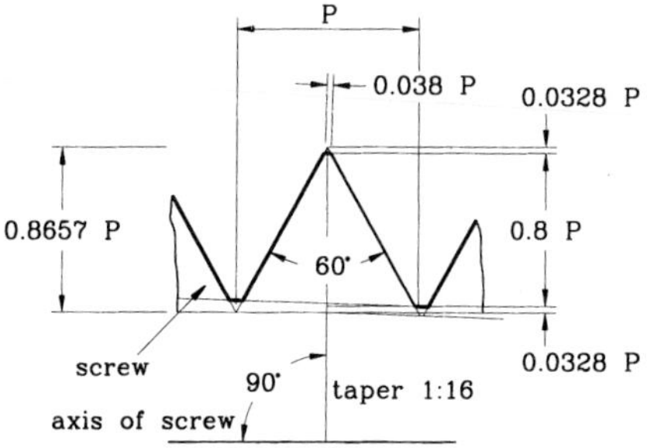

(e) NPT
(American National Pipe Thread)

FIGURE 4.5(e)

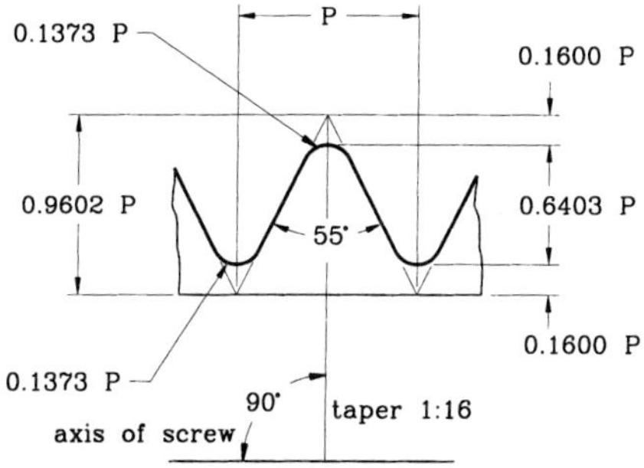

(f) BSPT
(British Standard Pipe Thread)

FIGURE 4.5(f)

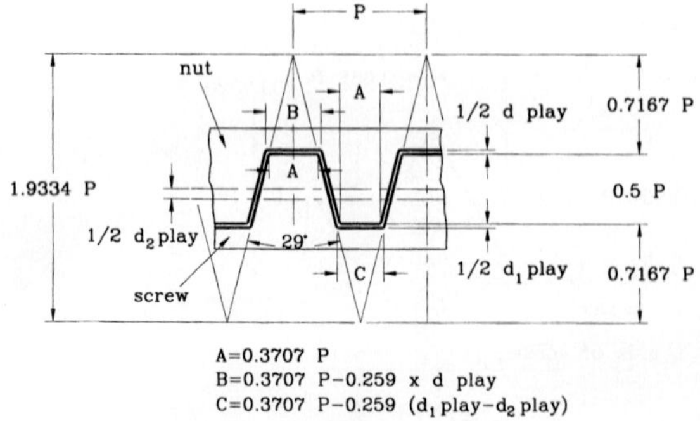

(g) Acme (29°)

FIGURE 4.5(g)

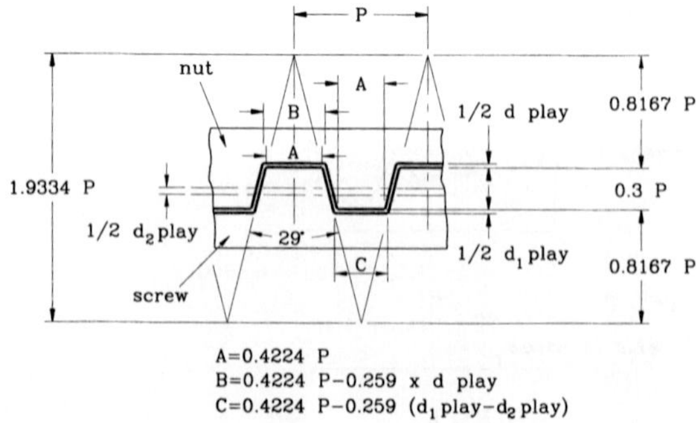

(h) Acme Stub (29°)

FIGURE 4.5(h)

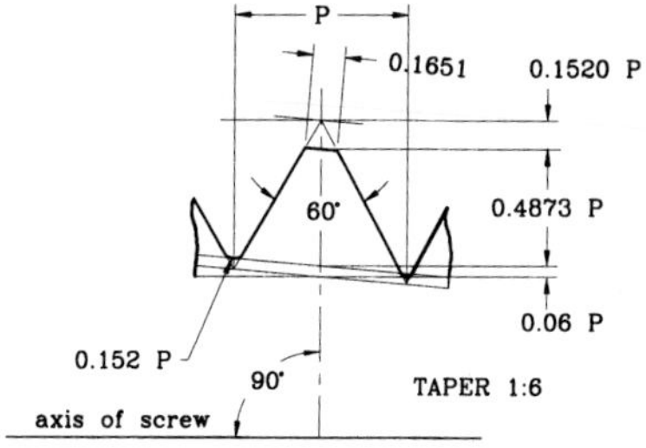

(i) API
Taper: 1:6 (V-0.38"R)

FIGURE 4.5(i)

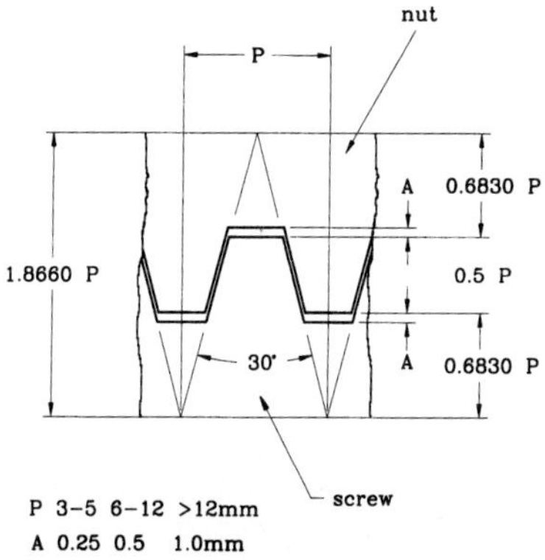

(j) TR DIN 103

FIGURE 4.5(j)

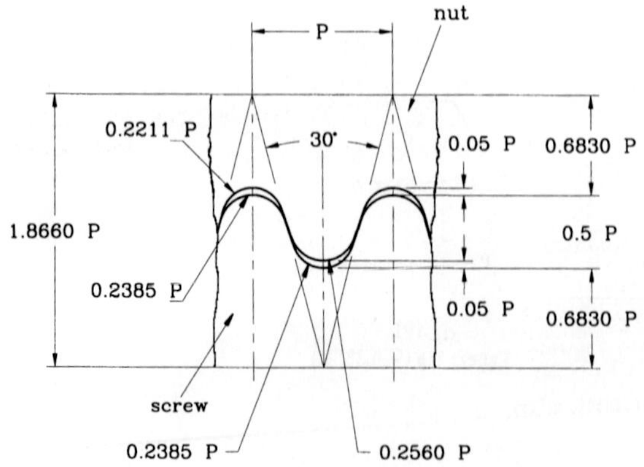

(k) RD DIN 405 (Round)

FIGURE 4.5(k)

The law was scheduled to take effect on May 25, 1997, but provisions allowed that date to be pushed back if the Secretary of Commerce determined that insufficient accredited laboratories exist to fulfill the requirements of the law.

Application. The Fastener Quality Act applies to any nut, bolt, screw, or stud 0.25 in or 5 mm in dia and larger which is through-hardened or is required to have a grade designation marking on the fastener. This includes the following specific commodities:

Inch series externally threaded products:

- Grades 5, 5.1, 5.2, 7, 8, 8.1, 8.2 per SAE J429, any commodity
- All standard products that accept a hex ("allen") key, including socket head cap screws, button socket heads, flat socket heads, socket set screws, and socket shoulder screws (stripper bolts)

Inch series internally threaded products:

- Nuts, grades 5, 8 per SAE J995
- Nuts, grades C, C3, D, DH, DH3 per ASTM A563

4.3 TABLE OF U.S. CUSTOMARY THREADS (SEE TABLE 4.1)

TABLE 4.1 Standard Limits of Size: Unified and American Screw Threads (0–80 through 1½–18)

Nominal size and threads per inch	Series designation	External									Internal						
		Class	Allowance, in	Major diameter limits, in			Pitch diameter limits, in			Minor diameter, in	Class	Minor diameter limits, in		Pitch diameter limits, in			Major diameter, in
				Max*	Min	Min[†]	Max*	Min	Tolerance			Min	Max	Min	Max	Tolerance	Min
No. 0–80 to ½–13																	
0–80	NF	2A	0.0005	0.0595	0.0563		0.0514	0.0496	0.0018	0.0442	2B	0.0465	0.0514	0.0519	0.0542	0.0023	0.0600
		3A	0.0000	0.0600	0.0568		0.0519	0.0506	0.0013	0.0447	3B	0.0465	0.0514	0.0519	0.0536	0.0017	0.0600
1–64	NC	2A	0.0006	0.0724	0.0686		0.0623	0.0603	0.0020	0.0532	2B	0.0561	0.0623	0.0629	0.0655	0.0026	0.0730
		3A	0.0000	0.0730	0.0692		0.0629	0.0614	0.0015	0.0538	3B	0.0561	0.0623	0.0629	0.0648	0.0019	0.0730
1–72	NF	2A	0.0006	0.0724	0.0689		0.0634	0.0615	0.0019	0.0554	2B	0.0580	0.0635	0.0640	0.0665	0.0025	0.0730
		3A	0.0000	0.0730	0.0695		0.0640	0.0626	0.0014	0.0560	3B	0.0580	0.0635	0.0640	0.0659	0.0019	0.0730
2–56	NC	2A	0.0006	0.0854	0.0813		0.0738	0.0717	0.0021	0.0635	2B	0.0667	0.0737	0.0744	0.0772	0.0028	0.0860
		3A	0.0000	0.0860	0.0819		0.0744	0.0728	0.0016	0.0641	3B	0.0667	0.0737	0.0744	0.0765	0.0021	0.0860
2–64	NF	2A	0.0006	0.0854	0.0816		0.0753	0.0733	0.0020	0.0662	2B	0.0691	0.0753	0.0759	0.0786	0.0027	0.0860
		3A	0.0000	0.0860	0.0822		0.0759	0.0744	0.0015	0.0668	3B	0.0691	0.0753	0.0759	0.0779	0.0020	0.0860
3–48	NC	2A	0.0007	0.0983	0.0938		0.0848	0.0825	0.0023	0.0727	2B	0.0764	0.0845	0.0855	0.0885	0.0030	0.0990
		3A	0.0000	0.0990	0.0945		0.0855	0.0838	0.0017	0.0734	3B	0.0764	0.0845	0.0855	0.0877	0.0022	0.0990
3–56	NF	2A	0.0007	0.0983	0.0942		0.0867	0.0845	0.0022	0.0764	2B	0.0797	0.0865	0.0874	0.0902	0.0028	0.0990
		3A	0.0000	0.0990	0.0949		0.0874	0.0858	0.0016	0.0771	3B	0.0797	0.0865	0.0874	0.0895	0.0021	0.0990
4–40	NC	2A	0.0008	0.1112	0.1061		0.0950	0.0925	0.0025	0.0805	2B	0.0849	0.0939	0.0958	0.0991	0.0033	0.1120
		3A	0.0000	0.1120	0.1069		0.0958	0.0939	0.0019	0.0813	3B	0.0849	0.0939	0.0958	0.0982	0.0024	0.1120
4–48	NF	2A	0.0007	0.1113	0.1068		0.0978	0.0954	0.0024	0.0857	2B	0.0894	0.0968	0.0985	0.1016	0.0031	0.1120
		3A	0.0000	0.1120	0.1075		0.0985	0.0967	0.0018	0.0864	3B	0.0894	0.0968	0.0985	0.1008	0.0023	0.1120
5–40	NC	2A	0.0008	0.1242	0.1191		0.1080	0.1054	0.0026	0.0935	2B	0.0979	0.1062	0.1088	0.1121	0.0033	0.1250
		3A	0.0000	0.1250	0.1199		0.1088	0.1069	0.0019	0.0943	3B	0.0979	0.1062	0.1088	0.1113	0.0025	0.1250
5–44	NF	2A	0.0007	0.1243	0.1195		0.1095	0.1070	0.0025	0.0964	2B	0.1004	0.1079	0.1102	0.1134	0.0032	0.1250
		3A	0.0000	0.1250	0.1202		0.1102	0.1083	0.0019	0.0971	3B	0.1004	0.1079	0.1102	0.1126	0.0024	0.1250
6–32	NC	2A	0.0008	0.1372	0.1312		0.1169	0.1141	0.0028	0.0989	2B	0.104	0.114	0.1177	0.1214	0.0037	0.1380
		3A	0.0000	0.1380	0.1320		0.1177	0.1156	0.0021	0.0997	3B	0.1040	0.1140	0.1177	0.1204	0.0027	0.1380
6–40	NF	2A	0.0008	0.1372	0.1321		0.1210	0.1184	0.0026	0.1065	2B	0.111	0.119	0.1218	0.1252	0.0034	0.1380
		3A	0.0000	0.1380	0.1329		0.1218	0.1198	0.0020	0.1073	3B	0.1110	0.1186	0.1218	0.1243	0.0025	0.1380

8–32	NC	2A	0.0009	0.1631	0.1571		0.1428	0.1399	0.0029	0.1248	2B	0.130	0.139	0.1437	0.1475	0.0038	0.1640
		3A	0.0000	0.1640	0.1580		0.1437	0.1415	0.0022	0.1257	3B	0.1300	0.1389	0.1437	0.1465	0.0028	0.1640
8–36	NF	2A	0.0008	0.1632	0.1577		0.1452	0.1424	0.0028	0.1291	2B	0.134	0.142	0.1460	0.1496	0.0036	0.1640
		3A	0.0000	0.1640	0.1585		0.1460	0.1439	0.0021	0.1299	3B	0.1340	0.1416	0.1460	0.1487	0.0027	0.1640
10–24	NC	2A	0.0010	0.1890	0.1818		0.1619	0.1586	0.0033	0.1379	2B	0.145	0.156	0.1629	0.1672	0.0043	0.1900
		3A	0.0000	0.1900	0.1828		0.1629	0.1604	0.0025	0.1389	3B	0.1450	0.1555	0.1629	0.1661	0.0032	0.1900
10–32	NF	2A	0.0009	0.1891	0.1831		0.1688	0.1658	0.0030	0.1508	2B	0.156	0.164	0.1697	0.1736	0.0039	0.1900
		3A	0.0000	0.1900	0.1840		0.1697	0.1674	0.0023	0.1517	3B	0.1560	0.1641	0.1697	0.1726	0.0029	0.1900
12–24	NC	2A	0.0010	0.2150	0.2078		0.1879	0.1845	0.0034	0.1639	2B	0.171	0.181	0.1889	0.1933	0.0044	0.2160
		3A	0.0000	0.2160	0.2088		0.1889	0.1863	0.0026	0.1649	3B	0.1710	0.1807	0.1889	0.1922	0.0033	0.2160
12–28	NF	2A	0.0010	0.2150	0.2085		0.1918	0.1886	0.0032	0.1712	2B	0.177	0.186	0.1928	0.1970	0.0042	0.2160
		3A	0.0000	0.2160	0.2095		0.1928	0.1904	0.0024	0.1722	3B	0.1770	0.1857	0.1928	0.1959	0.0031	0.2160
12–32	NEF	2A	0.0009	0.2151	0.2091		0.1948	0.1917	0.0031	0.1768	2B	0.182	0.190	0.1957	0.1998	0.0041	0.2160
		3A	0.0000	0.2160	0.2100		0.1957	0.1933	0.0024	0.1777	3B	0.1820	0.1895	0.1957	0.1988	0.0031	0.2160
¼–20	UNC	1A	0.0011	0.2489	0.2367		0.2164	0.2108	0.0056	0.1876	1B	0.196	0.207	0.2175	0.2248	0.0073	0.2500
		2A	0.0011	0.2489	0.2408	0.2367	0.2164	0.2127	0.0037	0.1876	2B	0.196	0.207	0.2175	0.2223	0.0048	0.2500
		3A	0.0000	0.2500	0.2419		0.2175	0.2147	0.0028	0.1887	3B	0.1960	0.2067	0.2175	0.2211	0.0036	0.2500
¼–28	UNF	1A	0.0010	0.2490	0.2392		0.2258	0.2208	0.0050	0.2052	1B	0.211	0.220	0.2268	0.2333	0.0065	0.2500
		2A	0.0010	0.2490	0.2425		0.2258	0.2225	0.0033	0.2052	2B	0.211	0.220	0.2268	0.2311	0.0043	0.2500
		3A	0.0000	0.2500	0.2435		0.2268	0.2243	0.0025	0.2062	3B	0.2110	0.2190	0.2268	0.2300	0.0032	0.2500
¼–32	NEF	2A	0.0010	0.2490	0.2430		0.2287	0.2255	0.0032	0.2107	2B	0.216	0.224	0.2297	0.2339	0.0042	0.2500
		3A	0.0000	0.2500	0.2440		0.2297	0.2273	0.0024	0.2117	3B	0.2160	0.2229	0.2297	0.2328	0.0031	0.2500
5/16–18	UNC	1A	0.0012	0.3113	0.2982		0.2752	0.2691	0.0061	0.2431	1B	0.252	0.265	0.2764	0.2843	0.0079	0.3125
		2A	0.0012	0.3113	0.3026	0.2982	0.2752	0.2712	0.0040	0.2431	2B	0.252	0.265	0.2764	0.2817	0.0053	0.3125
		3A	0.0000	0.3125	0.3038		0.2764	0.2734	0.0030	0.2443	3B	0.2520	0.2630	0.2764	0.2803	0.0039	0.3125
5/16–24	UNF	1A	0.0011	0.3114	0.3006		0.2843	0.2788	0.0055	0.2603	1B	0.267	0.277	0.2854	0.2925	0.0071	0.3125
		2A	0.0011	0.3114	0.3042		0.2843	0.2806	0.0037	0.2603	2B	0.267	0.277	0.2854	0.2902	0.0048	0.3125
		3A	0.0000	0.3125	0.3053		0.2854	0.2827	0.0027	0.2614	3B	0.2670	0.2754	0.2854	0.2890	0.0036	0.3125

TABLE 4.1 Standard Limits of Size: Unified and American Screw Threads (*Continued*)

Nominal size and threads per inch	Series designation	External									Internal						
		Class	Allowance, in	Major diameter limits, in			Pitch diameter limits, in			Minor diameter, in	Class	Minor diameter limits, in		Pitch diameter limits, in			Major diameter, in
				Max*	Min	Min†	Max*	Min	Tolerance			Min	Max	Min	Max	Tolerance	Min
								No. 0–80 to ½–13									
5/16–32	NEF	2A	0.0010	0.3115	0.3055		0.2912	0.2880	0.0032	0.2732	2B	0.279	0.286	0.2922	0.2964	0.0042	0.3125
		3A	0.0000	0.3125	0.3065		0.2922	0.2898	0.0024	0.2742	3B	0.2790	0.2847	0.2922	0.2953	0.0031	0.3125
		1A	0.0013	0.3737	0.3595		0.3331	0.3266	0.0065	0.2970	1B	0.307	0.321	0.3344	0.3429	0.0085	0.3750
3/8–16	UNC	2A	0.0013	0.3737	0.3643	0.3595	0.3331	0.3287	0.0044	0.2970	2B	0.307	0.321	0.3344	0.3401	0.0057	0.3750
		3A	0.0000	0.3750	0.3656		0.3344	0.3311	0.0033	0.2983	3B	0.3070	0.3182	0.3344	0.3387	0.0043	0.3750
		1A	0.0011	0.3739	0.3631		0.3468	0.3411	0.0057	0.3228	1B	0.330	0.340	0.3479	0.2553	0.0074	0.3750
3/8–24	UNF	2A	0.0011	0.3739	0.3667		0.3468	0.3430	0.0038	0.3228	2B	0.330	0.340	0.3479	0.3528	0.0049	0.3750
		3A	0.0000	0.3750	0.3678		0.3479	0.3450	0.0029	0.3239	3B	0.3300	0.3372	0.3479	0.3516	0.0037	0.3750
3/8–32	NEF	2A	0.0010	0.3740	0.3680		0.3537	0.3503	0.0034	0.3357	2B	0.341	0.349	0.3547	0.3591	0.0044	0.3750
		3A	0.0000	0.3750	0.3690		0.3547	0.3522	0.0025	0.3367	3B	0.3410	0.3469	0.3547	0.3580	0.0033	0.3750
		1A	0.0014	0.4361	0.4206		0.3897	0.3826	0.0071	0.3485	1B	0.360	0.376	0.3911	0.4003	0.0092	0.4375
7/16–14	UNC	2A	0.0014	0.4361	0.4258	0.4206	0.3897	0.3850	0.0047	0.3485	2B	0.360	0.376	0.3911	0.3972	0.0061	0.4375
		3A	0.0000	0.4375	0.4272		0.3911	0.3876	0.0035	0.3499	3B	0.3600	0.3717	0.3911	0.3957	0.0046	0.4375
		1A	0.0013	0.4362	0.4240		0.4037	0.3975	0.0062	0.3749	1B	0.383	0.395	0.4050	0.4131	0.0081	0.4375
7/16–20	UNF	2A	0.0013	0.4362	0.4281		0.4037	0.3995	0.0042	0.3749	2B	0.383	0.395	0.4050	0.4104	0.0054	0.4375
		3A	0.0000	0.4375	0.4294		0.4050	0.4019	0.0031	0.3762	3B	0.3830	0.3916	0.4050	0.4091	0.0041	0.4375
7/16–28	UNEF	2A	0.0011	0.4364	0.4299		0.4132	0.4096	0.0036	0.3926	2B	0.399	0.407	0.4143	0.4189	0.0046	0.4375
		3A	0.0000	0.4375	0.4310		0.4143	0.4116	0.0027	0.3937	3B	0.3990	0.4051	0.4143	0.4178	0.0035	0.4375
½–12	N	2A	0.0016	0.4984	0.4870		0.4443	0.4389	0.0054	0.3962	2B	0.410	0.428	0.4459	0.4529	0.0070	0.5000
		3A	0.0000	0.5000	0.4886		0.4439	0.4419	0.0040	0.3978	3B	0.4100	0.4223	0.4459	0.4511	0.0052	0.5000
		1A	0.0015	0.4985	0.4822		0.4485	0.4411	0.0074	0.4041	1B	0.417	0.434	0.4500	0.4597	0.0097	0.5000
½–13	UNC	2A	0.0015	0.4985	0.4876	0.4822	0.4485	0.4435	0.0050	0.4041	2B	0.417	0.434	0.4500	0.4565	0.0065	0.5000
		3A	0.0000	0.5000	0.4891		0.4500	0.4463	0.0037	0.4056	3B	0.4170	0.4284	0.4500	0.4548	0.0048	0.5000

½–20 to 1⅛–12

½–20	UNF	1A	0.0013	0.4987	0.4865		0.4662	0.4598	0.0064	0.4374	1B	0.446	0.457	0.4675	0.4759	0.0084	0.5000
		2A	0.0013	0.4987	0.4906		0.4662	0.4619	0.0043	0.4374	2B	0.446	0.457	0.4675	0.4731	0.0056	0.5000
		3A	0.0000	0.5000	0.4919		0.4675	0.4643	0.0032	0.4387	3B	0.4460	0.4537	0.4675	0.4717	0.0042	0.5000
½–28	UNEF	2A	0.0011	0.4989	0.4924		0.4757	0.4720	0.0037	0.4551	2B	0.461	0.470	0.4768	0.4816	0.0048	0.5000
		3A	0.0000	0.5000	0.4935		0.4768	0.4740	0.0028	0.4562	3B	0.4610	0.4676	0.4768	0.4804	0.0036	0.5000
9/16–12	UNC	1A	0.0016	0.5609	0.5437		0.5068	0.4990	0.0078	0.4587	1B	0.472	0.490	0.5084	0.5186	0.0102	0.5625
		2A	0.0016	0.5609	0.5495	0.5437	0.5068	0.5016	0.0052	0.4587	2B	0.472	0.490	0.5084	0.5152	0.0068	0.5625
		3A	0.0000	0.5625	0.5511		0.5084	0.5045	0.0039	0.4603	3B	0.4720	0.4843	0.5084	0.5135	0.0051	0.5625
9/16–18	UNF	1A	0.0014	0.5611	0.5480		0.5250	0.5182	0.0068	0.4929	1B	0.502	0.515	0.5264	0.5353	0.0089	0.5625
		2A	0.0014	0.5611	0.5524		0.5250	0.5205	0.0045	0.4929	2B	0.502	0.515	0.5264	0.5323	0.0059	0.5625
		3A	0.0000	0.5625	0.5538		0.5264	0.5230	0.0034	0.4943	3B	0.5020	0.5106	0.5264	0.5308	0.0044	0.5625
9/16–24	NEF	2A	0.0012	0.5613	0.5541		0.5342	0.5303	0.0039	0.5102	2B	0.517	0.527	0.5354	0.5405	0.0051	0.5625
		3A	0.0000	0.5625	0.5553		0.5354	0.5325	0.0029	0.5114	3B	0.5170	0.5244	0.5354	0.5392	0.0038	0.5625
⅝–11	UNC	1A	0.0016	0.6234	0.6052		0.5644	0.5561	0.0083	0.5119	1B	0.527	0.546	0.5660	0.5767	0.0107	0.6250
		2A	0.0016	0.6234	0.6113	0.6052	0.5644	0.5589	0.0055	0.5119	2B	0.527	0.546	0.5660	0.5732	0.0072	0.6250
		3A	0.0000	0.6250	0.6129		0.5660	0.5619	0.0041	0.5135	3B	0.5270	0.5391	0.5660	0.5714	0.0054	0.6250
⅝–12	N	2A	0.0016	0.6234	0.6120		0.5693	0.5639	0.0054	0.5212	2B	0.535	0.553	0.5709	0.5780	0.0071	0.6250
		3A	0.0000	0.6250	0.6136		0.5709	0.5668	0.0041	0.5228	3B	0.5350	0.5463	0.5709	0.5762	0.0053	0.6250
⅝–18	UNF	1A	0.0014	0.6236	0.6105		0.5875	0.5805	0.0070	0.5554	1B	0.565	0.578	0.5889	0.5980	0.0091	0.6250
		2A	0.0014	0.6236	0.6149		0.5875	0.5828	0.0047	0.5554	2B	0.565	0.578	0.5889	0.5949	0.0060	0.6250
		3A	0.0000	0.6250	0.6163		0.5889	0.5854	0.0035	0.5568	3B	0.5650	0.5730	0.5889	0.5934	0.0045	0.6250
⅝–24	NEF	2A	0.0012	0.6238	0.6166		0.5967	0.5927	0.0040	0.5727	2B	0.580	0.590	0.5979	0.6031	0.0052	0.6250
		3A	0.0000	0.6250	0.6178		0.5979	0.5949	0.0030	0.5739	3B	0.5800	0.5809	0.5979	0.6018	0.0039	0.6250
11/16–12	N	2A	0.0016	0.6859	0.6745		0.6318	0.6264	0.0054	0.5837	2B	0.597	0.615	0.6334	0.6405	0.0071	0.6875
		3A	0.0000	0.6875	0.6761		0.6334	0.6293	0.0041	0.5853	3B	0.5970	0.6085	0.6334	0.6387	0.0053	0.6875
11/16–24	NEF	2A	0.0012	0.6863	0.6791		0.6592	0.6552	0.0040	0.6352	2B	0.642	0.652	0.6604	0.6656	0.0052	0.6875
		3A	0.0000	0.6875	0.6803		0.6604	0.6574	0.0030	0.6364	3B	0.6420	0.6494	0.6604	0.6643	0.0039	0.6875
¾–10	UNC	1A	0.0018	0.7482	0.7288		0.6832	0.6744	0.0088	0.6255	1B	0.642	0.663	0.6850	0.6965	0.0115	0.7500
		2A	0.0018	0.7482	0.7353	0.7288	0.6832	0.6773	0.0059	0.6255	2B	0.642	0.663	0.6850	0.6927	0.0077	0.7500
		3A	0.0000	0.7500	0.7371		0.6850	0.6806	0.0044	0.6273	3B	0.6420	0.6545	0.6850	0.6907	0.0057	0.7500
¾–12	N	2A	0.0017	0.7483	0.7369		0.6942	0.6887	0.0055	0.6461	2B	0.660	0.678	0.6959	0.7031	0.0072	0.7500
		3A	0.0000	0.7500	0.7386		0.6959	0.6918	0.0041	0.6478	3B	0.6600	0.6707	0.6959	0.7013	0.0054	0.7500

TABLE 4.1 Standard Limits of Size: Unified and American Screw Threads (*Continued*)

Nominal size and threads per inch	Series designation	External									Internal						
				Major diameter limits, in			Pitch diameter limits, in					Minor diameter limits, in		Pitch diameter limits, in			Major diameter, in
		Class	Allowance, in	Max	Min	Min	Max	Min	Tolerance	Minor diameter, in	Class	Min	Max	Min	Max	Tolerance	Min
½–20 to 1⅛–12																	
¾–16	UNF	1A	0.0015	0.7485	0.7343		0.7079	0.7004	0.0075	0.6718	1B	0.682	0.696	0.7094	0.7192	0.0098	0.7500
		2A	0.0015	0.7485	0.7391		0.7079	0.7029	0.0050	0.6718	2B	0.682	0.696	0.7094	0.7159	0.0065	0.7500
		3A	0.0000	0.7500	0.7406		0.7094	0.7056	0.0038	0.6733	3B	0.6820	0.6908	0.7094	0.7143	0.0049	0.7500
¾–20	UNEF	2A	0.0013	0.7487	0.7406		0.7162	0.7118	0.0044	0.6874	2B	0.696	0.707	0.7175	0.7232	0.0057	0.7500
		3A	0.0000	0.7500	0.7419		0.7175	0.7142	0.0033	0.6887	3B	0.6960	0.7037	0.7175	0.7218	0.0043	0.7500
13⁄16–12	N	2A	0.0017	0.8108	0.7994		0.7567	0.7512	0.0055	0.7086	2B	0.722	0.740	0.7584	0.7656	0.0072	0.8125
		3A	0.0000	0.8125	0.8011		0.7584	0.7543	0.0041	0.7103	3B	0.7220	0.7329	0.7584	0.7638	0.0054	0.8125
13⁄16–16	UN	2A	0.0015	0.8110	0.8016		0.7704	0.7655	0.0049	0.7343	2B	0.745	0.759	0.7719	0.7782	0.0063	0.8125
		3A	0.0000	0.8125	0.8031		0.7719	0.7683	0.0036	0.7358	3B	0.7450	0.7533	0.7719	0.7766	0.0047	0.8125
13⁄16–20	UNEF	2A	0.0013	0.8112	0.8031		0.7787	0.7743	0.0044	0.7498	2B	0.758	0.770	0.7800	0.7857	0.0057	0.8125
		3A	0.0000	0.8125	0.8044		0.7800	0.7767	0.0033	0.7512	3B	0.7580	0.7662	0.7800	0.7843	0.0043	0.8125
⅞–9	UNC	1A	0.0019	0.8731	0.8523		0.8009	0.7914	0.0095	0.7368	1B	0.755	0.778	0.8028	0.8151	0.0123	0.8750
		2A	0.0019	0.8731	0.8592	0.8523	0.8009	0.7946	0.0063	0.7368	2B	0.755	0.778	0.8028	0.8110	0.0082	0.8750
		3A	0.0000	0.8750	0.8611		0.8028	0.7981	0.0047	0.7387	3B	0.7550	0.7681	0.8028	0.8089	0.0061	0.8750
⅞–12	N	2A	0.0017	0.8733	0.8619		0.8192	0.8137	0.0055	0.7711	2B	0.785	0.803	0.8209	0.8281	0.0072	0.8750
		3A	0.0000	0.8750	0.8636		0.8209	0.8168	0.0041	0.7728	3B	0.7850	0.7952	0.8209	0.8263	0.0054	0.8750
⅞–14	UNF	1A	0.0016	0.8734	0.8579		0.8270	0.8189	0.0081	0.7858	1B	0.798	0.814	0.8286	0.8392	0.0106	0.8750
		2A	0.0016	0.8734	0.8631		0.8270	0.8216	0.0054	0.7858	2B	0.798	0.814	0.8286	0.8356	0.0070	0.8750
		3A	0.0000	0.8750	0.8647		0.8286	0.8245	0.0041	0.7874	3B	0.7980	0.8068	0.8286	0.8339	0.0053	0.8750
⅞–16	UN	2A	0.0015	0.8735	0.8641		0.8329	0.8280	0.0049	0.7968	2B	0.807	0.821	0.8344	0.8407	0.0063	0.8750
		3A	0.0000	0.8750	0.8656		0.8344	0.8308	0.0036	0.7983	3B	0.8070	0.8158	0.8344	0.8391	0.0047	0.8750
⅞–20	UNEF	2A	0.0013	0.8737	0.8656		0.8412	0.8368	0.0044	0.8124	2B	0.821	0.832	0.8425	0.8482	0.0057	0.8750
		3A	0.0000	0.8750	0.8669		0.8425	0.8392	0.0033	0.8137	3B	0.8210	0.8287	0.8425	0.8468	0.0043	0.8750
15⁄16–12	UN	2A	0.0017	0.9358	0.9244		0.8817	0.8760	0.0057	0.8336	2B	0.847	0.865	0.8834	0.8908	0.0074	0.9375
		3A	0.0000	0.9375	0.9261		0.8834	0.8793	0.0011	0.8353	3B	0.8470	0.8575	0.8834	0.8889	0.0055	0.9375
5⁄16–16	UN	2A	0.0015	0.9360	0.9266		0.8954	0.8904	0.0050	0.8593	2B	0.870	0.884	0.8969	0.9034	0.0065	0.9375
		3A	0.0000	0.9375	0.9281		0.8969	0.8932	0.0037	0.8608	3B	0.8700	0.8783	0.8969	0.9013	0.0049	0.9375

15/16–20	UNEF	2A	0.0014	0.9361	0.9280		0.9036	0.8991	0.0045	0.8748	2B	0.883	0.895	0.9050	0.9109	0.0059	0.9375
		3A	0.0000	0.9375	0.9294		0.9050	0.9016	0.0034	0.8762	3B	0.8830	0.8912	0.9050	0.9094	0.0044	0.9375
1–8	UNC	1A	0.0020	0.9980	0.9755		0.9168	0.9067	0.0101	0.8446	1B	0.865	0.890	0.9188	0.9320	0.0132	1.0000
		2A	0.0020	0.9980	0.9830	0.9755	0.9168	0.9100	0.0068	0.8446	2B	0.865	0.890	0.9188	0.9276	0.0088	1.0000
		3A	0.0000	1.0000	0.9850		0.9188	0.9137	0.0051	0.8466	3B	0.8650	0.8797	0.9188	0.9254	0.0066	1.0000
1–12	UNF	1A	0.0018	0.9982	0.9810		0.9441	0.9353	0.0088	0.8960	1B	0.910	0.928	0.9459	0.9573	0.0114	1.0000
		2A	0.0018	0.9982	0.9868		0.9441	0.9382	0.0059	0.8960	2B	0.910	0.928	0.9459	0.9535	0.0076	1.0000
		3A	0.0000	1.0000	0.9886		0.9459	0.9415	0.0044	0.8978	3B	0.9100	0.9198	0.9459	0.9516	0.0057	1.0000
1–16	UN	2A	0.0015	0.9985	0.9891		0.9579	0.9529	0.0050	0.9218	2B	0.932	0.946	0.9594	0.9659	0.0065	1.0000
		3A	0.0000	1.0000	0.9906		0.9594	0.9557	0.0037	0.9233	3B	0.9320	0.9408	0.9594	0.9643	0.0049	1.0000
1–20	UNEF	2A	0.0014	0.9986	0.9905		0.9661	0.9616	0.0045	0.9373	2B	0.946	0.957	0.9675	0.9734	0.0059	1.0000
		3A	0.0000	1.0000	0.9919		0.9675	0.9641	0.0034	0.9387	3B	0.9460	0.9537	0.9675	0.9719	0.0044	1.0000
1 1/16–12	UN	2A	0.0017	1.0608	1.0494		1.0067	1.0010	0.0057	0.9586	2B	0.972	0.990	1.0084	1.0158	0.0074	1.0625
		3A	0.0000	1.0625	1.0511		1.0084	1.0042	0.0042	0.9603	3B	0.9720	0.9823	1.0084	1.0139	0.0055	1.0625
1 1/16–16	UN	2A	0.0015	1.0610	1.0516		1.0204	1.0154	0.0050	0.9843	2B	0.995	1.009	1.0219	1.0284	0.0065	1.0625
		3A	0.0000	1.0625	1.0531		1.0219	1.0182	0.0037	0.9853	3B	0.9950	1.0033	1.0219	1.0268	0.0049	1.0625
1 1/16–18	NEF	2A	0.0014	1.0611	1.0524		1.0250	1.0203	0.0047	0.9929	2B	1.002	1.015	1.0264	1.0326	0.0062	1.0625
		3A	0.0000	1.0625	1.0538		1.0264	1.0228	0.0036	0.9943	3B	1.0020	1.0105	1.0264	1.0310	0.0046	1.0625
1⅛–7	UNC	1A	0.0022	1.1228	1.0982		1.0300	1.0191	0.0109	0.9475	1B	0.970	0.998	1.0322	1.0463	0.0141	1.1250
		2A	0.0022	1.1228	1.1064	1.0982	1.0300	1.0228	0.0072	0.9475	2B	0.970	0.998	1.0322	1.0416	0.0094	1.1250
		3A	0.0000	1.1250	1.1086		1.0322	1.0268	0.0054	0.9497	3B	0.9700	0.9875	1.0322	1.0393	0.0071	1.1250
1⅛–8	N	2A	0.0021	1.1229	1.1079	1.1004	1.0417	1.0348	0.0069	0.9695	2B	0.900	1.015	1.0438	1.0528	0.0090	1.1250
		3A	0.0000	1.1250	1.1100		1.0438	1.0386	0.0052	0.9716	3B	0.9900	1.0047	1.0438	1.0505	0.0067	1.1250
1⅛–12	UNF	1A	0.0018	1.1232	1.1060		1.0691	1.0601	0.0090	1.0210	1B	1.035	1.053	1.0709	1.0826	0.0117	1.1250
		2A	0.0018	1.1232	1.1118		1.0691	1.0631	0.0060	1.0210	2B	1.035	1.053	1.0709	1.0787	0.0078	1.1250
		3A	0.0000	1.1250	1.1136		1.0709	1.0664	0.0045	1.0228	3B	1.0350	1.0448	1.0709	1.0768	0.0059	1.1250

1⅛–16 to 1½–18

1⅛–16	UN	2A	0.0015	1.1235	1.1141		1.0829	1.0779	0.0050	1.0468	2B	1.057	1.071	1.0844	1.0909	0.0065	1.1250
		3A	0.0000	1.1250	1.1156		1.0844	1.0807	0.0037	1.0483	3B	1.0570	1.0658	1.0844	1.0893	0.0049	1.1250
1⅛–18	NEF	2A	0.0014	1.1236	1.1149		1.0875	1.0828	0.0047	1.0554	2B	1.065	1.078	1.0889	1.0951	0.0062	1.1250
		3A	0.0000	1.1250	1.1163		1.0889	1.0853	0.0036	1.0568	3B	1.0650	1.0730	1.0889	1.0935	0.0046	1.1250
1 3/16–12	UN	2A	0.0017	1.1858	1.1744		1.1317	1.1259	0.0058	1.0836	2B	1.097	1.115	1.1334	1.1409	0.0075	1.1875
		3A	0.0000	1.1875	1.1761		1.1334	1.1291	0.0043	1.0853	3B	1.0970	1.1073	1.1334	1.1390	0.0056	1.1875
1 3/16	UN	2A	0.0015	1.1860	1.1766		1.1454	1.1403	0.0051	1.1093	2B	1.120	1.134	1.1469	1.1535	0.0066	1.1875
		3A	0.0000	1.1875	1.1781		1.1469	1.1431	0.0038	1.1108	3B	1.1200	1.1283	1.1469	1.1519	0.0050	1.1875
1 3/16–18	NEF	2A	0.0015	1.1860	1.1773		1.1499	1.1450	0.0049	1.1178	2B	1.127	1.140	1.1514	1.1577	0.0063	1.1875
		3A	0.0000	1.1875	1.1788		1.1514	1.1478	0.0036	1.1193	3B	1.1270	1.1355	1.1514	1.1561	0.0047	1.1875

TABLE 4.1 Standard Limits of Size: Unified and American Screw Threads (*Continued*)

Nominal size and threads per inch	Series designation	External									Internal						
		Class	Allowance, in	Major diameter limits, in			Pitch diameter limits, in			Minor diameter, in	Class	Minor diameter limits, in		Pitch diameter limits, in			Major diameter, in
				Max	Min	Min	Max	Min	Tolerance			Min	Max	Min	Max	Tolerance	Min
1⅛–16 to 1¾–12																	
1¼–7	UNC	1A	0.0022	1.2478	1.2232		1.1550	1.1439	0.0111	1.0725	1B	1.095	1.123	1.1572	1.1716	0.0144	1.2500
		2A	0.0022	1.2478	1.2314	1.2232	1.1550	1.1476	0.0074	1.0725	2B	1.095	1.123	1.1572	1.1668	0.0096	1.2500
		3A	0.0000	1.2500	1.2336		1.1572	1.1517	0.0055	1.0747	3B	1.0950	1.1125	1.1572	1.1644	0.0072	1.2500
1¼–8	N	2A	0.0021	1.2479	1.2329	1.2254	1.1667	1.597	0.0070	1.0945	2B	1.115	1.140	1.1688	1.1780	0.0092	1.2500
		3A	0.0000	1.2500	1.2350		1.1688	1.1635	0.0053	1.0966	3B	1.1150	1.1297	1.1688	1.1757	0.0069	1.2500
1¼–12	UNF	1A	0.0018	1.2482	1.2310		1.1941	1.1849	0.0092	1.1460	1B	1.160	1.178	1.1959	1.2079	0.0120	1.2500
		2A	0.0018	1.2482	1.2368		1.1941	1.1879	0.0062	1.1460	2B	1.160	1.178	1.1959	1.2039	0.0080	1.2500
		3A	0.0000	1.2500	1.2386		1.1959	1.1913	0.0046	1.1478	3B	1.1600	1.1698	1.1959	1.2019	0.0060	1.2500
1¼–16	UN	2A	0.0015	1.2485	1.2391		1.2079	1.2028	0.0051	1.1718	2B	1.182	1.196	1.2094	1.2160	0.0066	1.2500
		3A	0.0000	1.2500	1.2406		1.2094	1.2056	0.0038	1.1733	3B	1.1820	1.1908	1.2094	1.2144	0.0050	1.2500
1¼–18	NEF	2A	0.0015	1.2485	1.2398		1.2124	1.2075	0.0049	1.1803	2B	1.190	1.203	1.2139	1.2202	0.0063	1.2500
		3A	0.0000	1.2500	1.2413		1.2139	1.2103	0.0036	1.1818	3B	1.1900	1.1980	1.2139	1.2186	0.0047	1.2500
1⁵⁄₁₆–12	UN	2A	0.0017	1.3108	1.2994		1.2567	1.2509	0.0058	1.2086	2B	1.222	1.240	1.2584	1.2659	0.0075	1.3125
		3A	0.0000	1.3125	1.3011		1.2584	1.2541	0.0043	1.2103	3B	1.2220	1.2323	1.2584	1.2640	0.0056	1.3125
1⁵⁄₁₆–16	UN	2A	0.0015	1.3110	1.3016		1.2704	1.2653	0.0051	1.2343	2B	1.245	1.259	1.2719	1.2785	0.0066	1.3125
		5A	0.0000	1.3125	1.3031		1.2719	1.2681	0.0038	1.2358	3B	1.2450	1.2533	1.2719	1.2769	0.0050	1.3125
1⁵⁄₁₆–18	NEF	2A	0.0015	1.3110	1.3023		1.2749	1.2700	0.0049	1.2428	2B	1.252	1.265	1.2764	1.2827	0.0063	1.3125
		3A	0.0000	1.3125	1.3038		1.2764	1.2728	0.0036	1.2443	3B	1.2520	1.2605	1.2764	1.2811	0.0047	1.3125
1⅜–6	UNC	1A	0.0024	1.3726	1.3453		1.2643	1.2523	0.0120	1.1681	1B	1.195	1.225	1.2667	[‡]1.2822	[‡]0.0155	1.3750
		2A	0.0024	1.3726	1.3544	1.3453	1.2643	1.2563	0.0080	1.1681	2B	1.195	1.225	1.2667	1.2771	0.0104	1.3750
		3A	0.0900	1.3750	1.3568		1.2667	1.2607	0.0060	1.1705	3B	1.1950	1.2146	1.2667	1.2745	0.0078	1.3750
1⅜–8	N	2A	0.0022	1.3728	1.3578	1.3503	1.2916	1.2844	0.0072	1.2194	2B	1.240	1.265	1.2938	1.3031	0.0093	1.3750
		3A	0.0000	1.3750	1.3600		1.2938	1.2884	0.0054	1.2216	3B	1.2400	1.2547	1.2938	1.3008	0.0070	1.3750
1⅜–12	UNF	1A	0.0019	1.3731	1.3559		1.3190	1.3096	0.0094	1.2709	1B	1.285	1.303	1.3209	1.3332	0.0123	1.3750
		2A	0.0019	1.3731	1.3617		1.3190	1.3127	0.0063	1.2709	2B	1.285	1.303	1.3209	1.3291	0.0082	1.3750
		3A	0.0000	1.3750	1.3636		1.3209	1.3162	0.0047	1.2728	3B	1.2850	1.2948	1.3209	1.3270	0.0061	1.3750

1⅜–16	UN	2A	0.0015	1.3735	1.3641		1.3329	1.3278	0.0051	1.2968	2B	1.307	1.321	1.3344	1.3410	0.0066	1.3750
		3A	0.0000	1.3750	1.3656		1.3344	1.3306	0.0038	1.2983	3B	1.3070	1.3158	1.3344	1.3394	0.0050	1.3750
1⅜–18	NEF	2A	0.0015	1.3735	1.3648		1.3374	1.3325	0.0049	1.3053	2B	1.315	1.328	1.3389	1.3452	0.0063	1.3750
		3A	0.0000	1.3750	1.3663		1.3389	1.3353	0.0036	1.3068	3B	1.3150	1.3230	1.3389	1.3436	0.0047	1.3750
1$\frac{7}{16}$–12	UN	2A	0.0018	1.4357	1.4243		1.3816	1.3757	0.0059	1.3335	2B	1.347	1.365	1.3834	1.3910	0.0076	1.4375
		3A	0.0000	1.4375	1.4261		1.3834	1.3790	0.0044	1.3353	3B	1.3470	1.3573	1.3834	1.3891	.0057	1.4375
1$\frac{7}{16}$–16	UN	2A	0.0016	1.4359	1.4265		1.3953	1.3901	0.0052	1.3592	2B	1.370	1.384	1.3969	1.4037	0.0068	1.4375
		3A	0.0000	1.4375	1.4281		1.3969	1.3930	0.0039	1.3608	3B	1.3700	1.3783	1.3969	1.4020	0.0051	1.4375
1$\frac{7}{16}$–18	NEF	2A	0.0015	1.4360	1.4273		1.3999	1.3949	0.0050	1.3678	2B	1.377	1.390	1.4014	1.4079	0.0065	1.4375
		3A	0.0000	1.4375	1.4288		1.4014	1.3977	0.0037	1.3693	3B	1.3770	1.3855	1.4014	1.4062	0.0048	1.4375
1½–6	UNC	1A	0.0024	1.4976	1.4703		1.3893	1.3772	0.0121	1.2931	1B	1.320	1.350	1.3917	1.4075	0.0158	1.5000
		2A	0.0024	1.4976	1.4794	1.4703	1.3893	1.3812	0.0081	1.2931	2B	1.320	1.350	1.3917	1.4022	0.0105	1.5000
		3A	0.0000	1.5000	1.4818		1.3917	1.3856	0.0061	1.2955	3B	1.3200	1.3396	1.3917	1.3996	0.0079	1.5000
1½–8	N	2A	0.0022	1.4978	1.4828	1.4753	1.4166	1.4093	0.0073	1.3444	2B	1.365	1.390	1.4188	1.4283	0.0095	1.5000
		3A	0.0000	1.5000	1.4850		1.4188	1.4133	0.0055	1.3466	3B	1.3650	1.3797	1.4188	1.4259	0.0071	1.5000
1½–12	UNF	1A	0.0019	1.4981	1.4809		1.4440	1.4344	0.0096	1.3959	1B	1.410	1.428	1.4459	1.4584	0.0125	1.5000
		2A	0.0019	1.4981	1.4867		1.4440	1.4376	0.0064	1.3959	2B	1.410	1.428	1.4459	1.4542	0.0083	1.5000
		3A	0.0000	1.5000	1.4886		1.4459	1.4411	0.0048	1.3978	3B	1.4100	1.4198	1.4459	1.4522	0.0063	1.5000
1½–16	UN	2A	0.0016	1.4984	1.4890		1.4578	1.4526	0.0052	1.4217	2B	1.432	1.446	1.4594	1.4662	0.0068	1.5000
		3A	0.0000	1.5000	1.4906		1.4594	1.4555	0.0039	1.4233	3B	1.4320	1.4408	1.4594	1.4645	0.0051	1.5000
1½–18	NEF	2A	0.0015	1.4985	1.4898		1.4624	1.4574	0.0050	1.4303	2B	1.440	1.452	1.4639	1.4704	0.0065	1.5000
		3A	0.0000	1.5000	1.4913		1.4639	1.4602	0.0037	1.4318	3B	1.4400	1.4480	1.4639	1.4687	0.0048	1.5000

CHAPTER 5

FASTENING DEVICES—PROPERTIES AND DIMENSIONS

5.1 THREADED FASTENERS

Bolts and screws are the most commonly used types of fastening devices. The thread on a bolt or screw may be compared to an inclined plane wrapped around a cylinder, thus assuming the form of one of the basic machines. Many different thread form standards are used throughout the world, but this chapter addresses only the forms in common use, such as the 60° V thread form for the U.S. Customary (inch) system and the metric (ISO) standard. The other basic thread forms and their geometry are shown in Chap. 4, "Thread Systems."

Threaded fasteners have countless applications in industry, and in many cases, their performance is critical. Failure of a threaded fastener such as a heavily loaded bolt or bolts can cause severe injury, property damage, and death in critical applications. It is for these reasons that standards of performance have been devised and specified in the various American standards, such as those of the American National Standards Institute (ANSI), the American Society for Testing and Materials (ASTM), the Society of Automotive Engineers (SAE), the Industrial Fasteners Institute (IFI), the American Iron and Steel Institute (AISI), and others. In recent years, it has been discovered that "counterfeit" bolts are being imported on a large scale into the United States. False head markings and inferior steels have been found, and these have contributed to a number of failures in critical applications. This is a direct indication of improperly processed or poor-quality material. In other cases cited, failure was through the threaded section of such bolts or the result of cracking due to improper heat treatment combined with faulty material.

Not only are the counterfeit fasteners causing problems, but improper materials specifications also have been noted.

Threaded Fastener Dimensions

Dimensions of bolts, screws, nuts, and washers. Basic dimensions for the most commonly used American standard bolts are shown in Fig. 5.1. Basic dimensions for the most commonly used American standard machine screws are shown in Fig. 5.2. Basic dimensions for American standard miniature screws are shown in Fig. 5.3.

The dimensions shown in these figures are the most important ones needed for machining clearances and tool/design engineering purposes. The complete dimensional specifications and applicable tolerances for all standard hardware or fasteners should be obtained from the fastener handbooks distributed by the Industrial Fasteners Institute (IFI) (see Chap. 11).

The design specifications for strength of all types of threaded fasteners are shown in a subsection to follow. In addition, the normal tightening torques required for bolts and screws will be shown in another subsection to

	Hex Bolts				Hex Cap Screw (Finished Hex Bolts)				Round Head Square Neck Bolts				Countersunk Bolts			
Nominal size	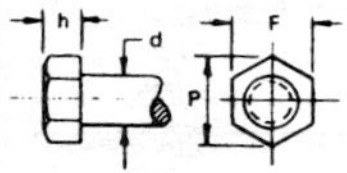				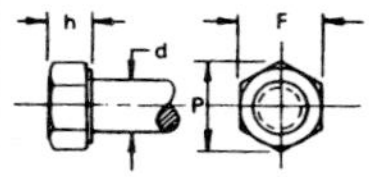				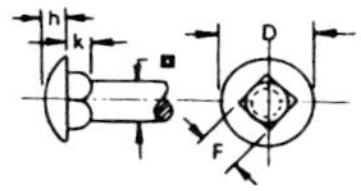				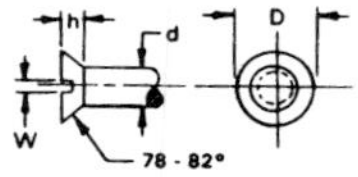			
Basic Dia.	d	F	P	h	d	F	P	h	D	F	h	k	d	D	h	W ♦
0.2500 - 1/4	0.260	0.438	0.505	0.188	0.2500	0.438	0.505	0.163	0.594	0.260	0.145	0.156	0.260	0.493	0.150	0.064
0.3125 - 5/16	0.324	0.500	0.577	0.235	0.3125	0.500	0.577	0.211	0.719	0.324	0.176	0.187	0.324	0.618	0.189	0.072
0.3750 - 3/8	0.388	0.562	0.650	0.268	0.3750	0.562	0.650	0.243	0.844	0.388	0.208	0.219	0.388	0.740	0.225	0.081
0.4375 - 7/16	0.452	0.625	0.722	0.316	0.4375	0.625	0.722	0.291	0.969	0.452	0.239	0.250	0.452	0.803	0.226	0.081
0.5000 - 1/2	0.515	0.750	0.866	0.364	0.5000	0.750	0.866	0.323	1.094	0.515	0.270	0.281	0.515	0.935	0.269	0.091
0.5625 - 9/16	—	—	—	—	0.5625	0.812	0.938	0.371	—	—	—	—	—	—	—	—
0.6250 - 5/8	0.642	0.938	1.083	0.444	0.6250	0.938	1.083	0.403	1.344	0.642	0.344	0.344	0.642	1.169	0.336	0.116
0.7500 - 3/4	0.768	1.125	1.299	0.524	0.7500	1.125	1.299	0.483	1.594	0.768	0.406	0.406	0.768	1.402	0.403	0.131
0.8750 - 7/8	0.895	1.312	1.516	0.604	0.8750	1.312	1.516	0.563	1.844	0.895	0.469	0.469	0.895	1.637	0.470	0.147
1.0000 - 1	1.022	1.500	1.732	0.700	1.0000	1.500	1.732	0.627	2.094	1.022	0.531	0.531	1.022	1.869	0.537	0.166
1.1250 - 1-1/8	1.149	1.688	1.949	0.780	1.1250	1.688	1.949	0.718	—	—	—	—	1.149	2.104	0.604	0.178
1.2500 - 1-1/4	1.277	1.875	2.165	0.876	1.2500	1.875	2.165	0.813	—	—	—	—	1.277	2.337	0.671	0.193
1.3750 - 1-3/8	1.404	2.062	2.382	0.940	1.3750	2.062	2.382	0.878	—	—	—	—	1.404	2.571	0.738	0.208
1.5000 - 1-1/2	1.531	2.250	2.598	1.036	1.5000	2.250	2.598	0.974	—	—	—	—	1.531	2.804	0.805	0.240
1.7500 - 1-3/4	1.785	2.625	3.031	1.196	1.7500	2.625	3.031	1.134	—	—	—	—	—	—	—	—
2.0000 - 2	2.039	3.000	3.464	1.388	2.0000	3.000	3.464	1.263	—	—	—	—	—	—	—	—

Note: ♦ Minimum dimensions of slot widths. All other tabulated values are maximum dimensions. ■ diameters same as hex bolts (d) of same basic size.

FIGURE 5.1 Dimensions for American standard bolts, in inches.

Nominal Size		Fillister Head			Binding Head			Pan Head			Countersunk			Undercut C'sunk		
		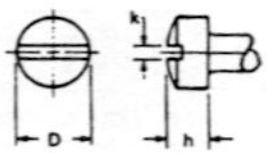			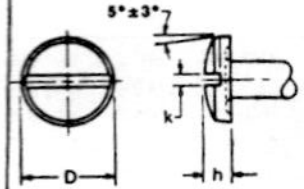			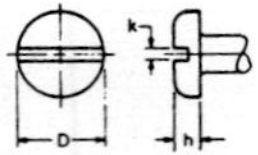			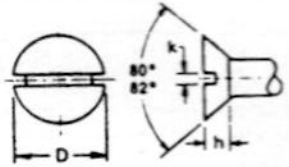			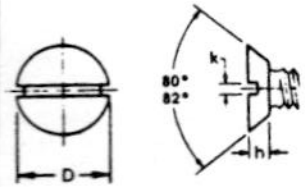		
Basic Dia.		D	h	k♦	D	h	k♦	D	h	k♦	D	h	k♦	D	h	k♦
0 -	0.0600	0.096	0.055	0.016	0.126	0.032	0.016	0.116	0.039	0.016	0.119	0.035	0.016	0.119	0.025	0.016
1 -	0.0730	0.118	0.066	0.019	0.153	0.041	0.019	0.142	0.046	0.019	0.146	0.043	0.019	0.146	0.031	0.019
2 -	0.0860	0.140	0.083	0.023	0.181	0.050	0.023	0.167	0.053	0.023	0.172	0.051	0.023	0.172	0.036	0.023
3 -	0.0990	0.161	0.095	0.027	0.208	0.059	0.027	0.193	0.060	0.027	0.199	0.059	0.027	0.199	0.042	0.027
4 -	0.1120	0.183	0.107	0.031	0.235	0.068	0.031	0.219	0.068	0.031	0.225	0.067	0.031	0.225	0.047	0.031
5 -	0.1250	0.205	0.120	0.035	0.263	0.078	0.035	0.245	0.075	0.035	0.252	0.075	0.035	0.252	0.053	0.035
6 -	0.1380	0.226	0.132	0.039	0.290	0.087	0.039	0.270	0.082	0.039	0.279	0.083	0.039	0.279	0.059	0.039
8 -	0.1640	0.270	0.156	0.045	0.344	0.105	0.045	0.322	0.096	0.045	0.332	0.100	0.045	0.332	0.070	0.045
10 -	0.1900	0.313	0.180	0.050	0.399	0.123	0.050	0.373	0.110	0.050	0.385	0.116	0.050	0.385	0.081	0.050
12 -	0.2160	0.357	0.205	0.056	0.454	0.141	0.056	0.425	0.125	0.056	0.438	0.132	0.056	0.438	0.092	0.056
1/4-	0.2500	0.414	0.237	0.064	0.525	0.165	0.064	0.492	0.144	0.064	0.507	0.153	0.064	0.507	0.107	0.064
5/16-	0.3125	0.518	0.295	0.072	0.656	0.209	0.072	0.615	0.178	0.072	0.635	0.191	0.072	0.635	0.134	0.072
3/8-	0.3750	0.622	0.335	0.081	0.788	0.253	0.081	0.740	0.212	0.081	0.762	0.230	0.081	0.762	0.161	0.081
7/16-	0.4375	0.625	0.368	0.081	—	—	—	0.863	0.247	0.081	0.812	0.223	0.081	0.812	0.156	0.081
1/2-	0.5000	0.750	0.412	0.091	—	—	—	0.987	0.281	0.091	0.875	0.223	0.091	0.875	0.156	0.091

Note: ♦ Minimum dimensions of slot widths. All other tabulated values are maximum dimensions.

FIGURE 5.2 Dimensions for standard machine screws.

Size	Thds/in	Basic dia.	Fillister			Binding			Pan			100° Flat Head		
			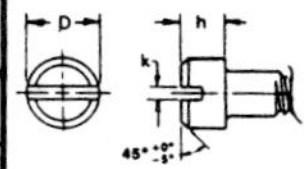			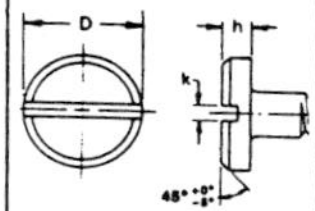			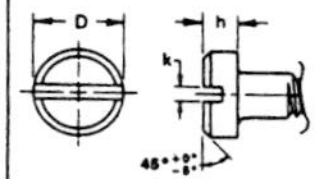			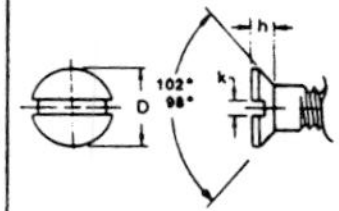		
			D	h	k♦	D	h	k♦	D	h	k♦	D	h	k♦
30 UNM	318	0.0118	0.021	0.012	0.003	—	—	—	0.025	0.010	0.003	0.023	0.007	0.003
40 UNM	254	0.0157	0.025	0.016	0.003	0.041	0.010	0.004	0.033	0.012	0.004	0.029	0.008	0.003
50 UNM	203	0.0197	0.033	0.020	0.004	0.051	0.012	0.005	0.041	0.016	0.005	0.037	0.011	0.004
60 UNM	169	0.0236	0.041	0.025	0.005	0.062	0.016	0.007	0.051	0.020	0.007	0.045	0.013	0.005
80 UNM	127	0.0315	0.051	0.032	0.007	0.082	0.020	0.008	0.062	0.025	0.008	0.056	0.016	0.007
100 UNM	102	0.0394	0.062	0.040	0.008	0.103	0.025	0.012	0.082	0.032	0.012	0.072	0.019	0.008
120 UNM	102	0.0472	0.082	0.050	0.012	0.124	0.032	0.015	0.103	0.040	0.015	0.092	0.025	0.012

Note: ♦ Minimum dimensions of slot widths. All other tabulated values are maximum dimensions.

FIGURE 5.3 Dimensions for standard miniature screws.

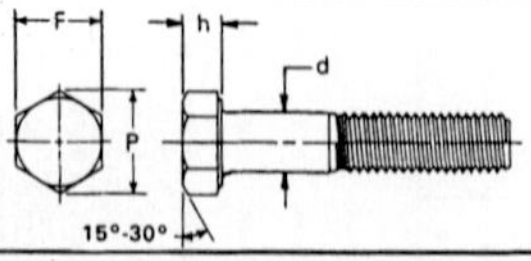

Diameter & thread pitch	d	F	P	h
M5 - 0.8	5.00	8.00	9.24	3.65
M6 - 1	6.00	10.00	11.55	4.15
M8 - 1.25	8.00	13.00	15.01	5.50
M10 - 1.50	10.00	16.00	18.48	6.63
M12 - 1.75	12.00	18.00	20.78	7.76
M14 - 2	14.00	21.00	24.25	9.09
M16 - 2	16.00	24.00	27.71	10.32
M20 - 2.5	20.00	30.00	34.64	12.88
M24 - 3	24.00	36.00	41.57	15.44
M30 - 3.5	30.00	46.00	53.12	19.48
M36 - 4	36.00	55.00	63.51	23.38

Note: Tabulated dimensions are in millimeters and are maximum values.

FIGURE 5.4 Dimensions for metric hex-cap screws.

follow (see Table 5.1). Calculation procedures for bolts and screws are shown in the *McGraw-Hill Machining and Metalworking Handbook,* 1999, and the *Handbook of Machining and Metalworking Calculations,* McGraw-Hill, 2001.

Figure 5.4 shows the basic dimensions for metric hex cap screws. Figure 5.5 shows the dimensions of American standard hex nuts and jamb nuts. Figure 5.6 shows the dimensions of metric standard nuts, type 1. Figure 5.7 shows the dimensions of American standard flat washers. Figure 5.8 shows dimensions of metric standard flat washers.

Figure 5.9 shows dimensions of American standard shoulder bolts and socket head cap screws, which are

Size	Flats	Points	Thickness
#0	5/32	0.180	0.050
#1	5/32	0.180	0.050
#2	3/16	0.217	0.066
#3	3/16	0.217	0.066
#4	1/4	0.289	0.098
#5	5/16	0.361	0.114
#6	5/16	0.361	0.114
#8	11/32	0.397	0.130
#10	3/8	0.433	0.130
1/4	7/16	0.505	0.226 0.163, jamb nut
5/16	1/2	0.577	0.273 0.195, jamb nut
3/8	9/16	0.650	0.337 0.227, jamb nut
7/16	11/16	0.794	0.385 0.260, jamb nut
1/2	3/4	0.866	0.448 0.323, jamb nut
5/8	15/16	1.083	0.559 0.387, jamb nut
3/4	1-1/8	1.299	0.665 0.446, jamb nut
7/8	1-5/16	1.516	0.776 0.510, jamb nut
1	1-1/2	1.732	0.887 0.575, jamb nut

FIGURE 5.5 Dimensions for American standard hex and jamb nuts.

widely used in tooling applications as well as in other mechanical design applications.

Standard shoulder screw mechanical data

Thread class	UNC-3A (ANSI/ASME B1.3) (No plating allowance is provided.)
Material	Alloys of chrome, nickel, molybdenum, or vanadium
Hardness	32 to 43 Rockwell C at the surface

Size	Flats	Points	Thickness
M1.6x0.35	3.20	3.70	1.30
M2x0.4	4.00	4.62	1.60
M2.5x0.45	5.00	5.77	2.00
M3x0.5	5.50	6.35	2.40
M3.5x0.6	6.00	6.93	2.80
M4x0.7	7.00	8.08	3.20
M5x0.8	8.00	9.24	4.70
M6x1	10.00	11.55	5.20
M8x1.25	13.00	15.01	6.80
M10x1.5	16.00	18.48	8.40
M12x1.75	18.00	20.78	10.80
M14x2	21.00	24.25	12.80
M16x2	24.00	27.71	14.80
M20x2.5	30.00	34.64	18.00
M24x3	36.00	41.57	21.50
M30x3.5	46.00	53.12	25.60
M36x4	55.00	63.51	31.00

NOTE: Tabulated dimensions are in millimeters. 1mm = 0.03937 inches.

FIGURE 5.6 Dimensions for metric standard nuts, hexagonal.

Size	Outside Diameter	Thickness
#0	3/16	0.028
#1	7/32	0.028
#2	1/4	0.036
#3	5/16	0.036
#4	3/8	0.045
#5	13/32	0.045
#6	7/16	0.045
#8	1/2	0.045
#10	9/16	0.045
¼	47/64	0.071
5/16	7/8	0.071
3/8	1	0.071
7/16	1-1/8	0.071
½	1-1/4	0.112
5/8	1-3/4	0.112
¾	2	0.112
7/8	2-1/4	0.174
1	2-1/2	0.174

FIGURE 5.7 Dimensions for American standard flat washers.

Dimensions for Metric Standard Flat Washers

Size	Series	Outside Diameter	Thickness
1.6	Narrow	4.00	0.70
	Regular	5.00	0.70
	Wide	6.00	0.90
2	Narrow	5.00	0.90
	Regular	6.00	0.90
	Wide	8.00	0.90
2.5	Narrow	6.00	0.90
	Regular	8.00	0.90
	Wide	10.00	1.20
3	Narrow	7.00	0.90
	Regular	10.00	1.20
	Wide	12.00	1.40
3.5	Narrow	9.00	1.20
	Regular	10.00	1.40
	Wide	15.00	1.75
4	Narrow	10.00	1.20
	Regular	12.00	1.40
	Wide	16.00	2.30
5	Narrow	11.00	1.40
	Regular	15.00	1.75
	Wide	20.00	2.30
6	Narrow	13.00	1.75
	Regular	18.80	1.75
	Wide	25.40	2.30
8	Narrow	18.80	2.30
	Regular	25.40	2.30
	Wide	32.00	2.80
10	Narrow	20.00	2.30
	Regular	28.00	2.80
	Wide	39.00	3.50
12	Narrow	25.40	2.80
	Regular	34.00	3.50
	Wide	44.00	3.50
14	Narrow	28.00	2.80
	Regular	39.00	3.50
	Wide	50.00	4.00
16	Narrow	32.00	3.50
	Regular	44.00	4.00
	Wide	56.00	4.60
20	Narrow	39.00	4.00
	Regular	50.00	4.60
	Wide	66.00	5.10
24	Narrow	44.00	4.60
	Regular	56.00	5.10
	Wide	72.00	5.60
30	Narrow	56.00	5.10
	Regular	72.00	5.60
	Wide	90.00	6.40
36	Narrow	66.00	5.60
	Regular	90.00	6.40
	Wide	110.00	8.50

Note: Tabulated dimensions are in millimeters.

FIGURE 5.8 Dimensions for metric standard flat washers.

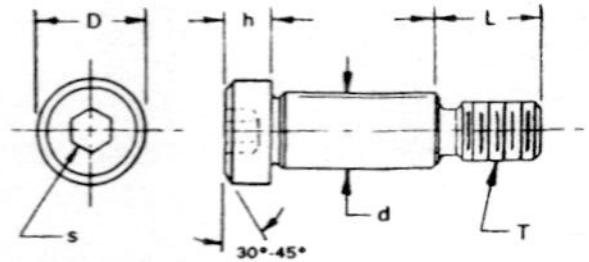

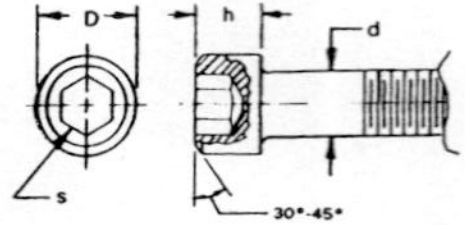

Nominal size		d	D	h	s	T	L	d	D	h	s
4	0.1120	—	—	—	—	—	—	0.1120	0.183	0.112	0.094
5	0.1250	—	—	—	—	—	—	0.1250	0.205	0.125	0.094
6	0.1380	—	—	—	—	—	—	0.1380	0.226	0.138	0.109
8	0.1640	—	—	—	—	—	—	0.1640	0.270	0.164	0.141
10	0.1900	—	—	—	—	—	—	0.1900	0.312	0.190	0.156
1/4	0.2500	0.2480	0.375	0.188	0.125	0.190-24	0.375	0.2500	0.375	0.250	0.188
5/16	0.3125	0.3105	0.438	0.219	0.156	0.250-20	0.438	0.3125	0.469	0.312	0.250
3/8	0.3750	0.3730	0.562	0.250	0.188	0.312-18	0.500	0.3750	0.562	0.375	0.312
7/16	0.4375	—	—	—	—	—	—	0.4375	0.656	0.438	0.375
1/2	0.5000	0.4980	0.750	0.312	0.250	0.375-16	0.625	0.5000	0.750	0.500	0.375
5/8	0.6250	0.6230	0.875	0.375	0.312	0.500-13	0.750	0.6250	0.938	0.625	0.500
3/4	0.7500	0.7480	1.000	0.500	0.375	0.625-11	0.875	0.7500	1.125	0.750	0.625
7/8	0.8750	—	—	—	—	—	—	0.8750	1.312	0.875	0.750
1	1.0000	0.9980	1.312	0.625	0.500	0.750-10	1.000	1.0000	1.500	1.000	0.750
1-1/8	1.1250	—	—	—	—	—	—	1.1250	1.688	1.125	0.875
1-1/4	1.2500	1.2480	1.750	0.750	0.625	0.875-9	1.125	1.2500	1.875	1.250	0.875
1-3/8	1.3750	—	—	—	—	—	—	1.3750	2.062	1.375	1.000
1-1/2	1.5000	1.4980	2.125	1.000	0.875	1.125-7	1.500	1.5000	2.250	1.500	1.000
1-3/4	1.7500	1.7480	2.375	1.125	1.000	1.250-7	1.750	1.7500	2.625	1.750	1.250
2	2.0000	1.9980	2.750	1.250	1.250	1.500-6	2.000	2.0000	3.000	2.000	1.500

Note: All tabulated dimensions are maximum values. See text for materials, hardness, etc.

FIGURE 5.9 American standard socket-head shoulder screws and socket-head cap screws.

Ultimate tensile strength	140,000 lb/in^2 based on minimum thread neck area
Shear strength	84,000 lb/in^2 in thread neck and shoulder areas

Figures 5.10 and 5.11 show the head marking standards, for SAE- and ASTM-graded bolts, together with the tensile strength requirements and proof load ratings. Figure 5.12 shows the head marking standard for metric bolts, together with the tensile strength requirements and proof load ratings.

Table 5.1 shows the tightening torque ranges for American Standard steel bolts for SAE grades 2, 5, and 8.

Set, self-tapping, thread-forming, and wood screws. Figures 5.13 and 5.14 show the American standard socket set screws with dimensions and applications data. The size of the hex wrench required for each set-screw size is also shown in Fig. 5.13 under Column "W." Figure 5.15 shows a chart of the standard self-tapping and thread-forming screws. Types AB, A, B, BF, and C are thread-forming, while types D, F, G, T, BF, and BT are thread-cutting or self-tapping. Type U is a spiral screw type that is driven or press-fit into the appropriate-sized hole. All thread forms shown are 60° V thread.

Holes for self-tapping and thread-forming screws. Extensive tables of recommended hole sizes for self-tapping and thread-forming screws are available from screw manufacturers. Because of the great variety of materials and thickness ranges possible for applying these types of screws, no tables are given here. In actual practice, you may measure the outside diameter of the

SAE bolt head markings	Diameter, in.	Tensile strength, psi	Proof load, psi	
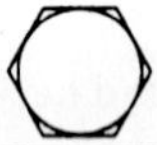	1/4 - 3/4 7/8 - 1 1/2	74,000 psi 60,000	55,000 psi 33,000	Grade 2, *low or medium-carbon steel
	1/4 - 1 1 1/8 - 1 1/2	120,000 105,000	85,000 74,000	Grade 5, *medium carbon steel (Q/T)†
	-	120,000	85,000	Grade 5.1, screws
	1/4 - 1	120,000	85,000	Grade 5.2, low-carbon martensitic steel (Q/T)
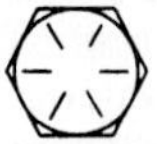	1/4 - 1 1/2	150,000	120,000	Grade 8, medium-carbon alloy steel (Q/T)
	1/4 - 1	150,000	120,000	Grade 8.2, low-carbon martensitic steel (Q/T)

*Highest usage grades †(Q/T) designates quenched and tempered

FIGURE 5.10 SAE bolt-head grade markings.

ASTM bolt head markings	Diameter, in.	Tensile strength, psi	Proof load, psi	
	1/4 - 4	A & B-60,000 min. B-100,000 max.		*A307 carbon steel*
	1/4 - 1 1 1/8 - 1 1/2 1 3/4 - 3	120,000 105,000 90,000	85,000 74,000 55,000	*A449 medium-carbon steel (Q/T)*
	1/2 - 1 1 1/8 - 1 1/2	120,000 105,000	85,000 74,000	*A325 Type 1 medium carbon steel (Q/T)*
	1/2 - 1 1 1/8 - 1 1/2	120,000 105,000	85,000 74,000	*A325 Type 2 low carbon martensitic steel (Q/T)*
	1/2 - 1 1 1/8 - 1 1/2	120,000 105,000	85,000 74,000	*A325 Type 3 weathering steel (Q/T)*
	1/2 - 1 1/2	150,000	120,000	*A490 alloy steel (Q/T)*
	1/4 - 4	150,000	120,000	*A354 grade BD alloy steel (Q/T)*

(Q/T) designates quenched and tempered

FIGURE 5.11 ASTM bolt-head grade markings.

Property Class	Size Range	Proof Strength Mpa (psi)	Tensile Strength Mpa (psi)	Material Type	Head Marking
4.6	M5 -M36	225 (32,625)	400 (58,000)	Low or Medium Carbon Steel	4.6
4.8	M1.6 - M16	310 (44,950)	420 (60,900)	Low or Medium Carbon Steel	4.8
5.8	M5 - M24	380 (55,100)	520 (75,400)	Low or Medium Carbon Steel	5.8
8.8	M16 - M36	600 (87,000)	830 (120,350)	Medium Carbon Steel, Q & T	8.8
9.8	M1.6 - M16	650 (94,250)	900 (130,500)	Medium Carbon Steel, Q & T	9.8
10.9	M5 - M36	830 (120,350)	1040 (150,800)	Low Carbon Martensitic Steel, Q&T	10.9
12.9	M1.6 - M36	970 (140,650)	1220 (176,900)	Alloy Steel, Q&T	12.9

Sources: Derived from SAE Standard J1199, and ASTM Standard F568. Q&T = quenched and tempered.
Note: the psi figures have been rounded to the nearest 25 psi increment. 145.0377 psi = 1 MPa

FIGURE 5.12 Metric bolt-head grade markings and strengths.

TABLE 5.1 Tightening Torque Ranges for American Standard Steel Bolts (Hardware, Dry, Nonlubricated, and Zinc Plated)

	SAE grade 2		SAE grade 5		SAE grade 8	
Bolt size	Tightening torque range, lb·ft	Clamp load range, lb	Tightening torque range, lb·ft	Clamp load range, lb	Tightening torque range, lb·ft	Clamp load range, lb
¼–20	5.7–4.3	1,813–1,360	9.1–6.9	2,926–2,195	12.9–9.7	4,134–3,101
¼–28	6.5–4.9	2,075–1,556	10.5–7.9	3,349–2,512	14.8–11.1	4,732–3,549
5⁄16–18	11.7–8.8	2,987–2,240	18.8–14.1	4,821–3,616	26.2–20.0	6,812–5,109
5⁄16–24	12.9–9.7	3,306–2,480	20.8–15.6	5,336–4,002	29.5–22.1	7,540–5,655
⅜–16	20.7–15.5	4,418–3,314	33.4–25.1	7,130–5,348	47.2–35.4	10,075–7,556
⅜–24	23.5–17.6	5,005–3,754	37.9–28.4	8,078–6,059	53.5–40.1	11,414–8,561
7⁄16–14	33.1–24.9	6,059–4,544	53.5–40.1	9,780–7,335	75.6–56.7	13,819–10,364
7⁄16–20	37.0–27.8	6,766–5,075	59.7–44.8	10,920–8,190	84.4–63.3	15,431–11,573
½–13	50.6–37.9	8,088–6,066	81.6–61.2	13,055–9,791	115.3–86.5	18,447–13,835
½–20	57.0–42.7	9,114–5,835	91.9–69.0	14,711–11,033	130.0–97.4	20,787–15,590
9⁄16–12	73.0–54.7	10,374–7,780	117.7–88.1	16,744–12,558	166.4–124.8	23,660–17,745
9⁄16–18	81.4–61.0	11,571–8,678	131.3–98.1	18,676–14,007	185.6–139.2	26,390–19,793
⅝–11	100.6–75.5	12,882–9,662	162.4–121.8	20.792–15,594	229.5–172.1	29,380–22,035
⅝–18	114–85.5	14,592–10,944	184–138	23,552–17,664	260.0–195.0	33,280–24,960
¾–10	178.5–133.9	19,038–14,279	288–216	30,728–23,046	407.1–305.3	43,420–35,368
¾–16	199–149.5	21,261–15,946	321.7–241.3	34,316–25,737	454.6–341.0	48,490–45,045
⅞–9	288–216	26,334–19,751	464.9–348.7	42,504–31,878	656.9–492.7	60,060–45,045
⅞–14	317–238	29,013–19,751	512.2–384.1	46,828–35,121	723.7–542.8	66,170–49,628
1–8	432–324	34,542–25,907	696.9–522.7	55,752–41,814	984.8–738.6	78,780–59,085
1–12	472–354	37,791–28,343	761.1–571.8	60,996–45,747	1077–808	86,190–64,643

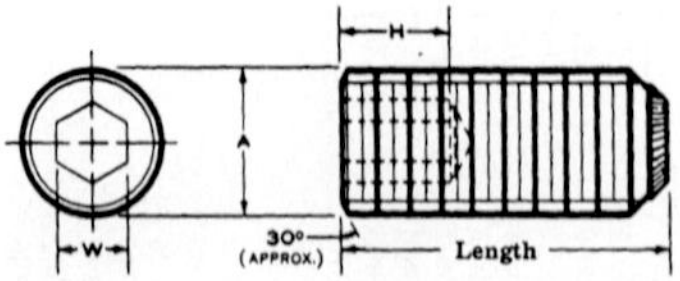

Dimensions

size	threads per inch UNC	threads per inch UNF	A max.	A min. UNC	A min. UNF	C max.	C min.	D max.	D min.	F max.	F min.	H min.	W nom.
#0	....	80	.0600	....	.0568	.033	.027	.040	.037	.017	.013	.022	.028
#1	64	72	.0730	.0692	.0695	.040	.033	.049	.045	.021	.017	.028	.035
#2	56	64	.0860	.0819	.0822	.047	.039	.057	.053	.024	.020	.028	.035
#3	48	56	.0990	.0945	.0949	.054	.045	.066	.062	.027	.023	.040	.050
#4	40	48	.1120	.1069	.1075	.061	.051	.075	.070	.030	.026	.040	.050
#5	40	44	.1250	.1199	.1202	.067	.057	.083	.078	.033	.027	.050	.0625
#6	32	40	.1380	.1320	.1329	.074	.064	.092	.087	.038	.032	.050	.0625
#8	32	36	.1640	.1580	.1585	.087	.076	.109	.103	.043	.037	.062	.0781
#10	24	32	.1900	.1828	.1840	.102	.088	.127	.120	.049	.041	.075	.0937
1/4	20	28	.2500	.2419	.2435	.132	.118	.156	.149	.0565	.0585	.100	.125
5/16	18	24	.3125	.3038	.3053	.172	.156	.203	.195	.082	.074	.125	.1562
3/8	16	24	.3750	.3656	.3678	.212	.194	.250	.241	.0987	.0887	.150	.1875
7/16	14	20	.4375	.4272	.4294	.252	.232	.296	.287	.114	.104	.175	.2187
1/2	13	20	.5000	.4891	.4919	.291	.270	.343	.334	.130	.120	.200	.250
9/16	12	18	.5625	.5511	.5538	.332	.309	.390	.379	.1456	.1356	.200	.250
5/8	11	18	.6250	.6129	.6163	.371	.347	.468	.456	.164	.148	.250	.3125
3/4	10	16	.7500	.7371	.7406	.450	.425	.562	.549	.1955	.1795	.300	.375
7/8	9	14	.8750	.8611	.8647	.530	.502	.656	.642	.2267	.2107	.400	.500
1	8	12	1.0000	.9850	.9886	.609	.579	.750	.734	.260	.240	.450	.5625
1 1/8	7	12	1.1250	1.1086	1.1136	.689	.655	.843	.826	.291	.271	.450	.5625
1 1/4	7	12	1.2500	1.2336	1.2386	.767	.733	.937	.920	.3225	.3025	.500	.625
1 3/8	6	12	1.3750	1.3568	1.3636	.848	.808	1.031	1.011	.3537	.3337	.500	.625
1 1/2	6	12	1.5000	1.4818	1.4886	.926	.886	1.125	1.105	.385	.365	.600	.750

FIGURE 5.13 Socket set-screw dimensions.

screw thread and the root diameter and then take the mean difference to find an approximate hole diameter for the particular screw and material combination. A few trial combinations will give you the exact drill size to use for your application.

Figure 5.16 shows American standard wood screws and their dimensions. Wood screws are used by pattern makers and in wooden form blocks for vacuum-forming equip-

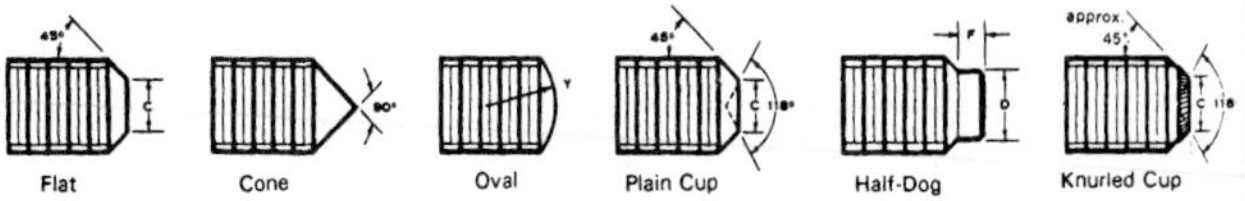

Application Data

Y	tap drill size			recommended seating torques - Pound-inches								
				regular and LOC-WEL screws			self-locking with NYLOK					
rad.	size	UNC	UNF	min. screw length	alloy steel	stain-less	min. screw length	alloy steel	stain-less	min. screw length	alloy steel	stain-less
.047	#0	....	1.25mm	3/32	.5	.4	3/32	.4	.3	1/8	.5	.4
.055	#1	1.5mm	1.5mm	3/32	1.5	1.2	1/8	1.2	1.0	5/32	1.5	1.2
.062	#2	#50	1.85mm	3/32	1.5	1.2	1/8	1.2	1.0	5/32	1.5	1.2
.078	#3	#46	2.1mm	3/32	5	4	5/32-3/16	4	3	7/32	5	4
.084	#4	2.3mm	#42	3/32	5	4	5/32-3/16	4	3	7/32	5	4
.093	#5	#37	#37	5/32	9	7	3/16	7	6	1/4	9	7
.109	#6	#33	#32	1/8	9	7	3/16	7	6	1/4	9	7
.125	#8	#29	3.5mm	1/8	20	16	1/4	16	13	5/16	20	16
.141	#10	#24	#20	5/32	33	26	1/4-5/16	28	22	3/8	33	26
.188	1/4	#6	5.5mm	3/16	87	70	1/4-7/16	52	42	1/2	87	70
.234	5/16	G	I	1/4	165	130	5/16-7/16	90	72	5/8	165	130
.281	3/8	O	8.6mm	5/16	290	230	3/8-5/8	200	160	3/4	290	230
.328	7/16	9.4mm	25/64	3/8	430	340	7/16-3/4	300	240	7/8	430	340
.375	1/2	27/64	11.5mm	1/2	620	500	1/2-3/4	500	400	7/8	620	500
.422	9/16	31/64	1/2	1/2	620	500	1/2-3/4	500	400	7/8	620	500
.468	5/8	17/32	14.5mm	9/16	1,225	980	5/8-7/8	980	780	1	1,225	980
.562	3/4	21/32	17.5mm	5/8	2,125	1,700	3/4-1 1/8	1,700	1,360	1 1/4	2,125	1,700
.656	7/8	49/64	20.5mm	3/4	5,000	4,000	3/4-1 1/4	3,650	2,920	1 5/16	5,000	4,000
.750	1	7/8	23.5	7/8	7,000	5,600	7/8-1 5/16	5,200	4,160	1 1/2	7,000	5,600
.844	1 1/8	25mm	1 3/64	1	7,000	5,600	....	....	....	....	....	....
.938	1 1/4	1 7/64	1 11/64	1 1/8	9,600	7,700	....	....	....	....	....	....
1.032	1 3/8	1 7/32	1 19/64	1 1/4	9,600	7,700	....	....	....	....	....	....
1.125	1 1/2	34mm	36mm	1 1/4	11,320	9,100	....	....	....	....	....	....

Note: Materials: High grade alloy steel, austenitic stainless steel.
Hardness: Rc 45-53 for alloy steel grades. Thread class: 3A.

FIGURE 5.14 Set-screw point types.

ment applications (vacuum-forming plastics). Head styles of the various machine screws are shown in Fig. 5.17.

Wrench clearances for design applications

An important point that is sometimes forgotten in basic mechanical design work is wrench clearances. Figures 5.18 and 5.19 show an extensive listing of required clearances

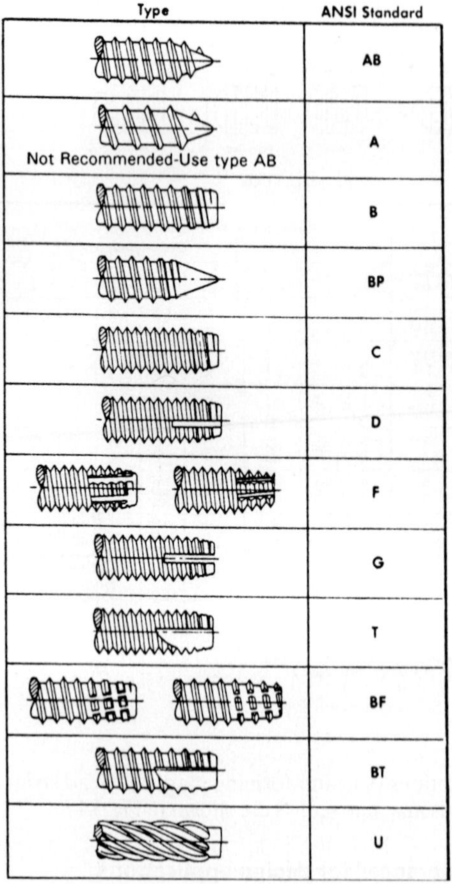

FIGURE 5.15 Self-tapping and thread-forming screw types.

Flat Head

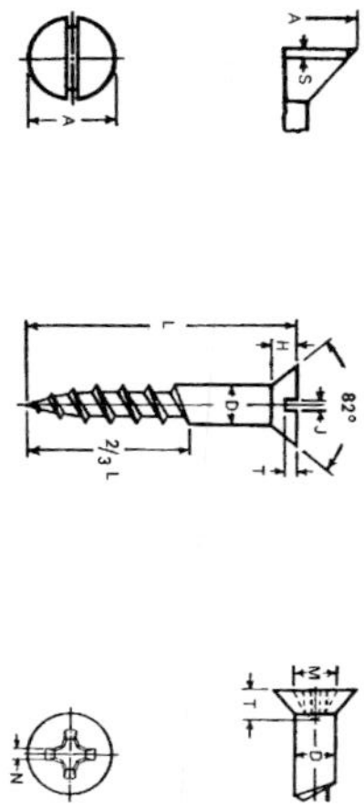

	D	A			S	H		J		T			D	M		T	N	
		SLOTTED											**PHILLIPS**					
Nominal Size	Basic Diameter of Screw	Head Diameter			Flat on Min Screw	Height of Head		Width of Slot		Depth of Slot		Number Threads per Inch	Basic Diameter of Screw	Diameter of Recess		Depth of Recess	Width of Recess	Driver Size
		Max Sharp	Min Sharp	Absolute Min with Max S		Max	Min	Max	Min	Max	Min			Max	Min	Max	Min	
0	0.060	0.119	0.105	0.099	0.002	0.035	0.026	0.023	0.016	0.015	0.010	32	0.060	0.069	0.056	0.043	0.014	0
1	0.073	0.146	0.130	0.123	0.003	0.043	0.033	0.026	0.019	0.019	0.012	28	0.073	0.077	0.064	0.051	0.015	0
2	0.086	0.172	0.156	0.147	0.003	0.051	0.040	0.031	0.023	0.023	0.015	26	0.086	0.102	0.089	0.063	0.017	1
3	0.099	0.199	0.181	0.171	0.004	0.059	0.048	0.035	0.027	0.027	0.017	24	0.099	0.107	0.094	0.068	0.018	1
4	0.112	0.225	0.207	0.195	0.004	0.067	0.055	0.039	0.031	0.030	0.020	22	0.112	0.128	0.115	0.089	0.018	1
5	0.125	0.252	0.232	0.220	0.005	0.075	0.062	0.043	0.035	0.034	0.022	20	0.125	0.154	0.141	0.086	0.027	2
6	0.138	0.279	0.257	0.244	0.005	0.083	0.069	0.048	0.039	0.038	0.024	18	0.138	0.174	0.161	0.106	0.029	2
7	0.151	0.305	0.283	0.268	0.005	0.091	0.076	0.048	0.039	0.041	0.027	16	0.151	0.189	0.176	0.121	0.031	2
8	0.164	0.332	0.308	0.292	0.006	0.100	0.084	0.054	0.045	0.045	0.029	15	0.164	0.204	0.191	0.136	0.032	2
9	0.177	0.358	0.334	0.316	0.006	0.108	0.091	0.054	0.045	0.049	0.032	14	0.177	0.214	0.201	0.146	0.033	2
10	0.190	0.385	0.359	0.340	0.007	0.116	0.098	0.060	0.050	0.053	0.034	13	0.190	0.258	0.245	0.146	0.034	3
12	0.216	0.438	0.410	0.389	0.008	0.132	0.112	0.067	0.056	0.060	0.039	11	0.216	0.283	0.270	0.171	0.036	3
14	0.242	0.491	0.461	0.437	0.009	0.148	0.127	0.075	0.064	0.068	0.044	10	0.242	0.303	0.290	0.191	0.039	3
16	0.268	0.544	0.512	0.485	0.010	0.164	0.141	0.075	0.064	0.075	0.049	9	0.268	0.327	0.314	0.216	0.045	3
18	0.294	0.597	0.563	0.534	0.011	0.180	0.155	0.084	0.072	0.083	0.054	8	0.294	0.378	0.365	0.230	0.062	4
20	0.320	0.650	0.614	0.582	0.012	0.196	0.170	0.084	0.072	0.090	0.059	8	0.320	0.393	0.380	0.245	0.065	4
24	0.372	0.756	0.716	0.679	0.013	0.228	0.198	0.094	0.081	0.105	0.069	7	0.372	0.424	0.411	0.276	0.069	4

FIGURE 5.16 American standard wood screw dimensions.

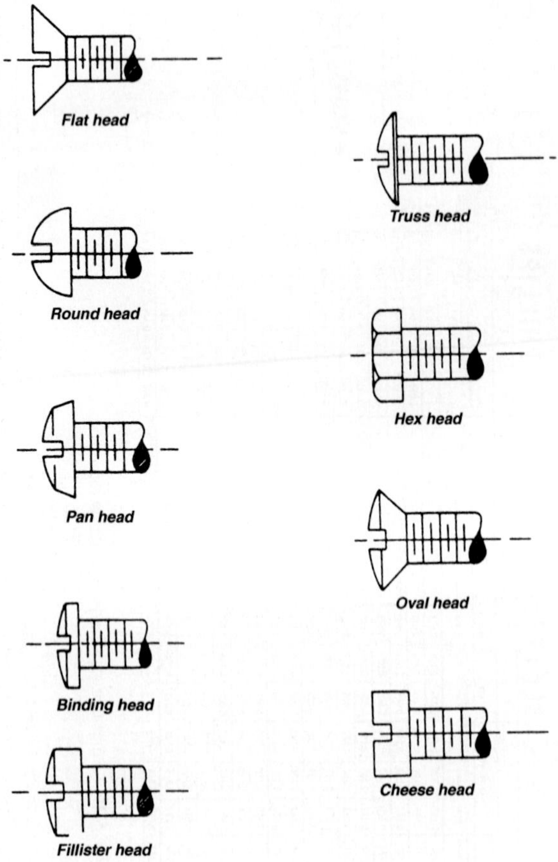

FIGURE 5.17 Head styles of standard machine screws.

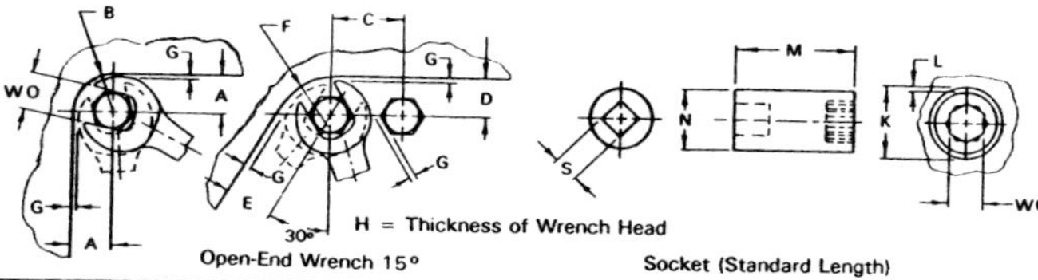

P = Torque that wrench will withstand, pound-inches

J = Torque that wrench will withstand, pound-inches

W.O.	Open End Wrench-15°											Socket Wrench (Standard Length)											
												S = 0.25			S = 0.375			S = 0.500			S = 0.750		
Wrench Opening	A min.	B max.	C min.	D min.	E min.	F max.	G ref.	H max.	J min.	K min.	L ref.	M max.	N max.	P min.	M max.	N max.	P min.	M max.	N max.	P min.	M max.	N max.	P min.
0.156	0.220	0.250	0.390	0.160	0.250	0.200	0.030	0.094	25	....	....	....	....	....	....	....	....	....	....	....	....	....	....
0.188	0.250	0.280	0.430	0.190	0.270	0.230	0.030	0.172	40	0.370	0.030	1.000	0.510	125	....	....	....	....	....	....	....	....	....
0.250	0.280	0.340	0.530	0.270	0.310	0.310	0.030	0.172	80	0.470	0.030	1.000	0.510	200	1.250	0.690	250	....	....	....	....	....	....
0.312	0.380	0.470	0.680	0.280	0.390	0.390	0.050	0.203	125	0.550	0.030	1.000	0.510	300	1.250	0.690	400	....	....	....	....	....	....
0.344	0.420	0.500	0.750	0.340	0.450	0.450	0.050	0.203	175	0.580	0.030	1.000	0.519	450	1.250	0.690	675	....	....	....	....	....	....
0.375	0.420	0.500	0.780	0.360	0.450	0.520	0.050	0.219	250	0.620	0.030	1.000	0.580	550	1.250	0.690	900	1.500	0.880	1600	....	....	....
0.438	0.470	0.590	0.890	0.420	0.520	0.640	0.050	0.250	375	0.750	0.030	1.000	0.683	550	1.250	0.880	1250	1.500	0.940	1700	....	....	....
0.500	0.520	0.640	1.000	0.470	0.580	0.660	0.050	0.266	490	0.810	0.030	1.000	0.692	600	1.250	0.880	1450	1.500	0.940	2000	....	....	....
0.562	0.590	0.770	1.130	0.520	0.660	0.700	0.050	0.297	700	0.870	0.030	....	....	....	1.250	0.932	1600	1.500	0.940	2700	....	....	....
0.594	0.640	0.830	1.210	0.530	0.700	0.700	0.050	0.344	800	0.920	0.030	....	....	....	1.250	0.963	1750	1.562	0.970	3000	....	....	....
0.625	0.640	0.830	1.230	0.550	0.700	0.700	0.050	0.344	935	0.960	0.030	....	....	....	1.250	0.995	2000	1.562	1.000	3600	....	....	....
0.688	0.770	0.920	1.470	0.660	0.880	0.800	0.080	0.375	1250	1.030	0.030	....	....	....	1.250	1.058	2000	1.562	1.066	4300	....	....	....
0.750	0.770	0.920	1.510	0.670	0.880	0.800	0.080	0.375	1500	1.120	0.030	....	....	....	1.250	1.120	2000	1.562	1.130	5000	....	....	....
0.781	0.830	0.950	1.550	0.690	0.890	0.840	0.080	0.375	1615	1.150	0.030	....	....	....	1.250	1.126	2000	1.625	1.130	5000	....	....	....
0.812	0.910	1.120	1.660	0.720	0.970	0.860	0.080	0.406	1710	1.200	0.030	....	....	....	1.250	1.213	2000	1.625	1.222	5000	....	....	....
0.875	0.970	1.150	1.810	0.800	1.060	0.910	0.080	0.438	2250	1.280	0.030	....	....	....	....	....	....	1.750	1.286	5000	....	....	....
0.938	0.970	1.150	1.850	0.810	1.060	0.950	0.080	0.438	2750	1.370	0.030	....	....	....	....	....	....	1.750	1.410	5000	....	....	....
1.000	1.050	1.230	2.00	0.880	1.160	1.060	0.080	0.500	3250	1.470	0.030	....	....	....	....	....	....	1.750	1.410	5000	....	....	....
1.062	1.090	1.250	2.100	0.970	1.200	1.200	0.080	0.500	3500	1.550	0.030	....	....	....	....	....	....	1.844	1.505	5000	....	....	....
1.125	1.140	1.370	2.210	1.000	1.270	1.230	0.080	0.500	4000	1.610	0.030	....	....	....	....	....	....	1.939	1.567	5000	....	....	....
1.250	1.270	1.420	2.440	1.080	1.390	1.310	0.080	0.562	5250	1.890	0.030	....	....	....	....	....	....	2.000	1.723	5000	2.375	1.855	7250
1.312	1.390	1.690	2.630	1.170	1.520	1.340	0.080	0.562	6000	1.980	0.030	....	....	....	....	....	....	....	....	....	2.500	1.920	8000
1.438	1.470	1.720	2.800	1.250	1.590	1.340	0.090	0.641	7500	2.140	0.030	....	....	....	....	....	....	....	....	....	2.625	2.075	9550
1.500	1.470	1.720	2.840	1.270	1.590	1.450	0.090	0.641	8250	2.200	0.030	....	....	....	....	....	....	....	....	....	2.625	2.170	10450
1.625	1.560	1.880	3.100	1.380	1.750	1.560	0.090	0.641	9000	2.390	0.030	....	....	....	....	....	....	....	....	....	2.750	2.325	11750

Note: P = Torque that wrench will withstand, pound-inches. No allowances have been made for torque devices. H = wrench head thickness. J = Torque that wrench will withstand, pound-inches. Tabulated dimensions are in inches.

FIGURE 5.18 Wrench clearance chart. Clearances for open-end wrench 15° and socket wrench (standard length) (derived from SAE Aeronautical Standard).

Wrench Clearances for Standard Box Wrenches - 12 Point Type
(Derived from SAE Aeronautical Standard)

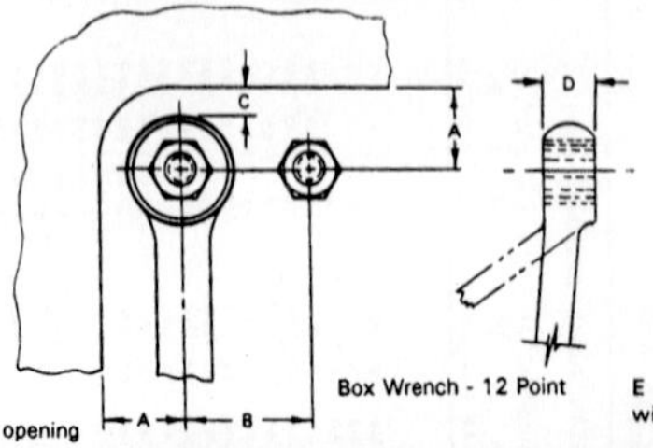

W.O.	A min.	B min.	C ref.	D max.	E min.
0.156	0.190	0.280	0.030	0.156	100
0.188	0.200	0.309	0.030	0.172	150
0.250	0.270	0.410	0.030	0.250	150
0.312	0.300	0.480	0.030	0.281	210
0.344	0.300	0.500	0.030	0.281	250
0.375	0.340	0.560	0.030	0.344	370
0.438	0.400	0.650	0.030	0.359	650
0.500	0.450	0.740	0.030	0.375	1020
0.562	0.500	0.830	0.030	0.406	1200
0.594	0.530	0.870	0.030	0.469	1200
0.625	0.560	0.920	0.030	0.469	2000
0.688	0.590	0.990	0.030	0.531	2300
0.750	0.660	1.090	0.030	0.594	2600
0.781	0.690	1.140	0.030	0.594	2600
0.812	0.720	1.190	0.030	0.594	3000
0.875	0.750	1.260	0.030	0.594	3300
0.938	0.780	1.320	0.030	0.656	4100
1.000	0.810	1.390	0.030	0.718	4900
1.062	0.840	1.450	0.030	0.781	5400
1.125	0.950	1.600	0.030	0.844	5900
1.250	0.980	1.700	0.030	0.875	7200
1.312	1.090	1.850	0.030	0.906	8000
1.438	1.220	2.050	0.030	1.000	8400
1.500	1.270	2.140	0.030	1.062	10450
1.625	1.340	2.280	0.030	1.156	11750

FIGURE 5.19 Box wrench clearance chart.

for open-end, box, and socket wrenches. Allowances should be made in initial design work to allow for more clearance than shown in the figures when possible. Tabulated dimensions shown in the figures are minimum allowables. No allowances have been made for torquing devices, since these are extensive and variable according

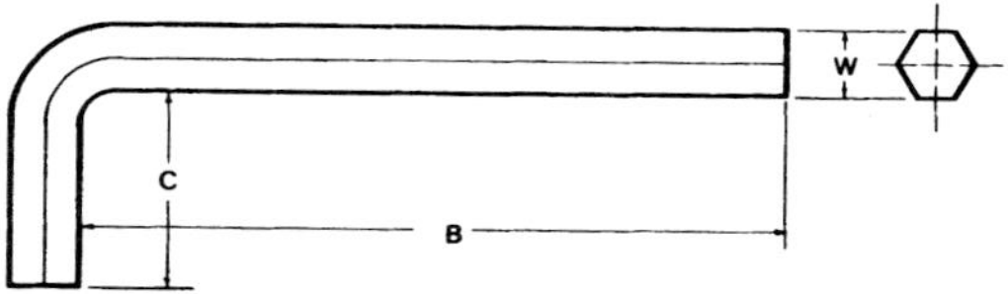

dash no.	key size W max.	key size W min.	B nom. short arm	B nom. long arm	C nom.	torsional shear strength Lb-ins min.	torsional yield Lb-ins min.
1	.028	.0275	1.219	2.594	.219	1.2	1.1
2	.035	.0345	1.219	2.672	.344	2.4	2.1
3	.050	.049	1.656	2.844	.531	7	6.0
4	1/16	.0615	1.750	3.000	.562	12	11
5	5/64	.0771	1.875	3.188	.609	26	23
6	3/32	.0927	2.000	3.375	.656	46	40
7	7/64	.1077	2.125	3.562	.703	73	63
8	1/8	.1235	2.250	3.750	.750	108	94
9	9/64	.1391	2.375	3.960	.796	154	134
10	5/32	.1547	2.500	4.125	.844	210	183
11	3/16	.1860	2.750	4.500	.938	364	317
12	7/32	.2172	3.000	4.875	1.031	580	502
13	1/4	.2480	3.250	5.250	1.125	860	750
14	5/16	.3110	3.750	6.000	1.250	1,685	1,465
15	3/8	.3730	4.250	6.750	1.375	2,900	2,520
16	7/16	.4355	4.750	7.500	1.500	4,400	3,860
17	1/2	.4975	5.250	8.250	1.625	6,600	5,750
18	9/16	.5600	5.750	9.000	1.750	9,200	8,000
19	5/8	.6225	6.250	9.750	1.875	12,650	11,000
20	3/4	.7470	7.250	11.250	2.125	20,800	18,100
21	7/8	.8720	8.250	12.750	2.375	29,200	25,400
22	1	.9970	9.250	14.250	2.625	43,700	38,000
23	1 1/4	1.243	11.250	....	3.000	71,900	62,500
24	1 1/2	1.493	13.250	....	3.500	124,000	108,000
25	1 3/4	1.743	15.250	....	4.000	198,000	172,000
26	2	1.993	17.250	....	4.500	276,000	240,000

FIGURE 5.20 Hex-key table, Allen wrenches. Dimensions and properties of hex keys (Allen wrenches, inches).

Maximum Width Across Flats of Nut	Wrench Opening	
	Minimum	Maximum
5/32	0.158	0.163
3/16	0.190	0.195
7/32	0.220	0.225
1/4	0.252	0.257
9/32	0.283	0.288
5/16	0.316	0.322
11/32	0.347	0.353
3/8	0.378	0.384
7/16	0.440	0.446
1/2	0.504	0.510
9/16	0.566	0.573
5/8	0.629	0.636
11/16	0.692	0.699
3/4	0.755	0.763
13/16	0.818	0.826
7/8	0.880	0.888
15/16	0.944	0.953
1	1.006	1.015
1-1/16	1.068	1.077
1-1/8	1.132	1.142
1-1/4	1.257	1.267
1-5/16	1.320	1.331
1-3/8	1.383	1.394
1-7/16	1.446	1.457
1-1/2	1.508	1.520
1-5/8	1.634	1.646
1-11/16	1.696	1.708
1-13/16	1.822	1.835
1-7/8	1.885	1.898
2	2.011	2.025
2-1/16	2.074	2.088
2-3/16	2.200	2.215
2-1/4	2.262	2.277
2-3/8	2.388	2.404
2-7/16	2.450	2.466
2-9/16	2.576	2.593
2-5/8	2.639	2.656
2-3/4	2.766	2.783
2-13/16	2.827	2.845
2-15/16	2.954	2.973
3	3.016	3.035

Note: Wrenches are marked with the "Nominal size of wrench," which is equal to the basic or maximum width across the flats of the corresponding nut.

FIGURE 5.21 Wrench openings for American standard nuts.

FIGURE 5.22 Samples of various hardware items.

FIGURE 5.23 Flanged "whiz" bolts, grades 5 and 8. An excellent lock-bolt design.

to manufacturer. Figure 5.20 shows the dimensions and mechanical properties of hex keys (Allen wrenches).

Figure 5.21 shows the wrench openings for American standard nuts. Figures 5.22 and 5.23 illustrate typical hardware components, including some of the screws, bolts, nuts, washers, pins, and special fasteners that are used throughout industry. The numbers and styles of standard hardware components are enormous, and new components are being developed constantly. Hardware manufacturers produce catalogs that show the dimensions and mechanical properties of their particular fasteners or hardware components. The fastener handbooks available from the Industrial Fasteners Institute (IFI) are among the most complete and technically correct for both the inch series and metric components.

5.2 PIN DIMENSIONS—CLEVIS, COTTER, SPRING, SPIRAL, TAPER, DOWEL, AND GROOVED

The styles, nominal sizes, and dimensions of the various pins used in industrial applications are shown in the following figures. Dimensions for clevis pins are shown in Figure 5.24. Cotter pin dimensions are shown in Figure 5.25. Dimensions for spring pins, often called roll-pins or split-pins, are shown in Figure 5.26. Spiral spring pin dimensions are shown in Figure 5.27. Standard taper pin size numbers and dimensions are shown in Figure 5.28. Dowel pins and dimensions are shown in Figure 5.29. Grooved pin styles and dimensions are shown in Figure 5.30. Standard sizes and lengths of grooved pins are shown in Figure 5.31. The recommended hole sizes for inserting or installing grooved pins are shown in Figure 5.32, and the minimum double-shear loads allowed for

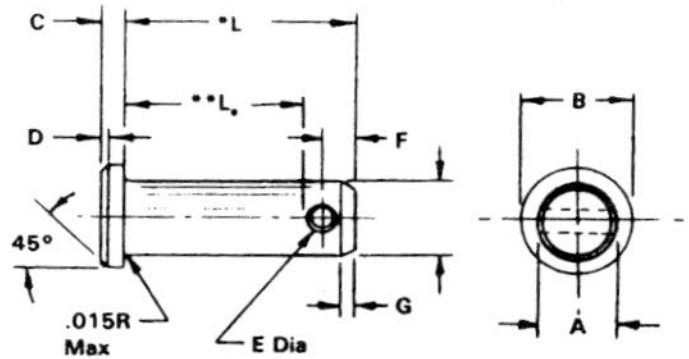

	A		B	C	D	E				
Basic Pin Diameter	Shank Diameter Max	Min	Head Diameter Max	Head Height Max	Head Chamfer ± 0.01	Hole Diameter Max	Min	F	G	Cotter Pin Size
0.188	0.186	0.181	0.32	0.07	0.02	0.088	0.073	0.09	0.055	1/16
0.250	0.248	0.243	0.38	0.10	0.03	0.088	0.073	0.09	0.055	1/16
0.312	0.311	0.306	0.44	0.10	0.03	0.119	0.104	0.12	0.071	3/32
0.375	0.373	0.368	0.51	0.13	0.03	0.119	0.104	0.12	0.071	3/32
0.438	0.436	0.431	0.57	0.16	0.04	0.119	0.104	0.12	0.071	3/32
0.500	0.496	0.491	0.63	0.16	0.04	0.151	0.136	0.15	0.089	1/8
0.625	0.621	0.616	0.82	0.21	0.06	0.151	0.136	0.15	0.089	1/8
0.750	0.746	0.741	0.94	0.26	0.07	0.182	0.167	0.18	0.110	5/32
0.875	0.871	0.866	1.04	0.32	0.09	0.182	0.167	0.18	0.110	5/32
1.000	0.996	0.991	1.19	0.35	0.10	0.182	0.167	0.18	0.110	5/32

*L = Total length under head
**L_e = Length of effective grip
Effective grip lengths must be selected from manufacturer's catalogs.

FIGURE 5.24 Standard clevis-pin dimensions, inches.

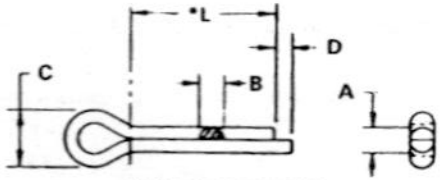

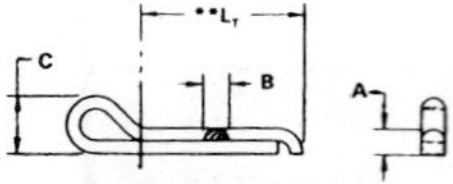

Basic Pin Diameter	Total Shank Diameter A	Wire Width Maximum B	Head Diameter Minimum C	Extended Prong Length, Minimum D	Recommended Hole Size
0.031	0.032	0.032	0.06	0.01	0.047
0.047	0.048	0.048	0.09	0.02	0.062
0.062	0.060	0.060	0.12	0.03	0.078
0.078	0.076	0.076	0.16	0.04	0.094
0.094	0.090	0.090	0.19	0.04	0.109
0.109	0.104	0.104	0.22	0.05	0.125
0.125	0.120	0.120	0.25	0.06	0.141
0.141	0.134	0.134	0.28	0.06	0.156
0.156	0.150	0.150	0.31	0.07	0.172
0.188	0.176	0.176	0.38	0.09	0.203
0.219	0.207	0.207	0.44	0.10	0.234
0.250	0.225	0.225	0.50	0.11	0.266
0.312	0.280	0.280	0.62	0.14	0.312
0.375	0.335	0.335	0.75	0.16	0.375
0.438	0.406	0.406	0.88	0.20	0.438
0.500	0.473	0.473	1.00	0.23	0.500
0.625	0.598	0.598	1.25	0.30	0.625
0.750	0.723	0.723	1.50	0.36	0.750

* L = Length
** L_T = Total length
Allow extra length for spreading and securing
Available lengths to be selected from manufacturer's catalogs.

FIGURE 5.25 Cotter-pin dimensions, inches.

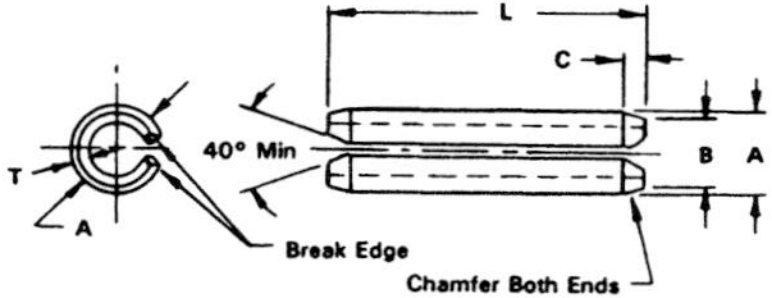

	A		B	C	T					
Basic Pin Diameter	Pin Diameter Max	Min	Chamfer Dia.	Chamfer Lth.	Stock Thickness	Hole Diameter Recommended Max	Min	Double Shear Load, Lb. ♦ AISI 1070 1095 & 420	AISI 302	Beryllium Copper
0.062	0.069	0.066	0.059	0.028	0.012	0.065	0.062	425	350	270
0.078	0.086	0.083	0.075	0.032	0.018	0.081	0.078	650	550	400
0.094	0.103	0.099	0.091	0.038	0.022	0.097	0.094	1,000	800	660
0.125	0.135	0.131	0.122	0.044	0.028	0.129	0.125	2,100	1,500	1,200
0.141	0.149	0.145	0.137	0.044	0.028	0.144	0.140	2,200	1,600	1,400
1.156	0.167	0.162	0.151	0.048	0.032	0.160	0.156	3,000	2,000	1,800
0.188	0.199	0.194	0.182	0.055	0.040	0.192	0.187	4,400	2,800	2,600
0.219	0.232	0.226	0.214	0.065	0.048	0.224	0.219	5,700	3,550	3,700
0.250	0.264	0.258	0.245	0.065	0.048	0.256	0.250	7,700	4,600	4,500
0.312	0.328	0.321	0.306	0.080	0.062	0.318	0.312	11,500	7,100	6,800
0.375	0.392	0.385	0.368	0.095	0.077	0.382	0.375	17,600	10,000	10,100
0.438	0.456	0.448	0.430	0.095	0.077	0.445	0.437	20,000	12,000	12,200
0.500	0.521	0.513	0.485	0.110	0.094	0.510	0.500	25,800	15,500	16,800
0.625	0.650	0.640	0.608	0.125	0.125	0.636	0.625	46,000	18,800	.....
0.750	0.780	0.769	0.730	0.150	0.150	0.764	0.750	66,000	23,200	.....

Length L, is selected from manufacturer's catalogs. ♦ Other materials may be available.

FIGURE 5.26 Spring-pin dimensions, slotted type, inches.

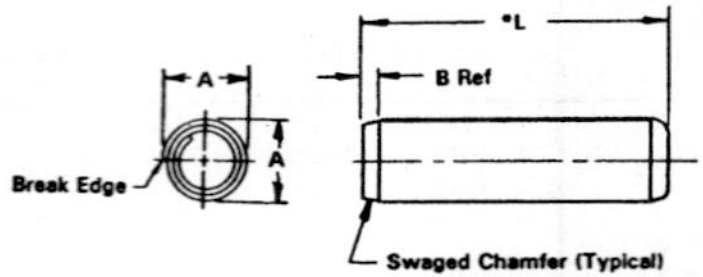

	A			B								
Basic Pin Diameter	Pin Diameter Std Duty Max	Pin Diameter Heavy Duty Max	Pin Diameter Light Duty Max	Chamfer Length Ref.	Hole Size Max	Hole Size Min	Double Shear Load, Min., Lb ♦ Std Duty 1070, 1095 and 420	Std Duty 302	Heavy Duty 1070, 1095 and 420	Heavy Duty 302	Light Duty 1070, 1095 and 420	Light Duty 302
0.031	0.035	-----	-----	0.024	0.032	0.031	75	60	-----	-----	-----	-----
0.047	0.052	-----	-----	0.024	0.048	0.046	170	140	-----	-----	-----	-----
0.062	0.072	0.070	0.073	0.028	0.065	0.061	300	250	450	350	-----	135
0.078	0.088	0.086	0.089	0.032	0.081	0.077	475	400	700	550	-----	225
0.094	0.105	0.103	0.106	0.038	0.097	0.093	700	550	1,000	800	375	300
0.109	0.120	0.118	0.121	0.038	0.112	0.108	950	750	1,400	1,125	525	425
0.125	0.138	0.136	0.139	0.044	0.129	0.124	1,250	1,000	2,100	1,700	675	550
0.156	0.171	0.168	0.172	0.048	0.160	0.155	1,925	1,550	3,000	2,400	1,100	875
0.188	0.205	0.202	0.207	0.055	0.192	0.185	2,800	2,250	4,400	3,500	1,500	1,200
0.219	0.238	0.235	0.240	0.065	0.224	0.217	3,800	3,000	5,700	4,600	2,100	1,700
0.250	0.271	0.268	0.273	0.065	0.256	0.247	5,000	4,000	7,700	6,200	2,700	2,200
0.312	0.337	0.334	0.339	0.080	0.319	0.308	7,700	6,200	11,500	9,200	4,440	3,500
0.375	0.403	0.400	0.405	0.095	0.383	0.370	11,200	9,000	17,600	14,000	6,000	5,000
0.438	0.469	0.466	0.471	0.095	0.446	0.431	15,200	13,000	22,500	18,000	8,400	6,700
0.500	0.535	0.532	0.537	0.110	0.510	0.493	20,000	16,000	30,000	24,000	11,000	8,800
0.625	0.661	0.658	-----	0.125	0.635	0.618	31,000	25,000	46,000	37,000	-----	-----
0.750	0.787	0.784	-----	0.150	0.760	0.743	45,000	36,000	66,000	53,000	-----	-----

* Length L is selected from manufacturer's catalogs. ♦ Other materials may be available.

FIGURE 5.27 Spiral-spring-pin dimensions, inches.

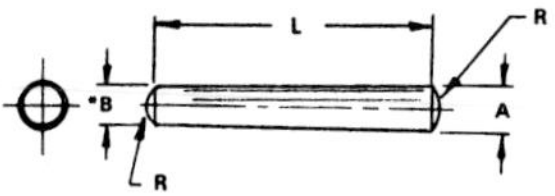

Pin Size Number and Basic Pin Dia		A Major Diameter (Large end) Commercial Class Max	Min	Precision Class Max	Min	R End Crown Radius Maximum
7/0	0.0625	0.0638	0.0618	0.0635	0.0625	0.072
6/0	0.0780	0.0793	0.0773	0.0790	0.0780	0.088
5/0	0.0940	0.0953	0.0933	0.0950	0.0940	0.104
4/0	0.1090	0.1103	0.1083	0.1100	0.1090	0.119
3/0	0.1250	0.1263	0.1243	0.1260	0.1250	0.135
2/0	0.1410	0.1423	0.1403	0.1420	0.1410	0.151
0	0.1560	0.1573	0.1553	0.1570	0.1560	0.166
1	0.1720	0.1733	0.1713	0.1730	0.1720	0.182
2	0.1930	0.1943	0.1923	0.1940	0.1930	0.203
3	0.2190	0.2203	0.2183	0.2200	0.2190	0.229
4	0.2500	0.2513	0.2493	0.2510	0.2500	0.260
5	0.2890	0.2903	0.2883	0.2900	0.2890	0.299
6	0.3410	0.3423	0.3403	0.3420	0.3410	0.351
7	0.4090	0.4103	0.4083	0.4100	0.4090	0.419
8	0.4920	0.4933	0.4913	0.4930	0.4920	0.502
9	0.5910	0.5923	0.5903	0.5920	0.5910	0.601
10	0.7060	0.7073	0.7053	0.7070	0.7060	0.716
11	0.8600	0.8613	0.8593	.....	.....	0.870
12	1.0320	1.0333	1.0313	.....	.....	1.042
13	1.2410	1.2423	1.2403	.....	.....	1.251
14	1.5210	1.5223	1.5203	.....	.....	1.531

* B dimension varies per length. Length L to be selected from manufacturer's catalogs.

FIGURE 5.28 Taper-pin dimensions, inches.

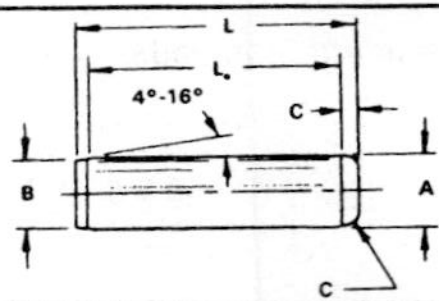

Nominal Diameter	Pin Diameter - A Standard Series Basic	Max	Min	Oversize Series Basic	Max	Min	B Point Dia Max	Min	C Crown Radius Max	Min	Double Shear Load, Lb Carbon or Alloy Steel
0.0625	0.0627	0.0628	0.0626	0.0635	0.0636	0.0634	0.058	0.048	0.020	0.008	800
0.0781	0.0783	0.0784	0.0782	0.0791	0.0792	0.0790	0.074	0.064	0.026	0.010	1,240
0.0938	0.0940	0.0941	0.0939	0.0948	0.0949	0.0947	0.089	0.079	0.031	0.012	1,800
0.1250	0.1252	0.1253	0.1251	0.1260	0.1261	0.1259	0.120	0.110	0.041	0.016	3,200
0.1562	0.1564	0.1565	0.1563	0.1572	0.1573	0.1571	0.150	0.140	0.052	0.020	5,000
0.1875	0.1877	0.1878	0.1876	0.1885	0.1886	0.1884	0.180	0.170	0.062	0.023	7,200
0.2500	0.2502	0.2503	0.2501	0.2510	0.2511	0.2509	0.240	0.230	0.083	0.031	12,800
0.3125	0.3127	0.3128	0.3126	0.3135	0.3136	0.3134	0.302	0.290	0.104	0.039	20,000
0.3750	0.3752	0.3753	0.3751	0.3760	0.3761	0.3759	0.365	0.350	0.125	0.047	28,700
0.4375	0.4377	0.4378	0.4376	0.4385	0.4386	0.4384	0.424	0.409	0.146	0.055	39,100
0.5000	0.5002	0.5003	0.5001	0.5010	0.5011	0.5009	0.486	0.471	0.167	0.063	51,000
0.6250	0.6252	0.6253	0.6251	0.6260	0.6261	0.6259	0.611	0.595	0.208	0.078	79,800
0.7500	0.7502	0.7503	0.7501	0.7510	0.7511	0.7509	0.735	0.715	0.250	0.094	114,000
0.8750	0.8752	0.8753	0.8751	0.8760	0.8761	0.8759	0.860	0.840	0.293	0.109	156,000
1.0000	1.0002	1.0003	1.0001	1.0010	1.0011	1.0009	0.980	0.960	0.333	0.125	204,000

Note: Sizes 0.0781 and 0.1562 diameter not recommended for new design.
L = Total pin length; L_e = Length of engagement.
Dowel pins listed are available in nominal lengths from 0.1875 to 6.000 inches.
Consult the manufacturer's catalogs for available lengths.

FIGURE 5.29 Dowel-pin dimensions, hardened and ground machine type, inches.

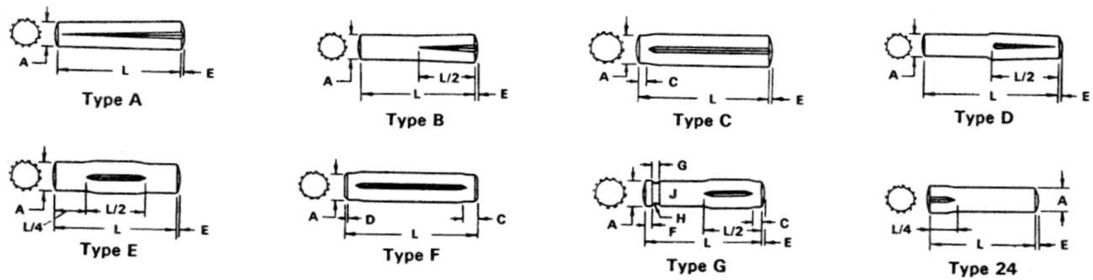

Basic Pin Diameter	A Max	C Ref.	D Min.	E Max	F Min.	G Max.	H Ref.	J Max
0.0312	0.0312	0.015	.....	.....	.....	.....	.....	.....
0.0469	0.0469	0.031	.....	.....	.....	.....	.....	.....
0.0625	0.0625	0.031	0.016	0.0115	.....	.....	.....	.....
0.0781	0.0781	0.031	0.016	0.0137	.....	.....	.....	.....
0.0938	0.0938	0.031	0.016	0.0141	0.028	0.041	0.016	0.067
0.1094	0.1094	0.031	0.016	0.0160	0.028	0.041	0.016	0.082
0.1250	0.1250	0.031	0.016	0.0180	0.059	0.041	0.031	0.088
0.1563	0.1563	0.062	0.031	0.0220	0.059	0.057	0.031	0.109
0.1875	0.1875	0.062	0.031	0.0230	0.059	0.057	0.031	0.130
0.2188	0.2188	0.062	0.031	0.0270	0.091	0.072	0.047	0.151
0.2500	0.2500	0.062	0.031	0.0310	0.091	0.072	0.047	0.172
0.3125	0.3125	0.094	0.047	0.0390	0.122	0.104	0.062	0.214
0.3750	0.3750	0.094	0.047	0.0440	0.122	0.135	0.062	0.255
0.4375	0.4375	0.094	0.047	0.0520	0.185	0.135	0.094	0.298
0.5000	0.5000	0.094	0.047	0.0570	0.185	0.135	0.094	0.317

FIGURE 5.30 Grooved-pin dimensions, type and dimensions, inches.

Nominal Length	Nominal Size														
	1/32	3/64	1/16	5/64	3/32	7/64	1/8	5/32	3/16	7/32	1/4	5/16	3/8	7/16	1/2
1/8	Y	Y	Y												
1/4	Y	Y	Y	Y	Y	Y	Y								
3/8	Y	Y	Y	Y	X	X	X	X	X						
1/2	Y	Y	Y	Y	X	X	X	X	X	X	X				
5/8		Y	Y	Y	X	X	X	X	X	X	X	X			
3/4			Y	Y	X	X	X	X	X	X	X	X	X		
7/8			Y	Y	X	X	X	X	X	X	X	X	X	X	
1			Y	Y	X	X	X	X	X	X	X	X	X	X	X
1-1/4					X	X	X	X	X	X	X	X	X	X	X
1-1/2							X	X	X	X	X	X	X	X	X
1-3/4								X	X	X	X	X	X	X	X
2								X	X	X	X	X	X	X	X
2-1/4									X	X	X	X	X	X	X
2-1/2										X	X	X	X	X	X
2-3/4										X	X	X	X	X	X
3										X	X	X	X	X	X
3-1/4											X	X	X	X	X
3-1/2												X	X	X	X
3-3/4													X	X	X
4													X	X	X
4-1/4													X	X	X
4-1/2														X	X

Note: Carbon steel pins are normally available in the marked sizes by X and Y. X designates all types of pins; Y designates all types except type G. Other lengths may be available from different manufacturers.

FIGURE 5.31 Sizes and lengths of standard grooved pins.

Recommended Hole Sizes For Grooved Pins			
Nominal Pin Size	Drill Size	Hole Diameter	
		Max	Min
1/32	1/32	0.0324	0.0312
3/64	3/64	0.0482	0.0469
1/16	1/16	0.0640	0.0625
5/64	5/64	0.0798	0.0781
3/32	3/32	0.0956	0.0938
7/64	7/64	0.1113	0.1094
1/8	1/8	0.1271	0.1250
5/32	5/32	0.1587	0.1563
3/16	3/16	0.1903	0.1875
7/32	7/32	0.2219	0.2188
1/4	1/4	0.2534	0.2500
5/16	5/16	0.3166	0.3125
3/8	3/8	0.3797	0.3750
7/16	7/16	0.4428	0.4375
1/2	1/2	0.5060	0.5000

FIGURE 5.32 Hole sizes for grooved pins.

Nominal Pin Size	Double Shear Load, Min, lb			
	Material			
	Low Carbon Steel	Alloy Steel (Rockwell C45 to 50)	Corrosion Resistant Steel	Brass
1/32	104	202	143	64
3/64	220	430	300	136
1/16	402	785	540	250
5/64	624	1,215	860	386
3/32	896	1,750	1,240	555
7/64	1,222	2,380	1,685	757
1/8	1,600	3,115	2,200	990
5/32	2,494	4,860	3,440	1,540
3/16	3,588	6,990	4,960	2,220
7/32	4,884	9,520	6,760	3,020
1/4	6,380	12,430	8,840	3,950
5/16	9,970	19,420	13,750	6,170
3/8	11,620	27,950	19,800	9,050
7/16	15,820	38,060	27,000	12,100
1/2	20,600	49,700	35,200	15,800

FIGURE 5.33 Double-shear loads for grooved pins, lbs.

the different sizes of grooved pins using different metals and alloys, are shown in Figure 5.33. Select the material that is most suitable for the intended application of the grooved pin from Figure 5.33.

5.3 SET-COLLARS, SHIMS, AND ARBOR SPACERS

Standard sizes of O.D., I.D., and thickness for set-collars are shown in Figure 5.34. Set-collars are produced in solid, clamp types, and split types, in steel, stainless steel, or aluminum alloy. Clamp types and split types are made with smooth bores (I.D.), or threaded bores or I.D.

Shims and arbor spacer sizes and thicknesses are shown in Figure 5.35. These useful mechanical components have many uses in tooling applications and mechanism assemblies. They are usually applied as spacer washers and spacer shims on shafts or in bolted assemblies. The normal material from which they are produced is carbon-steel, although other materials are available, such as stainless steel, brass, copper, or aluminum.

5.4 RETAINING RINGS AND X-WASHERS

Styles and types of retaining rings are shown in Figure 5.36. A retaining ring interchangeability chart is shown in Figure 5.37. If you need a Waldes type 5100 external ring for a military application, the chart lists this ring as type 16624. Replacability between different manufacturers for most styles are also shown in this chart. The groove diameters and groove widths for the different types and sizes of retaining rings are shown in the Waldes Truarc Retaining Ring catalog, available from

DIMENSIONS			
BORE	O. D.	WIDTH	SET SCREW
3/16	7/16	1/4	8-32
1/4	1/2	9/32	10-32
5/16	5/8	11/32	10-32
3/8	3/4	3/8	1/4-20
7/16	7/8	7/16	1/4-20
1/2	1"	7/16	1/4-20
9/16	1"	7/16	1/4-20
5/8	1-1/8	1/2	5/16-18
11/16	1-1/4	9/16	5/16-18
3/4	1-1/4	9/16	5/16-18
13/16	1-5/16	9/16	5/16-18
7/8	1-1/2	9/16	5/16-18
15/16	1-5/8	9/16	5/16-18
1"	1-5/8	5/8	5/16-18
1-1/16	1-3/4	5/8	5/16-18
1-1/8	1-3/4	5/8	5/16-18
1-3/16	2"	11/16	3/8-16
1-1/4	2"	11/16	3/8-16
1-5/16	2-1/8	11/16	3/8-16
1-3/8	2-1/8	3/4	3/8-16
1-7/16	2-1/4	3/4	3/8-16
1-1/2	2-1/4	3/4	3/8-16
1-9/16	2-1/2	13/16	3/8-16
1-5/8	2-1/2	13/16	3/8-16
1-11/16	2-1/2	13/16	3/8-16
1-3/4	2-3/4	7/8	1/2-13
1-13/16	2-3/4	7/8	1/2-13
1-7/8	2-3/4	7/8	1/2-13
1-15/16	3"	7/8	1/2-13
2"	3"	7/8	1/2-13
2-1/16	3"	7/8	1/2-13
2-1/8	3"	7/8	1/2-13
2-3/16	3-1/4	15/16	1/2-13
2-1/4	3-1/4	15/16	1/2-13
2-5/16	3-1/4	15/16	1/2-13
2-3/8	3-1/4	15/16	1/2-13
2-7/16	3-1/2	1"	1/2-13
2-1/2	3-1/2	1"	1/2-13
2-9/16	3-3/4	1-1/8	1/2-13
2-5/8	3-3/4	1-1/8	1/2-13
2-11/16	4"	1-1/8	1/2-13
2-3/4	4"	1-1/8	1/2-13
2-13/16	4-1/4	1-1/8	1/2-13
2-7/8	4-1/4	1-1/8	1/2-13
2-15/16	4-1/4	1-1/8	1/2-13
3"	4-1/4	1-1/8	1/2-13

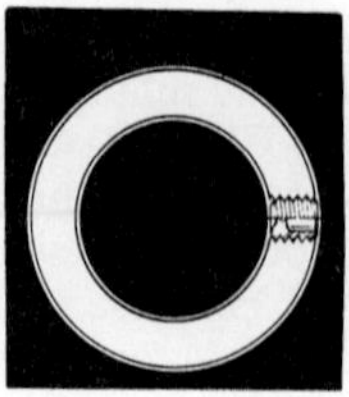

FIGURE 5.34 Set-collars (*Ruland, Inc.*).

Sizes and Thicknesses of Steel Shims - (Inches)

ID	OD	Thickness Ranges
0.375	0.625	0.001 0.002 0.003 0.004 0.005 0.006 0.007 0.008 0.010 0.012 0.015 0.020 0.025 0.031 0.047
0.500	0.750	0.0062 0.093 0.125 (These thicknesses available in all ID and OD sizes)
0.625	1.000	
0.750	1.125	
0.875	1.375	
1.000	1.500	
1.125	1.625	
1.250	1.750	
1.375	1.875	
1.500	2.125	
1.750	2.750	
2.000	2.750	

Sizes and Thicknesses of Steel Arbor Spacers - (Inches)

ID	OD	Thickness Ranges
0.500	0.750	0.001 0.002 0.003 0.004 0.005 0.006 (These thicknesses available for diameters listed)
0.625	1.000	
0.750	1.125	
0.875	1.375	
1.000	1.500	
1.250	1.750	
1.500	2.125	
2.000	2.250	
0.750	1.125	0.007 0.008 0.010 0.012 0.015 0.020 (These thicknesses available for diameters listed)
0.875	1.375	
1.000	1.500	
1.250	1.750	
1.500	2.125	
2.000	2.750	
1.000	1.500	0.025 0.031 0.047 0.062 0.093 0.125 (These thicknesses available for diameters listed)
1.250	1.750	
1.500	2.125	
2.000	2.750	

FIGURE 5.35 Steel shims and arbor spacers.

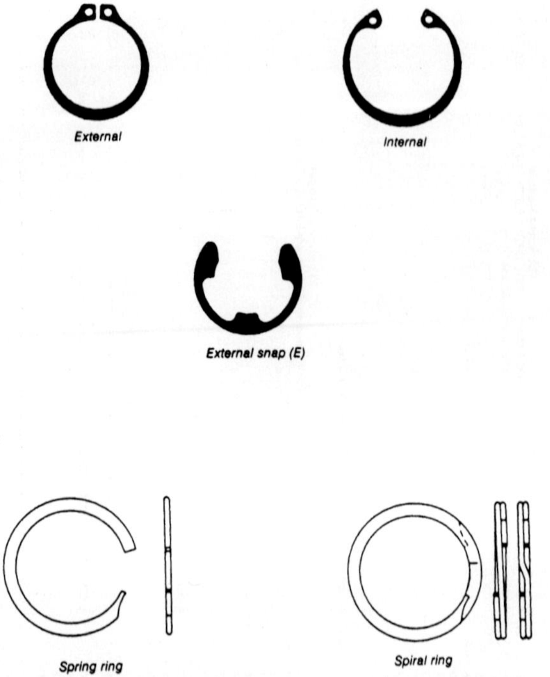

FIGURE 5.36 Retaining-ring styles.

	Roto Clip	Waldes	I.R.R.	Anderton	Mil Standard
	E	5133	1000	1500	16633
	BE	5131	1001	1501	16634
	RE	5144	1200	1540	3215
	C	5103	2000	1800	16632
	HO	5000	3000	1300	16625
	BHO	5001	3001	1301	16629
	VHO	5002	—	1302	16631
	SH	5100	3100	1400	16624
	BSH	5101	3101	1401	16628
	VSH	5102	—	1402	16630
	HOI	5008	4000	1308	16627
	SHI	5108	4100	1408	16626
	SHF	5555	7100	1440	90707
	SHR	5160	—	1460	3217
	SHM	5560	—	—	—

FIGURE 5.37 Retaining-ring interchangeability chart.

the manufacturer. The groove diameter and groove width for these types of rings must be very accurate, and dimensioned per the catalog tables. Failure to obtain the correct ring installation dimensions for the groove diameter and width for each particular ring can result in ring failure or pull-out under high mechanical loads.

Special types of retaining rings called X-Washers are shown in Figure 5.38. The A dimension is the groove diameter for the shaft, and the B dimension is the washer thickness. The width of the groove should be 0.003 to 0.005 in wider than the washer thickness or B dimension. The X-Washer is easily installed with a standard set of pliers. If the X-Washer is removed, a new washer should be installed in place of the removed washer. The X-Washer is installed into a shaft groove by squeezing the open *ears* together with a set of pliers. They are removed by squeezing the opposite set of *ears* together with pliers. The X-Washers are available in zinc plated carbon-steel. These types of retaining rings can be used only for over-shaft applications, or as external rings. A typical application of the X-Washers on an assembled product is shown in Figure 5.39, where their locations are indicated by the X marks and arrows in the photograph.

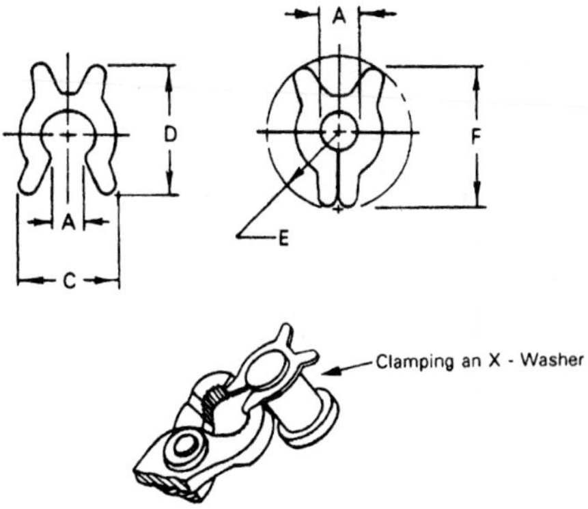

X - Washer Dimensions

A	B*	C	D	E	F
.086	.025	.320	.406	.210	.406
.098	.055	.364	.490	.297	.475
.130	.055	.430	.575	.359	.556
.164	.065	.523	.687	.422	.665
.190	.065	.593	.745	.437	.730
.222	.075	.622	.776	.469	.775
.256	.075	.698	.905	.500	.890
.285	.075	.822	.986	.563	.984
.317	.089	.872	1.100	.609	1.078
.347	.089	.948	1.190	.688	1.188
.381	.089	1.060	1.297	.797	1.281

NOTE: B* = thickness, inches

FIGURE 5.38 X-washer dimensional sizes. X washers are often called split washers.

FIGURE 5.39

CHAPTER 6

MACHINING AND MACHINE TOOL APPLICATIONS

Machine tools and cutting tools have advanced in great strides in the past 50 years. What at one time was a difficult operation on a machine tool such as a lathe or milling machine has become commonplace and greatly simplified with the advent of microprocessor controls, advanced positioning and control techniques, and modern cutting-tool materials. Chapter 1 described some of the modern equipment that is standard in many companies throughout the nation and the world. This chapter discusses other types of equipment, controls, and cutting tools together with their applications to machining and other metalworking procedures.

Over the years, we have progressed in our cutting tools from high-carbon steel, high-speed steel, cobalt matrix, and solid carbides to cemented or sintered carbides, ceramics, silicon nitride, cubic boron nitride, and special cutting tool coatings such as titanium nitride, titanium carbide, aluminum oxide, and others. The advantages of these new technologies in cutting tools have been increased production rates coupled with cost savings. Cutting tool technology is changing constantly, with new types of cutters, mills, drills, reamers, etc. being introduced to the market at a steady pace.

This chapter will show the latest types of cutting tools together with the important speeds and feeds recommended for them. Also shown are the typical speeds and feeds for the popular and widely used high-speed steel (HSS) and cobalt-matrix tools. The speeds and feeds tables for the various materials and machining methods are provided by the tables, figures, and charts contained in this chapter.

A typical machined detail part is shown in Fig. 6.1. Producing this part required stock cutoff, drilling, turning, boring, counterboring, broaching, facing, end milling, deburring, and finally, zinc plating. A typical part such as that shown in Fig. 6.1 can now be produced more rapidly than was possible in the past and at relatively less cost due to the advances in cutting and machine tool technologies.

6.1 TURNING AND BORING

Turning is the machining process used to generate external, cylindrical forms by removing material, usually with a single-point cutting tool. *Boring* is essentially internal turning to generate internal shapes. The common turning machines include engine lathes, single-spindle automatic lathes, horizontal turret lathes, automatic screw machines, Swiss-type automatic screw machines, multiple-spindle automatic bar and chucking machines, and computer-controlled automatic turning centers. The single-point tool is moved parallel to the machine spindle for straight or contour turning of the outside diameter and turning or boring of an internal surface. Form tools, both flat and circular, were at one time fed into the workpiece to produce the desired contour on the part.

FIGURE 6.1 Machined steel part, zinc plated.

Knurling produces a controlled rough surface pattern on the periphery of the part, either diamond shaped or straight, in coarse, medium, or fine texture.

Turning and Boring Tool Materials

High-speed steel (HSS), cast nonferrous alloys (Stellite and Tantung), and cemented carbides (sintered carbides) are among the most widely used turning, milling, and boring tool bit materials. The more advanced materials used for turning, boring, and milling operations include the following families:

- Tungsten carbides, coated (aluminum oxide and titanium nitride) and uncoated
- Cermets (*cer*amics with *met*allic binders)—titanium carbide and titanium nitride with a metallic binder
- Ceramics—alumina base (aluminum oxide) and silicon nitride base ceramics
- Polycrystallines—polycrystalline diamond and cubic boron nitride

Popular Grades of HSS and Tantung (Cast-Alloy) Tool Bits

HSS and Cobalt Ground Tool Bits

M-2 HSS: General-purpose tool bit material. Works extremely well on mild steel, alloys, and tool steels and is an excellent finishing tool. Hardness: Rockwell C63 to C65.

5% cobalt: For heavy cuts on castings and forgings. A T-4 type of tungsten high-speed steel which withstands higher heat than M-2 types. Hardness: Rockwell C64 to C66.

10% cobalt: A superior type of high-speed steel used for heavy cuts in hard materials. Hardness: Rockwell C64.5 to C67.5.

Cobalt M-34: For heavy cuts on castings and forgings. A T-4 type of tungsten HSS which withstands higher heat than M-2 types.

Cobalt M-43: A superior type of HSS which is excellent for high-heat applications and for heavy cuts in hard materials.

Tantung Tool Bits. *Tantung G:* A cast-alloy cutting tool material composed of chromium, tungsten, columbium, and carbon in a cobalt matrix. These materials have the ability to retain cutting hardness at temperatures up to 1500°F. As a cutting tool, Tantung G is excellent for all turning, boring, facing, milling, and cutoff applications on most metal and nonmetallic materials. Performs best at cutting speeds of 100 to 250 surface feet per minute (sfpm). Hardness: Rockwell C60 to C63 and transverse rupture strength of 300,000 psi, minimum.

Note: Typical modern high-speed turning and boring tool materials and insert forms are shown following the speeds and feeds tables in this section.

Turning Tool Types, Terms and Definitions, and Grinding

Single-Point Tools. A typical set of turning tools of HSS is shown in Fig. 6.2, together with a straight-shank lathe toolholder. The unground and ground tool bits are shown under the toolholder, followed by the different configurations of basic cutting bits. The cutting bit types are as follows:

A Left-hand turning tool

B Round-nose turning tool

C Right-hand turning tool

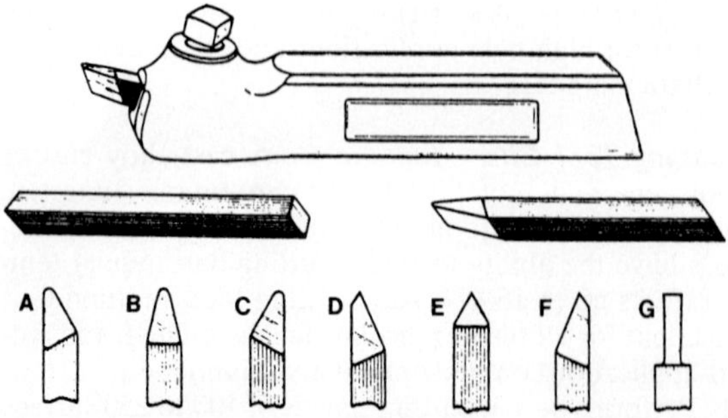

FIGURE 6.2 Standard lathe cutting tools, HSS (high-speed steel).

D	Left-hand facing tool
E	Threading tool
F	Right-hand facing tool
G	Cutoff tool

Applications for the cutting tools are shown in Fig. 6.3 and include turning, facing, threading, cutoff, boring, and inside threading. The important geometric form angles relating to single-point cutting tools which are ground from HSS and cast-alloy bit-stock materials (Tantung, Tantung G, Stellite, etc.) are shown in Fig. 6.4.

Tool Nomenclature

A single-point tool contains several geometric elements which are classified and defined by the American National Standards Institute (ANSI) B94.50-1975 (R1986),

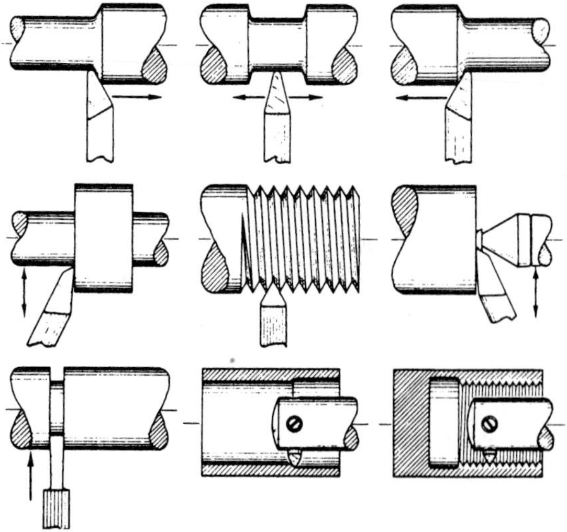

FIGURE 6.3 Turning operations.

Single-Point Cutting Tools, Basic Nomenclature and Definitions for. The basic definitions may be summarized as follows:

The *size* of a tool of square or rectangular section is expressed by giving, in the order named, the width of the shank *w,* the height of shank *h,* and the total tool length *l* in inches, such as 0.75 × 1.50 × 8 in.

The *shank* is that part of the tool on one end of which the point is formed or the tip or bit is supported. It is supported on the tool post of the machine.

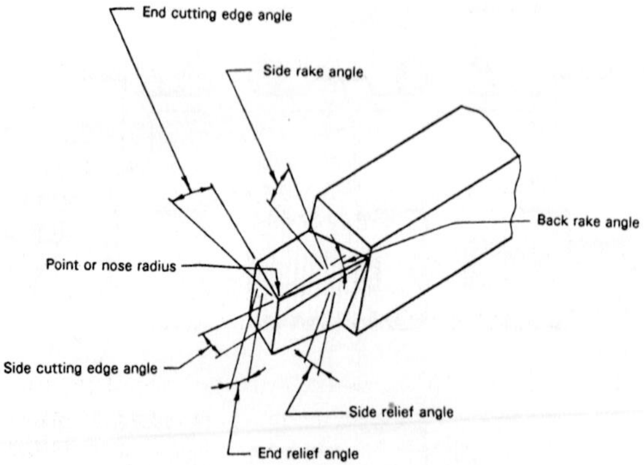

FIGURE 6.4 Form angles of single-point cutting tool.

The *base* is that surface of the shank which bears against the support and takes the tangential pressure of the cut.

The *heel* consists of the areas adjacent to the intersection of the base and flank.

The *face* is the surface on which the chip apparently impinges as it is separated from the workpiece.

A typical single-point carbide-tipped turning tool is shown in Fig. 6.4, and the angles shown are *normal,* that is, taken with reference to the cutting edges, since these are the ones specified in grinding a single-point tool. If a land is used, the rake of the land is given, followed by the rake of the face, such as –3 (12). This indicates a land rake of 3° negative with a positive side rake of 12°.

Tool Angles (see Fig. 6.5)

The tool angles shown in Fig. 6.5 are defined as follows:

Back-rake angle: The angle between the face of the tool and the base of the shank or holder. This is usually measured in a plane through the side-cutting edge and at right angles to the base. It is positive if the face slopes downward from the point toward the shank, tending to reduce the included angle of the tool point. It is negative if the face slopes upward toward the shank.

Side-rake angle: The angle between the face of the tool and the base of the shank or holder. It is usually measured in a plane perpendicular to the base and to the side-cutting edge and hence is *normal side rake.*

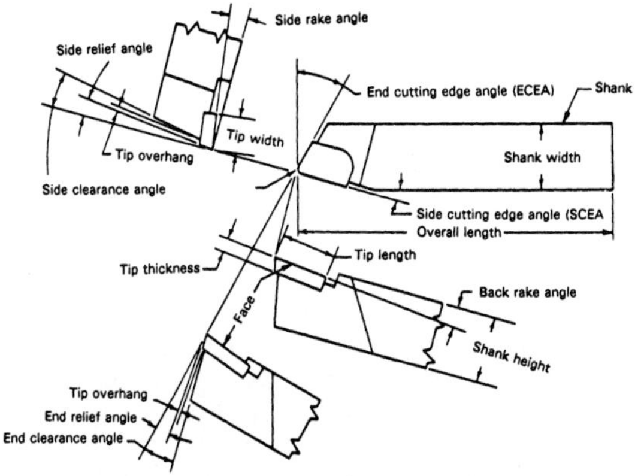

FIGURE 6.5 Tool angles and definitions.

Side-relief angle: The angle between the portion of the side flank immediately below the side-cutting edge and a line drawn through this cutting edge perpendicular to the base. It is usually measured in a plane at right angles to the side flank and hence is *normal side relief.*

End-relief angle: The angle between the portion of the end flank immediately below the end-cutting edge and a line drawn through that cutting edge perpendicular to the base. It is usually measured in a plane at right angles to the end flank and hence is *normal end relief.*

Clearance angle: The angle between a plane perpendicular to the base of the tool or holder and that portion of the flank immediately below the relieved flank. Side-clearance angle is measured in the plane of the back-rake angle. The clearance angle is greater than its corresponding relief angle, except when only one plane exists on the flank, in which case the clearance and relief angles coincide.

Side-cutting-edge angle: The angle between the straight side-cutting edge and the side of the tool shank or holder.

End-cutting-edge angle: The angle between the cutting edge on the end of the tool and a line at right angles to the side of the tool shank.

Note: As the setting of the tool or holder is changed in relation to the workpiece, the *effective working angles* will no longer agree with the tool angles.

Tool character. An abbreviated, convenient designation system for specifying the angles of single-point cutting tools is shown in Fig. 6.6. These angles are considered *normal* as used in most tool specifications and in grinding operations.

Working Angles. Working angles are the angles between the tool and the workpiece. These angles depend on the shape of the tool as well as on its position relative to the workpiece. These angles are defined on page 6-12.

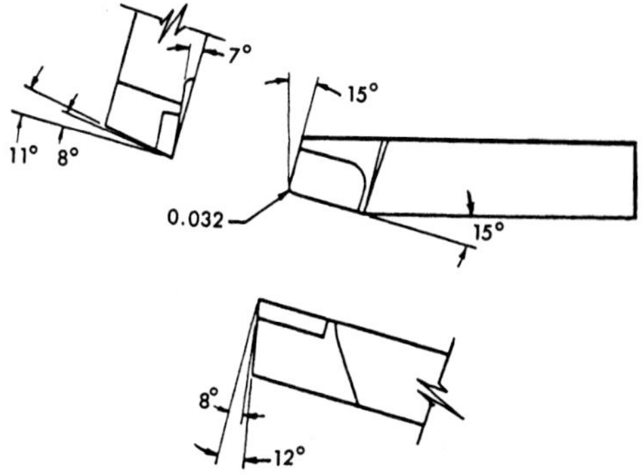

Tool Character 0^P 7^N 8^N $(11)^N$ 8^N $(12)^N$ 15 15 0.032

Tool Character	
Back rake angle, parallel	0^P
Side rake angle, normal	7^N
End relief angle, normal	8^N
End clearance angle, normal	$(11)^N$
Side relief angle, normal	8^N
Side clearance angle, normal	$(12)^N$
End cutting edge angle	15
Side cutting edge angle	15
Nose radius	0.032

FIGURE 6.6 Tool character number system.

Setting angle: The angle made by the straight portion of the shank of a tool or holder with the machined portion of the workpiece, commonly 90°.

Entering angle: The angle that the side-cutting edge of a tool makes with the machined surface of the workpiece, which is 90° in the case of a tool with 0° side-cutting-edge angle effective.

True rake angle: The slope of the tool face toward the tool base, from the active cutting edge in the direction of chip flow.

Cutting angle: The angle between the face of the tool and a tangent to the machined surface at the point of action. This angle is equal to 90° minus the true rake angle.

Lip angle: The included angle of the tool material between the face and the relieved flank. According to the direction of measurement, it may represent the *end lip angle, side lip angle,* or *true lip angle.*

Working relief angle: The angle between the relieved ground flank of the tool and a line tangent to the machined surface passing through the active cutting edge.

Working end-cutting-edge angle: The angle between the end-cutting edge and a plane tangent to the machined surface at the point of cutting.

Selection of Tool Geometry

Rake Angles. The rake-angle combination varies greatly according to tool material, workpiece material, and cost. Thermally efficient cutting is improved with positive rake angles or at least positive rake angles with respect to chip flow or true rake. It should be understood that a high positive rake results in a more fragile cutting edge and must often be compromised for tool durability. As cutting speeds increase, the rake angle has less effect on tool pressures. Therefore, in using tool materials which can be operated at higher cutting speeds, it becomes possible to use less positive rake angles or even negative rakes to increase tool strength and economize on tool maintenance. Prismatic-insert and throw-away-insert tools are greatly simplified in use and maintenance by using negative rake angles, although more cutting power and increased cutting forces are required.

Combining negative back rake with positive side rake allows safer cutting through slots or keyways, thus placing the initial impact loads on a portion of the cutting edge removed from the nose of the tool bit.

Relief Angles. Side and end relief angles vary between 5 and 15° for cutting metals and may run higher for some of the nonmetallic materials. Increased relief angles reduce cutting forces and result in a cleaner cut on low-tensile-strength materials. Reduced relief angles give more support to the cutting edge of the tool and are indicated for cutting high-strength metals. A wear land of excessive width may be the result of too small a relief angle, and tool breakage downward from the cutting edge may result from too large a relief angle.

Cutting-Edge Angles. The shape of the workpiece often determines the side-cutting-edge angle. Side-cutting-edge angles of approximately 15° will reduce the cutting power requirement, increase tool life, and aid in controlling the chip. A greater angle increases the chip thickness or makes an increase in feed possible. A decreased side-cutting-edge angle reduces the force at right angles to the workpiece and is required on thin-wall or long, slender workpieces.

End-cutting-edge angles of 6 to 15° are common. An angle of less than 6° may cause excessive forces normal to the work surface and lead to rapid dulling of the cutting tool. Angles greater than 15° may weaken the tool point but are used for tracing tools that must "in-feed" while operating. In all cases, the end-cutting-edge angle should be a minimum of 4° in relation to the surface of the workpiece.

A modification of the end-cutting-edge angle to 0° for a distance a little greater than the feed will produce a *drag,* or flattening, effect on the tool feed mark. If this modification is used, it must have a relief angle under it.

Nose Radius. A small nose radius reduces cutting forces and is indicated on long, slender parts or thin-walled sections. Large nose radii make the tool stronger and are indicated for roughing operations. Large radii are also generally used for machining cast iron and similar materials which produce a crumbling chip.

Chipbreaker. A *chipbreaker* is a small step or groove in the face of a tool or a separate piece attached to the tool or toolholder that causes the chip to break into small sections or curl. The three common types of chipbreakers are those which are ground into the tool point, those which are attached to the tool point, and those which are preformed into an insert. Figure 6.7 shows a typical removable insert toolholder and support for an

FIGURE 6.7 Typical removable insert tool holder and support for engine lathe.

engine lathe. The removable cutting insert shown mounted on the end of the toolholder has a chipbreaker groove formed into its perimeter.

Cutting Angles for Single-Point Cutting Tools (HSS, Cast-Alloy, and Carbide)

Cutting Angles for High-Speed-Steel Cutting Tools. Figure 6.8 shows the recommended cutting angles for high-speed-steel (HSS) lathe tool bits. Figure 6.9 shows the recommended cutting angles for cast-alloy (Stellite, Tantung/Tantung G, etc) lathe tool bits. Figure 6.10 shows the recommended cutting angles for carbide lathe bit tools.

Material	Side-relief angle	Front-relief angle	Back-rake angle	Side-rake angle
High-speed, alloy and high-carbon tool steels and stainless steels	7 - 9 (8)	6 - 8 (8)	0 - 7 (0)	8 - 10 (8)
SAE steels:				
1020, 1035, 1040	8 - 10 (8)	8 - 10 (8)	0 - 12 (0)	8 - 12 (8)
1045, 1095	7 - 9 (8)	8 - 10 (8)	0 - 12 (0)	8 - 12 (8)
1112, 1120	7 - 9 (8)	7 - 9 (8)	0 - 14 (0)	10 - 14 (10)
1314, 1315	7 - 9 (8)	7 - 9 (8)	0 - 14 (0)	10 - 16 (10)
1335	7 - 9 (8)	7 - 9 (8)	0 - 14 (0)	10 - 16 (10)
3115, 3120, 3130	7 - 9 (8)	7 - 9 (8)	0 - 10 (0)	8 - 12 (8)
3135, 3140	7 - 9 (8)	7 - 9 (8)	0 - 10 (0)	8 - 10 (8)
3250, 4140, 4340	7 - 9 (8)	7 - 9 (8)	0 - 8 (0)	8 - 10 (8)
6140, 6145	7 - 9 (8)	7 - 9 (8)	0 - 8 (0)	8 - 10 (8)
Aluminum & alloys	12 - 14 (14)	(14)	(0)	(15)
Phenol formaldahyde (Bakelite)	(14)	(14)	(0)	(10)
Brass, free-cutting	10 - 12 (10)	8 - 10 (10)	(0)	1 - 8 (8)
Bronzes-cast, red, yellow	8 - 10 (10)	8 - 10 (10)	(0)	-4 to +6 (+6)
Bronze, free-cutting	8 - 10 (10)	8 - 10 (10)	(0)	2 - 6 (6)
Phosphor-bronze, hard	8 - 10 (10)	6 - 10 (10)	(0)	0 - 6 (6)
Cast iron, gray	8 - 10 (8)	6 - 8 (8)	0 - 5 (0)	8 - 12 (8)
Copper	12 - 14 (12)	12 - 14 (12)	0 - 16 (0)	12 - 20 (12)
Copper alloys:				
Hard	8 - 10 (10)	6 - 10 (10)	(0)	0 - 8 (8)
Soft	10 - 12 (12)	8 - 12 (12)	0 - 2 (0)	0 - 10 (10)
Fiber	14 - 16 (14)	12 - 14 (14)	0 - 2 (0)	0 - 10 (10)
Formica	14 - 16 (14)	10 - 14 (14)	0 - 16 (0)	10 - 12 (12)
Nickel iron	14 - 16 (14)	10 - 14 (14)	0 - 8 (0)	12 - 15 (15)
Micarta	14 - 16 (14)	10 - 14 (14)	0 - 16 (0)	10 - 15 (15)
Monel and nickel	14 - 16 (14)	12 - 14 (14)	0 - 10 (0)	12 - 15 (15)
Nickel silvers	10 - 14 (14)	10 - 14 (14)	0 - 10 (0)	0 - 10 (10)
Rubber, hard and plastics	18 - 20 (20)	14 - 20 (20)	(0)	0 - 20 (20)

NOTE: Angles in parentheses are recommended as a starting point.

FIGURE 6.8 Recommended cutting angles for HSS single-point cutting tools for lathe; tabular data.

Material	Back-rake angle	Side-rake angle	Side-relief angle	Front-relief angle	Side-cutting-edge angle	End-cutting-edge angle
Steel	8 - 20 **	8 - 20 **	6	6	10	15
Cast steel	8	8	6	6	10	15
Cast iron	0	4	6	6	10	15
Bronze	4	4	6	6	10	10
Stainless steel	8 - 20 **	8 - 20 **	6	6	10	15

* Stellite 98M-2 turning tools & Tantung/Tantung "G"
** Angles depend on grade and type of steel. Soft materials require more positive rake than hard materials.
Boring tools use the same rake but greater relief to clear the workpiece.

FIGURE 6.9 Recommended cutting angles for cast-alloy tools; tabular data in degrees.

Material	Normal end relief	Normal side relief	Normal back rake	Normal side rake
Aluminum & magnesium alloys	6 - 10 (10)	6 - 10 (10)	0 to 10 (10)	10 to 20 (15)
Copper	6 - 8 (8)	6 - 8 (8)	0 to 4 (0)	15 to 20 (15)
Brass and bronze	6 - 8 (8)	6 - 8 (8)	0 to -5 (0)	+8 to -5 (+8)
Cast iron	5 - 8 (6)	5 - 8 (6)	0 to -7 (0)	+6 to -7 (-6)
Low carbon steels to SAE 1020	5 - 10 (6)	5 - 10 (6)	0 to -7 (0)	+6 to -7 (+6)
Carbon steels, SAE 1025 and above	5 - 8 (6)	5 - 8 (6)	0 to -7 (0)	+6 to -7 (+6)
Alloy steels	5 - 8 (6)	5 - 8 (6)	0 to -7 (0)	+6 to -7 (+6)
Free machining steels SAE 1100 and 1300	5 - 10 (6)	5 - 10 (6)	0 to -7 (0)	+6 to -7 (+6)
Stainless steels, austenitic	5 - 10 (6)	5 - 10 (6)	0 to -7 (0)	+6 to -7 (+6)
Stainless steels, hardenable grades	5 - 8 (6)	5 - 8 (6)	0 to -7 (0)	+6 to -7 (+6)
High nickel alloys (Monel, Inconel, etc.)	5 - 10 (8)	5 - 10 (8)	0 to -3 (0)	+6 to +10 (+10)
Titanium alloys	5 - 8 (6)	5 - 8 (6)	0 to -5 (0)	+6 to -5 (+6)

NOTE: Angles in parantheses are recommended as a starting point.

FIGURE 6.10 Recommended angles for carbide tools; tabular data.

Grinding/Sharpening of HSS, Cast-Alloy, and Carbide Tool Bits

Proper grinding and sharpening of standard tool-bit materials should produce a surface condition that promotes a substantial increase in the number of workpieces produced per grind. First, grind the tool bit to the proper angles using an aluminum oxide wheel of 40 to 60 grit, followed by a 320-grit aluminum oxide wheel. The second step should only remove a minimal amount of material, on the order of 0.0005 to 0.001 in. If hand honing of the tool is attempted, a hard Arkansas or medium India stone should be used. An experienced operator is required for hand honing tool bits.

Speeds, Cuts, and Feeds

HSS, Cast-Alloy, and Carbide Tools (see Fig. 6.11). The surface speed (sfpm), depth of cut (in), and feed (ipr) for various materials using HSS, cast-alloy, and carbide cutting tools are shown in Fig. 6.11. In all cases, especially where combinations of values are selected that have not been used previously on a given machine, the selected values should have their required horsepower or kilowatts calculated. Use the approximate calculations shown previously, or use one of the machining calculators available from the cutting tool manufacturers. The method indicated earlier for calculating the required horsepower gives a conservative value that is higher than the actual power required. In any event, on a manually controlled machine, the machinist or machine operator will know if the selected speed, depth of cut, and feed are more than the given machine can tolerate, and can make

Class	Material or SAE No.	Cutting Tool Material	Depth of Cut: 0.005-0.015 Feed: 0.002-0.005	Depth of Cut: 0.015-0.094 Feed: 0.005-0.015	Depth of Cut: 0.094-0.187 Feed: 0.015-0.030	Depth of Cut: 0.187-0.375 Feed: 0.030-0.050	Depth of Cut: 0.375-0.750 Feed: 0.050-0.090
Free-cutting steels	1112, 11L17	HSS	....	250 - 350	175 - 250	80 - 150	55 - 75
	1120	Cast-alloys	....	425 - 550	315 - 400	215 - 300	100 -210
	1315, etc.	Sintered carbides	750 - 1,500	600 - 750	450 - 600	350 - 450	175 - 350
Carbon & low-alloy steels	1010, 1020	HSS	....	225 - 300	150 - 200	75 - 125	45 - 65
	1025, etc.	Cast-alloys	....	375 - 500	275 - 350	180 - 250	100 - 175
		Sintered carbides	700 - 1,200	550 - 700	400 - 550	300 - 400	150 - 300
Medium alloy steels	1030	HSS	....	200 - 275	125 - 175	70 - 120	40 - 60
	1050	Cast-alloys	....	325 - 400	230 - 300	160 - 225	80 - 150
		Sintered carbides	600 - 1,000	450 - 600	350 - 450	250 - 350	125 - 250
High-alloy steels	1060	HSS	....	175 - 250	125 - 175	65 - 100	35 - 55
	1095	Cast-alloys	....	250 - 350	200 - 250	150 - 200	65 - 150
	1350	Sintered carbides	500 - 750	400 - 500	300 - 400	200 - 300	100 -200
Chromium-nickel steels	3120, 3450	HSS	....	150 - 200	100 - 125	50 - 75	30 - 50
	5140, 52100	Cast-alloy	....	230 - 315	165 - 225	110 - 160	55 - 110
		Sintered carbides	425 - 550	325 - 425	250 - 325	175 - 250	75 - 175
Molybdenum steels	4130	HSS	....	160 - 210	110 - 140	60 - 80	35 - 55
	4615	Cast-alloys	....	250 - 325	160 - 225	120 - 150	65 - 100
		Sintered carbides	475 - 650	350 - 475	275 - 350	200 - 275	100 - 200
Chrome, vanadium and stainless steels	6120	HSS	....	100 - 150	80 - 100	50 - 75	30 - 50
	6150, 18Cr-5Ni	Cast-alloys	....	210 - 250	170 - 200	110 - 165	55 - 100
	6195	Sintered carbides	375 - 500	300 - 375	250 - 300	175 - 250	75 - 175
Tungsten steels	7260	HSS	....	120 - 150	75 - 120	40 - 75	25 - 40
	18-4-1 annealed	Cast-alloys	....	130 - 175	110 - 130	80 - 100	35 - 80
		Sintered carbides	325 - 400	250 - 325	200 - 250	150 - 200	50 - 150
Special steels	12-14% Mn.	HSS	....	....	....	....	....
		Cast-alloys	....	....	....	....	....
		Sintered carbides	200 - 250	125 - 200	75 - 125	50 - 75	....

Special steels	Si. elect., sheet ingot iron, etc.	HSS	400 - 500	300 - 400	200 - 300	150 - 200	....
		Cast-alloys	....	500 - 600	350 - 450	250 - 300	....
		Sintered carbide	1,000 - 1,200	800 - 1,000	600 - 800	500 - 600	....
Cast iron	Soft gray	HSS	....	120 - 150	90 - 120	75 - 90	35 - 75
		Cast-alloys	....	225 - 300	160 - 220	125 - 160	70 - 125
		Sintered carbides	450 - 600	350 - 450	250 - 350	200 - 250	100 - 200
	Medium and malleable	HSS	....	120 - 150	90 - 120	60 - 90	30 - 60
		Cast-alloys	....	100 - 225	150 - 190	120 - 150	60 - 120
		Sintered carbides	350 - 450	250 - 350	200 - 250	150 - 200	75 - 150
	Hard alloys	HSS	....	90 - 125	60 - 90	40 - 60	20 - 40
		Cast-alloys	....	120 - 170	80 - 120	55 - 80	35 - 55
		Sintered carbides	250 - 300	150 - 250	100 - 150	75 - 100	50 - 75
	Chilled	HSS	10 - 15	....	....	....	....
		Cast-alloys	....	....	....	....	....
		Sintered carbides	30 - 50	10 - 30	....	....	....
Copper base alloys	Leaded, free - cutting, soft brass and bronze	HSS	....	300 - 400	225 - 300	150 - 255	100 - 150
		Cast-alloys	....	500 - 600	400 - 500	325 - 400	200 - 325
		Sintered carbides	1,000 - 1,250	800 - 1,000	650 - 800	500 - 650	300 - 500
	Normal brass, bronze low alloy	HSS	....	275 - 350	225 - 275	150 - 225	100 - 150
		Cast-alloys	....	375 - 425	325 - 375	250 - 325	175 - 250
		Sintered carbides	700 - 800	600 - 700	500 - 600	400 - 500	200 - 400
	Tough copper, high tin & alum. bronzes, gilding.	HSS	....	100 - 150	75 - 100	50 - 75	35 - 50
		Cast-alloys	....	225 - 300	180 - 225	125 - 180	75 - 125
		Sintered carbides	500 - 600	400 - 500	300 - 400	200 - 300	100 - 200
Light alloys	Magnesium	HSS	500 - 750	350 - 500	275 - 350	200 - 275	125 - 200
		Cast-alloys	700 - 1,000	500 - 700	400 - 500	300 - 400	200 - 300
		Sintered carbides	1,250 - 2,000	800 - 1,250	600 - 800	500 - 600	300 - 500
	Aluminum	HSS	350 - 500	225 - 350	150 - 225	100 - 150	50 - 100
		Cast-alloys	450 - 650	300 - 450	225 - 300	150 - 225	75 - 150
		Sintered carbides	700 - 1,000	450 - 700	300 - 450	200 - 300	100 - 200
Titanium	Pure & low alloys	HSS	....	100 - 160	70 - 110	50 - 75	....
		Cast-alloys	....	165 - 375	110 - 250	75 - 165	....
		Sintered carbides	550 - 900	375 - 600	250 - 400	165 - 265	....

FIGURE 6.11 Recommended cuts, feeds, and speeds for metals and nonmetallic turning (for HSS, cast-alloys, and carbide tools).

	Alpha alloys	HSS	....	30 - 75	20 - 50	....	....
		Cast-alloys	....	75 - 110	50 - 75	....	....
		Sintered carbides	165 - 450	110 - 300	75 - 200	50 - 135	....
	Beta alloys	HSS	....	30 - 40	20 - 25	....	....
		Cast-alloys	....	40 - 90	25 - 60	....	....
		Sintered carbides	125 - 225	90 - 150	60 - 100	40 - 70	....
Plastics	Thermoplastic, thermosetting	HSS	....	....	....	....	....
		Cast-alloys	....	....	....	....	....
		Sintered carbides	650 - 1,000	400 - 650	250 - 400	150 - 250	....
Abrasives	Glass, hard rubber, green ceramics, marble.	HSS	....	....	....	....	....
		Cast-alloys	....	....	....	....	....
		Sintered carbides	150 - 250	75 - 150	....	....	....

NOTE: It is possible that a combination of speeds, feeds and cuts may be so selected for a given application, that a higher horsepower may be required than is available at the lathe spindle. In all cases, especially where combinations of values are selected that have not been used previously on a given machine, the selected values should have their required horsepower calculated. See the subsection (Horsepower requirements) for calculating required horsepower when speed, feed and cut are given.
Values of depth of cut are in inches; feeds are given in ipr (inches per revolution) and speed is given in sfpm (surface feet per minute).

(Tabular values are in sfpm).

FIGURE 6.11 *(Cont.)* Recommended cuts, feeds, and speeds for metals and nonmetallic turning (for HSS, cast-alloys, and carbide tools).

corrections accordingly. On CNC/DNC-controlled automatic turning centers and other automatic machines, the cutting parameters must be selected carefully, with the machine operator carefully watching the first trial program run so that he or she may intervene if problems of overloading or tool damage occur.

Metal Removal Rates

When the feeds, cuts, and speeds are selected from Fig. 6.11, the metal removal rates may be found from Fig. 6.12.

Machining Insert Material Types

The following describes currently available machining insert materials and their applications. This information, together with the machining charts and remedies for cutting or machining problems appearing in this section of the *Handbook*, will enable you to select the proper type of insert and insert material for specific machining applications and requirements.

Uncoated Carbide. This tungsten carbide-based material with a cobalt binder, frequently alloyed with combinations of titanium and tantalum carbides, is available in all ANSI and ISO grade groups for metal-cutting applications. These inserts cut all materials, including irons, steels, and super alloys; aluminum and brass alloys; and nonmetallics, including plastics, phenolics, and composites. Applications range from heavy roughing to finishing on almost all machining operations.

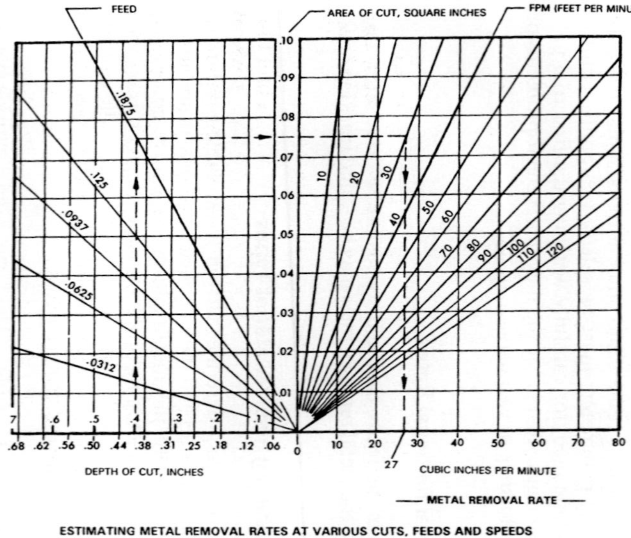

FIGURE 6.12 Metal-removal rate (mrr) chart.

CVD-Coated Carbide. This insert material is available in a complete range of grades, with substrates engineered for maximum compatibility with coatings. Coatings include:

- Titanium carbide (TiC)
- Titanium nitride (TiN)
- Composite ceramic (Al_2O_3)
- Multiple coatings

These grades are excellent for machining most materials—including cast irons, steels, stainless steels, and some super alloys—in turning, boring, and milling operations.

PVD-Coated Carbide. Inserts coated with a thin film (2–4 microns) of titanium nitride (TiN) by physical vapor deposition (PVD) are characterized by good wear-resistance, low coefficient of friction, and resistance to edge buildup. PVD (TiN) coatings are applied at temperatures substantially below those used for chemical vapor deposition (CVD), the process most commonly used for coating carbide cutting tools. The result is less reaction with the substrate and thus better edge-strength retention. Low coating temperatures allow greater control over deposition rate, and result in closer tolerances and more grain-size control.

Silicon Nitride. This hot-pressed monolithic silicon nitride-based ceramic is designed to do one thing exceptionally well: cut cast-iron materials. It offers outstanding toughness and thermal resistance, and handles turning, boring, and milling operations at speeds from

350 to 5,000 sfm. Because silicon nitride retains its hardness, toughness, and abrasion resistance at elevated temperatures, it can provide excellent tool life at these speeds.

PCBN. Polycrystalline cubic boron nitride inserts, available in both tipped and solid forms, are designed to machine ferrous workpiece materials in the 40- to 70-RC range. PCBN is significantly harder than cemented carbide or ceramic materials and is the only material that combines a high degree of toughness with exceptional hot-hardness. PCBN excels in turning of hardened steels, chilled and and gray cast irons, and Ni-Hard materials.

PCD. Polycrystalline diamond inserts provide premium performance when cutting highly abrasive, low tensile-strength, nonferrous, nonmetallic, and aluminum composite work materials. The exceptionally hard, wear-resistant diamond cutting tip retains its sharpness even in demanding applications, running up to 7500 sfm and beyond. As a result, PCD inserts provide up to one hundred times the tool life of carbide. Excellent for roughing and high-speed finishing on turning, boring, and milling operations.

Coated Carbide. Coated carbide grades represent the latest development in advanced coating technology and offer unique substrates for turning and milling. This combination allows Spectra-coated inserts to provide increased productivity at higher cutting speeds with greater wear-resistance over a broad range of applications and work materials. Applications range from heavy roughing to precision finishing on work materials including all steels, stainless steels, many super alloys, and alloyed cast irons.

Cermet. This type includes composite titanium carbide and titanium nitride-based materials designed for light roughing to high-speed finishing in turning, boring, and milling operations. These grades show excellent results on steels, stainless steels, cast irons, and aluminum. They are capable of running at moderate to high speeds and are best used with light feeds and depths of cut. They feature excellent abrasion resistance and thermal properties, and lubricity for crater resistance.

Ceramic. Metal-oxide composite with edge strength well suited for materials with hardness from 35RC to 65RC, including some hardened steels, irons, and some super alloys. Ceramic inserts are available in a wide range of shapes, sizes, and geometries.

Material Characteristics and Insert-Tool Application Considerations

The following listings provide extensive information relating to insert-tool machining of various materials. Each item describes machining characteristics for the specific material, notes various conditions that may be encountered while machining it, and provides remedies for correcting the conditions or problems listed. Figure 6.13 is also included for milling recommendations on different milling operations.

Low-Carbon and Free-Machining Steels. 1100 series, 1008, 1010, 1018, and 1200 series, leaded steels, and steels with free-machining additives.

-GENERAL APPLICATIONS FOR CUTTING CONDITIONS-		
CONDITION	**-SPEED-**	**-FEED-**
Roughing	↓	↑
Finishing	↑	↓
End Milling	↑	↓
Slotting	↑	↓
Hard Material	↓	→
Soft Material	↑	↑
Scale	↓	↑
Tool Life	↓	→
Heavy d.o.c.	↓	↓

Higher- ↑

Lower- ↓

Same- →

FIGURE 6.13 Chart for general cutting conditions on milling machine operations.

Material Characteristics. Low-carbon, soft and gummy; difficult chip control; torn finish is common; high-speed capability. Free-machining, easy to machine; high-speed capability; high depths of cut achievable; easy chip control.

Tool-Application Considerations

Condition: Torn finish (built-up edge)

Remedy: Increase speed; use cermets; use PVD-coated grade with sharp edge; use positive-rake inserts; increase coolant concentration.

Condition: Difficult chip control

Remedy: Increase speed; change lead angle to redirect chip away from workpiece.

Condition: Burrs and sharp edges

Remedy: Change cutting path; use PVD-coated grade with sharp edge; use positive-rake or high-positive-rake inserts.

Medium/High-Carbon Steels and Alloy Steels. 1045, 1085, 1541, 1561, 1572, and 4000 series, 4300 series, 5100 series, 8600 series, 52100, and 300M.

Material Characteristics. Higher chrome, nickel, and molybdenum content, work hardening. Higher carbon content, abrasive but achieves surface finish easier. Higher nickel content, more carbon and alloy, more difficult to machine.

Tool-Application Considerations

Condition: Crater wear

Remedy: Use aluminum oxide–coated grade; reduce speed; use more free-cutting chip control; use cermets; increase coolant pressure; reduce feed.

Condition: Flank wear

Remedy: Use TiC/alumina ceramics; use aluminum oxide–coated grade; increase feed; increase coolant concentration; increase depth of cut.

Condition: Thermal deformation

Remedy: Use more wear-resistant grade; reduce speed; reduce feed.

Tool Steels

Material Characteristics. Abrasive, work hardening, and tough, hard-to-break chips.

Tool-Application Considerations

Condition: Flank wear

Remedy: Use aluminum oxide–coated grade; reduce speed; increase feed.

Condition: Crater wear

Remedy: Use aluminum oxide–coated grade; use cermets; reduce speed; reduce feed.

Condition: Dull surface finish

Remedy: Increase speed; use positive-rake inserts (PVD-coated); use cermets; increase coolant concentration.

Gray Cast Iron. Gray, classes 20, 25, 30, 45, and 60, high tensile.

Material Characteristics. Abrasive, potential scale and inclusions, tendency to break out during exit from cut, potential for chatter on thin-wall sections.

Tool-Application Considerations

Condition: Excessive edge wear

Remedy: Use harder grade or aluminum oxide–

coated grades; use ceramics; increase feed to reduce in-cut time.

Condition: Chipping

Remedy: Increase lead angle; use stronger grade; use prehoned inserts or inserts with edge preparation.

Condition: Workpiece break-out

Remedy: Use sharp inserts or PVD-coated grades; use lower feed rate during exit; increase speed; increase lead angle; use positive geometries for finish cuts.

Condition: Workpiece chatter

Remedy: Use sharp inserts or PVD-coated grades; use positive-rake inserts; increase lead angle; increase feed to stabilize workpiece.

Ductile Cast Iron. Nodular/ductile, malleable.

Material Characteristics. Abrasive, difficult to machine, good surface finishes difficult to achieve.

Tool-Application Considerations

Condition: Edge wear

Remedy: Use harder grades; apply aluminum oxide–coated grades; consider ceramics and cermets for finishing; increase feed to reduce in-cut time.

Condition: Crater wear

Remedy: Apply aluminum oxide–coated grades; apply geometries that are free-cutting; lower speed (sfpm).

Condition: Torn or dull surface finish

Remedy: Apply ceramics, cermets, or PVDs at higher speeds and lower feeds; increase lead angle; increase coolant concentration.

Hardened Irons

Material Characteristics. Highly abrasive, high cutting forces.

Tool-Application Considerations

Condition: Excessive edge wear

Remedy: Apply alumina-based ceramics or cubic boron nitride (CBN) grades; feed above 0.004 ipr; increase lead angles or use round inserts; check coolant application.

Condition: Catastrophic breakage

Remedy: Increase lead angles or use round inserts; use large nose radii; repair toolholder; reduce feeds.

Austenitic Stainless Steels. 200 and 300 series.

Material Characteristics. Work hardens rapidly; small depths of cut are difficult; usually abrasive rather than hard; tough and stringy chips.

Tool-Application Considerations

Condition: Depth-of-cut notch

Remedy: Increase lead angle; use tougher grade; feed over 0.005 ipr.

Condition: Built-up edge

Remedy: Increase speed; use cermets; use positive-rake inserts (PVD-coated).

Condition: Workpiece glazing

Remedy: Increase depth of cut; reduce nose radius; use sharp PVD-coated grades.

Condition: Dull surface finish

Remedy: Increase speed; reduce feed; increase positive rake.

Ferritic/Martensitic Stainless Steels. 400 series and precipitation-hardening PH grades.

Material Characteristics. Brittle, stringy chips; high cutting forces; high work hardening (especially PH stainless grades).

Tool-Application Considerations

Condition: Built-up edge

Remedy: Increase speed; use cermets; use positive-rake inserts (PVD-coated); increase coolant concentration.

Condition: Workpiece glazing

Remedy: Increase depth of cut; reduce nose radius; use positive-rake inserts (PVD-coated); increase coolant concentration.

Condition: Depth-of-cut notch

Remedy: Increase lead angle; use tougher grade; feed over 0.005 ipr.

Condition: Torn or dull surface finish

Remedy: Increase speed; reduce feed; use positive-rake inserts (PVD-coated); increase coolant concentration.

High-Temperature Alloys. Iron, nickel, and cobalt base.

Material Characteristics. Relatively poor tool life; small depths of cut are difficult; work hardens rapidly; usually abrasive rather than hard; tough and stringy chips.

Tool-Application Considerations

Condition: Depth-of-cut notch

Remedy: Increase lead angle; use tougher grade; use 0.015-in or greater depth of cut; feed over 0.005 ipr; increase coolant concentration.

Condition: Built-up edge

Remedy: Increase speed; change cutting tool material; use positive-rake inserts (PVD-coated); do not overrun cutting edge; increase coolant concentration.

Condition: Workpiece glazing

Remedy: Increase depth of cut; increase feed rate; reduce nose radius; use positive-rake inserts (PVD-coated).

Condition: Torn or dull surface finish

Remedy: Increase speed; reduce feed; use positive-rake inserts (PVD-coated); increase coolant concentration.

Titanium Alloys. Pure, alpha, alpha-beta, and beta alloys.

Material Characteristics. Relatively poor tool life; chips tend to gall and weld to cutting edge; requires lower cutting speeds than steels of equal hardness; usually abrasive; tough and stringy; low thermal conductivity.

Tool-Application Considerations

Condition: Depth-of-cut notch

Remedy: Increase lead angle; use 0.015-in or greater depth of cut; remove embrittled surface from mill stock; feed over 0.005 ipr; increase coolant concentration.

Condition: Built-up edge

Remedy: Keep a sharp edge; use positive-rake inserts (PVD-coated); increase coolant concentration.

Condition: Workpiece glazing

Remedy: Increase depth of cut; reduce nose radius; use positive-rake inserts (PVD-coated); index cutting edge.

Condition: Torn or dull surface finish

Remedy: Index cutting edge; reduce feed; use positive-rake inserts (PVD-coated); increase speed; increase coolant concentration.

High-Silicon and Free-Machining Aluminum

Free-machining aluminum. 1000, 1100, 1200, and 1300 series; 2011 through 2024 series; 3000 series. 1050 to 1350: Almost pure. 2011 to 2024: 2024 is popular (screw-machine products, aircraft parts); contains copper, manganese, and magnesium.

Material Characteristics. Soft and gummy.

Tool-Application Considerations

Condition: Built-up edge

Remedy: Increase speed; use positive-rake inserts (PVD-coated); use coolants designed for machining aluminum; J polish top of insert.

Condition: Torn or dull surface finish

Remedy: Increase speed; reduce feed; use positive-rake inserts (PVD-coated); use coolant designed for aluminum; use polycrystalline diamond tools; J polish top of insert.

High-Silicon Aluminum. 4000, 5000, 6000, and 7000 series (7000 series uses zinc, zirconium, and titanium as alloying additives). 4000 series: Contains silicon for heat resistance and wear. 5000 series: Contains chromium. 6000 series: 6061 typical, contains silicon, copper, magnesium, and chromium. 7000 series: Contains copper, magnesium, chromium, and zinc (aircraft parts).

Material Characteristics. Abrasive and tough.

Tool-Application Considerations

Condition: Edge wear

Remedy: Use coated grade; use polycrystalline diamond tools; increase coolant concentration.

Condition: Built-up edge/surface finish

Remedy: Increase speed; increase coolant concentration.

Nonferrous Alloys and Nonmetallics

Nonferrous Metals. Free-machining and high-strength copper alloys, brass, zinc, and magnesium.

Material Characteristics. Easily machined; high machining speeds; positive chip control with magnesium to reduce fire hazards.

Caution: Fire hazard is present when machining magnesium at high speeds.

Nonmetallics. Nylons, acrylics, phenolics, and resin materials.

Material Characteristics. Easily machined; high machining speeds; machine with polycrystalline diamonds whenever possible; center height of cutting is important.

Tool-Application Considerations

> *Condition:* Built-up edge
>
> *Remedy:* Increase speed; use positive-rake inserts; J polish top of inserts.
>
> *Condition:* Poor surface finish
>
> *Remedy:* Increase speed; use polycrystalline diamond inserts; use positive-rake inserts; J polish top of inserts; use positive chip control.

Relation of Speed to Feed. The following general rules apply to most turning and boring operations:

- If the tool shows a built-up edge, increase feed or increase speed.
- If the tool shows excessive cratering, reduce feed or reduce speed.
- If the tool shows excessive edge wear, increase feed or reduce speed.

Caution: The productivity settings from the machining calculators and any handbook speed and feed tables are suggestions and guides only. *A safety hazard may exist* if the user calculates or uses a table-selected machine setting *without also considering* the machine power and the condition, size, strength, and rigidity of the workpiece, machine, and cutting tools.

Turning, Boring, and Threading Inserts. Figure 6.14 shows a sample of some of the typical cutting inserts available today. The versatile 80° diamond inserts are shown in the figure as "DNGA," "DNGZ" and "DNGP." Various types and styles of toolholders for the inserts are produced by the insert manufacturers to hold the disposable inserts rigidly in position during the cutting operation. Fig. 6.15 shows a modern toolholder with a multipurpose, double-ended cutting insert. With the one insert, finishing, facing, and cutoff operations may be performed according to the orientation angle of the toolholder in relation to the workpiece.

ANSI and ISO Classification Systems for Turning, Boring, and Milling Inserts

Removable turning, boring, and milling inserts are classified and coded according to the (JIC) and the Interna-

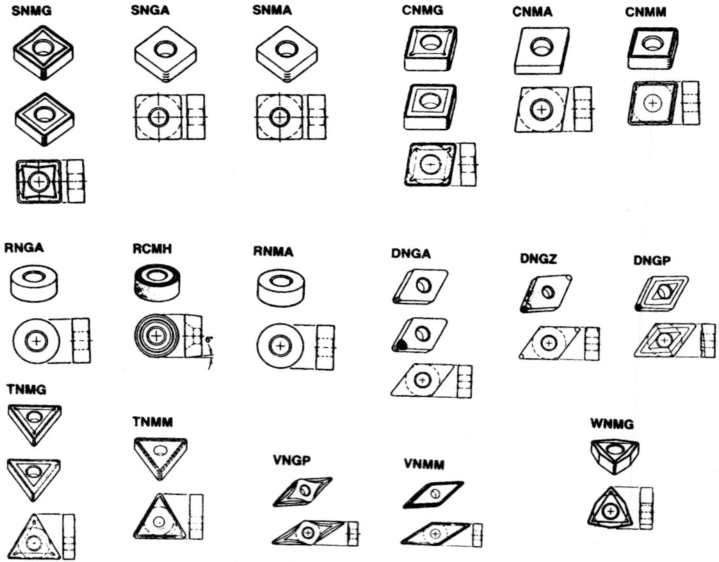

FIGURE 6.14 Samples of typical cutting inserts and their code identification letters.

FIGURE 6.15 Modern tool holder with multipurpose, double-ended cutting insert.

tional Standards Organization (ISO) classification systems shown in Figs. 6.16 and 6.17, respectively. Sample drawings of various types of inserts are shown in Fig. 6.14. The identification code is listed at the top right of each insert and may be referenced back to the ANSI identification system.

Thread Turning

Thread-turning inserts are available in different styles or types for turning external and internal thread systems such as UN series, 60° metric, Whitworth (BSW), Acme, ISO, American buttress, etc. Figure 6.18 shows some of the typical thread-cutting inserts.

Threading Operations. Prior to cutting (turning) any particular thread, the following should be determined:

Code	Application	Carbide Characteristics
C-1	Roughing	Medium-high shock resistance Medium-low wear resistance
C-2	General purpose	Medium shock resistance Medium wear resistance
C-3	Finishing	Medium-low shock resistance Medium-high wear resistance
C-4	Precision finishing	Low shock resistance High wear resistance
C-5	Roughing	High resistance to cutting temperatures Shock and cutting load Medium wear resistance
C-6	General purpose	Medium-high shock resistance Medium wear resistance Medium shock resistance
C-7	Finishing	Medium shock resistance Medium wear resistance
C-8	Precision finishing	Very high wear resistance Low shock resistance

Note: Hardness increases from top to bottom; toughness increases from bottom to top

FIGURE 6.16 JIC classification system for inserts.

- Machining toward the spindle (standard helix)
- Machining away from the spindle (reverse helix)
- Helix angle
- Insert and toolholder
- Insert grade
- Speed (sfpm)
- Number of thread passes
- Method of in-feed

Main Machining Group	Color Marking	Application Group	Operations and Working conditions
P Group: steel cast steel, long chip malleables	Blue	P01	High precision turning and boring. High cutting speeds. Good surface finish and vibration-free machining
		P10	Turning, thread cutting and milling. High cutting speeds. Small/medium chip cross-section
		P20	Turning, milling, medium cutting speeds and medium chip cross-section. Planing - small chip.
		P30	Turning, milling, planing. Medium to low cutting speeds. Medium/large chip cross-section. Unfavorable conditions
		P40	Planing, turning, milling, shaping. Low cutting speeds. Large chip cross-section. High rake angles. Unfavorable conditions
		P50	Where highest demands are made on carbide toughness. Turning, planing and shaping. Low cutting speeds. Large chips and high rakes. Unfavorable conditions.
M Group: steel, cast steel austenitic manganese steels, cast iron alloys, austenitic steels, malleable cast iron, free-cutting steels.	Yellow	M10	Turning. Medium-high cutting speeds. Small to medium chip cross-section.
		M20	Turning, milling. Medium cutting speeds and medium chip cross-section.
		M30	Turning, milling, planing. Medium cutting speeds. Medium/large chip cross-section.
		M40	Turning, form turning, parting off and recessing - for automatics.
K Group: cast iron, chilled cast irons, short chip malleables hardened steels, nonferrous metals and nonmetallic materials.	Red	K01	Turning, precision turning/boring. Finish milling and scraping.
		K10	Turning, milling, boring, countersinking. reaming, scraping and broaching.
		K20	Turning, milling, planing, countersinking. scraping, reaming and broaching under tougher conditions than K10.
		K30	Turning, milling, planing, shaping under unfavorable conditions. High rakes.
		K40	Turning, milling, planing, shaping under unfavorable conditions. High rakes.

Note: Cutting speed and wear resistance increase from bottom to top; feed and carbide toughness increase from top to bottom.

FIGURE 6.17 ISO carbide classification system for inserts.

Cutting Procedures for External and Internal Threads: Machine Setups. Figure 6.19 illustrates the methods for turning the external thread systems (standard and reverse helix). Figure 6.20 illustrates the methods for turning the internal thread systems (standard and reverse helix).

Problems in thread cutting

Problem	Possible remedy
Burr on crest of thread	1. Increase surface feet per minute (rpm). 2. Use positive rake. 3. Use full-profile insert (NTC type).
Poor tool life	1. Increase surface feet per minute (rpm). 2. Increase chip load. 3. Use more wear-resistant tool.
Built-up edge	1. Increase surface feet per minute (rpm). 2. Increase chip load. 3. Use positive rake, sharp tool. 4. Use coolant or increase concentration.
Torn threads on workpiece	1. Use neutral rake. 2. Alter infeed angle. 3. Decrease chip load. 4. Increase coolant concentration. 5. Increase surface feet per minute (rpm).

In-Feed Methods for Thread Cutting

0° In-Feed Angle. When the in-feed angle is 0°, cutting occurs on both sides of the thread form, which places all the cutting edge in the cut and helps protect the edge from chipping. *Disadvantages:* May produce difficult-to-handle chip; the tip of the tool may chip when cutting high-tensile-strength materials; burring is increased; and chatter problems may develop.

Large In-Feed Angle (15 to 30°). Cutting with the leading edge of the tool produces an easy-to-handle chip, and reduced burr problems occur on the trailing edge of the tool. *Disadvantages:* Trailing edge of tool may drag and tend to chip; torn or poor surface occurs when cutting soft, gummy materials such as low-carbon

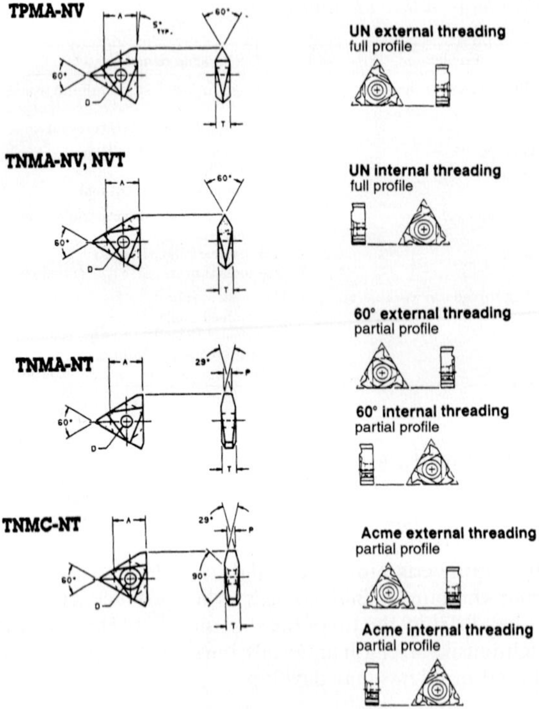

FIGURE 6.18 Typical thread-cutting inserts.

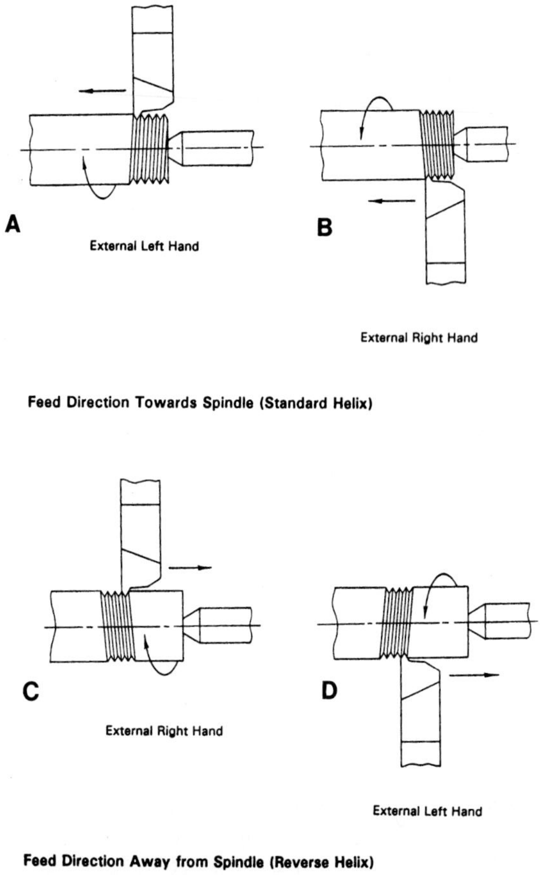

FIGURE 6.19 Methods for turning the external thread systems (standard and reverse helix).

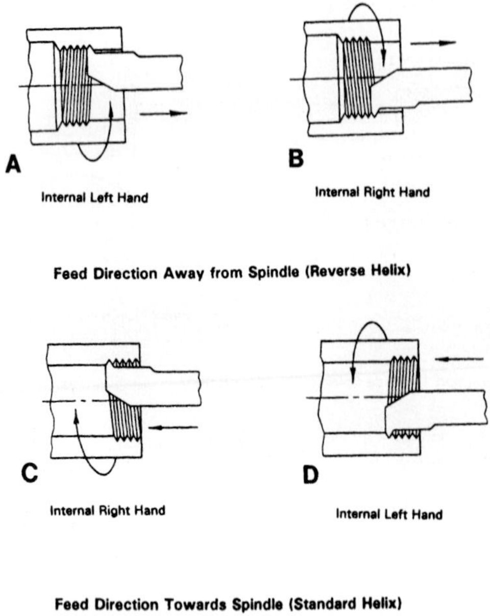

FIGURE 6.20 Methods for turning the internal thread systems (standard and reverse helix).

steels (1020, etc.) and low-alloy aluminums or pure aluminum.

Small In-Feed Angle (5 to 10°). Tool cuts from both sides and is protected from chipping; chip is easy to handle. *Disadvantages:* Similar to 0° in-feed, but less in magnitude because cutting forces are equalized and chip is easy to control.

Chemical Composition						Comparative Physical Properties *			
AISI - SAE Grade	Tungsten	Chromium	Vanadium	Molybdenum	Cobalt	Red Hardness	Abrasion Resistance	Edge Strength	Grinding Ability
M - 1	1.5	4.0	1.0	8.5	-	10	10	18	18
M - 2	6.0	4.0	2.0	5.0	-	10	11	17	17
M - 3-1	6.0	4.0	2.4	6.0	-	10	14	12	8
M - 3-2	6.0	4.0	3.0	6.0	-	10	17	11	6
M - 4	5.5	4.5	4.0	4.5	-	10	18	8	5
M - 7	1.7	4.0	2.0	8.75	-	10	12	17	14
M - 10	-	4.0	2.0	8.0	-	10	10	17	17
M - 15	6.5	4.5	5.0	3.0	5.0	13	19	9	2
M - 30	2.0	4.0	1.0	8.0	5.0	13	10	14	14
M - 34	2.0	4.0	2.0	8.5	8.0	15	11	14	14
M - 36	6.0	4.0	2.0	6.0	9.0	16	11	14	14
M - 42	1.5	3.75	1.15	9.5	8.0	17	19	15	14
T - 1	18.0	4.0	1.0	-	-	10	10	16	18
T - 2	18.0	4.0	2.0	-	-	10	11	15	17
T - 3	18.0	4.0	3.0	-	-	10	14	8	7
T - 4	18.0	4.0	1.0	-	5.0	13	10	11	14
T - 5	18.0	4.0	2.0	-	8.0	15	11	11	14
T - 6	22.0	4.5	1.5	-	12.0	20	10	11	14
T - 15	13.0	4.5	5.0	-	5.0	17	20	9	2

* Comparative Range of Physical Properties:
0 - 4 = Poor; 5 - 8 = Fair; 9 - 12 = Good; 13 - 16 = Very Good; 17 - 20 = Excellent

FIGURE 6.21 Classifications of high-speed steels (HSS).

Kenna-metal	IMCO	Adamas	Carboloy	Carmet	New-comber	Sandvic	SECO	Firth Sterling	TRW	Valenite	VR/ Wesson	Walmet
Uncoated Grades												
K1	IC-10	B	44A	CA-3	N-10			H-6	CQ-12	VC-1	VR-54	WA-1
K68	MG-10	PWX	820	CA-310		H10	G-27	H-17	CQ-22	VC-27	RAM-1	WA-10
K68	IC-20	A	883	CA-4	N-22	H20	H-13	H-21	CQ-2	VC-2	2A5	WA-2
K313	IC-40	AAA	999	CA-8	N-40	H05		HF	CQ-4	VC-4		WA-4
K420	IC-50	434	390	CA-740	N-50		S6	T-04	CY-12		2A7	WA-54
K420	IC-55	499	370	CA-720	N-55	S6	S6	T-12	CY-16	VC-5	VR-75	WA-5
K2884	IC-67	548		CA-717	N-60	SIP	S2	T-25	CY-14	VC-6	VR-73	WA-47
K313	IC-70	490	350		N-70	SIP	SIG	T-25	CY-31	VC-8		WA-73
Coated Grades												
KC730	IC-202	ACT-2	523	CA9443	NT-2	GC315	TP-15	TC	027	VN2	630	P2
KC810	IC-552	ACT-5	550	CA9720	NT-5	GC1025	TP-35	HN	715	VN5	660	P5
KC850	IC-672	ACT-7	518		NT-8		TP-25	TC1	714		670	P47
KC910	IC-554	ROXIDE	570		NA02	GC015		CC46	918	V01		A60
KC950	IC-674		545			GC415		CC44		V05	680	A62

NOTE: This grade selection chart is intended to be used as a guide to grade selection of various manufacturers. It indicates those grades having similar characteristics under most general conditions. It is not intended to imply that all comparable grades are exact duplicates in physical and metallurgical properties, or that they perform exactly the same on all applications.

FIGURE 6.22 Carbide grade insert comparison chart for different manufacturereers.

Classification of High-Speed Steel (HSS), and Modern Carbide Insert Comparison Chart

Figure 6.21 shows the classification, chemical composition, and comparative physical properties of high-speed steels. A carbide cutting grade comparison chart for different manufacturers is shown in Fig. 6.22.

Modern Turning Centers

The modern DNC (direct numerical control) turning center has replaced many types of older machine tools such as the automatic screw machines, small turret lathes, and profiling equipment. Figure 6.23 shows a typical automatic CNC turning center, and Fig. 6.24 shows a close-up view of the CNC controller.

FIGURE 6.23 A modern CNC controlled turning center.

FIGURE 6.24 Close-up view of the CNC control panel for the machine shown in Fig. 6.23.

6.2 MILLING

Milling is a machining process for generating machined surfaces by removing a predetermined amount of material progressively from the workpiece. The milling process employs relative motion between the workpiece and the rotating cutting tool to generate the required surfaces. In some applications the workpiece is stationary and the cutting tool moves, while in others the cutting tool and the workpiece are moved in relation to each other and to the machine. A characteristic feature of the milling process is that each tooth of the cutting tool takes a portion of the stock in the form of small, individual chips.

Typical cutting tool types for milling-machine operations are shown in Fig. 6.25.

Milling Methods

- Peripheral milling (slab milling)
- Face milling and straddle milling
- End milling
- Single-piece milling
- String or "gang" milling
- Slot milling
- Profile milling
- Thread milling
- Worm milling
- Gear milling

Modern milling machines have many forms, but the most common types are shown in Fig. 6.26. The well-

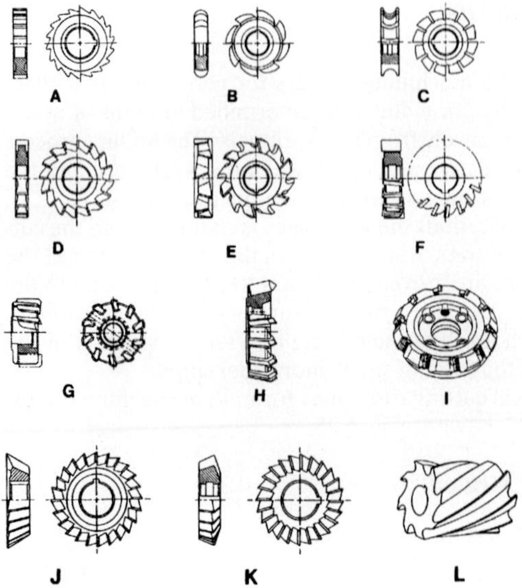

Milling Cutter Styles - High-Speed Steel and Carbide Insert

A	Disk type milling cutter	G	Face milling cutter
B	Convex half-round milling cutter	H	Face milling head
C	Concave half-round milling cutter	I	Double-angle carbide insert milling cutter
D	Three-side milling cutter	J	Single-angle milling cutter
E	Staggered-tooth milling cutter	K	Double-angle milling cutter
F	Inserted blade milling cutter	L	Left hand slab milling cutter

FIGURE 6.25 Various cutting tool types for the milling machine.

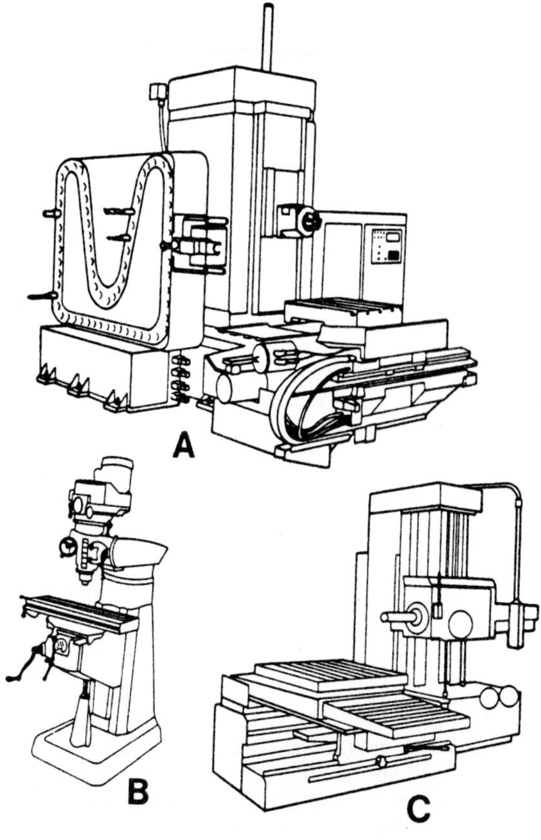

FIGURE 6.26 Typical modern milling machines.

known and highly popular Bridgeport-type milling machine is shown in Fig. 6.27. The Bridgeport machine is often used in tool and die making operations and in model shops, where prototype work is done. The great stability and accuracy of the Bridgeport makes this machine popular with many experienced machinists and die makers. The Bridgeport shown in Fig. 6.27 is equipped with digital sensing controls and read-out panel, reading to ±0.0005 in.

Many of the better grades of aluminum alloys, such as 2024, 6061, and 7075, can be considered free-machining and are often mill cut without coolants and at high surface speeds (see speeds and feeds section). When cutting solutions and coolants are used for turning or milling aluminum alloys, special formulations are available and may be used (see coolants and cutting solutions section).

The modern machining center is being used to replace the conventional milling machine in many industrial applications. Figure 1.2*a* shows a machining center, with its control panel at the right side of the machine. Machines such as these generally cost $250,000 or more depending on the accessories and auxiliary equipment obtained with the machine. These machines are the modern "workhorses" of industry and cannot remain idle for long periods owing to their cost. As described in Chap. 1, these machines are computer-controlled and make their own tool changes automatically during ongoing machining operations. One needs to see these machines in actual operation to appreciate the great speed and accuracy with which they perform their programmed (CNC) machining functions.

The modern machining center may be equipped for three-, four-, or five-axis operation. The normal or com-

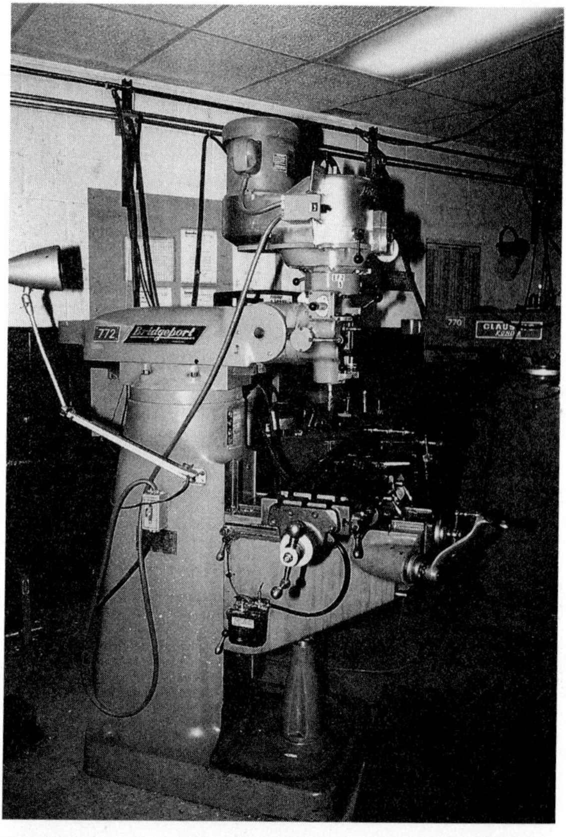

FIGURE 6.27 A Bridgeport milling machine with digital control panel (accuracy of ±0.0005 in).

mon operations usually call for three-axis machining, while more involved machining procedures require four- or even five-axis operation. Three-axis operation consists of *x* and *y* table movements and *z*-axis vertical spindle movements. The four-axis operation includes the addition of spindle rotation with three-axis operation. Five-axis operation includes a horizontal fixture for rotating the workpiece on a horizontal axis at a predetermined speed (rpms), together with the functions of the four-axis machine. This allows all types of screw threads to be machined on the part and other operations such as producing a worm for worm-gear applications, segment cuts, arcs, etc. Very complex parts may be mass produced economically on a three-, four-, or five-axis machining center, all automatically, using CNC (computer numerical control).

The control panels on these machining centers contain a microprocessor that is, in turn, controlled by a host computer, generally located in the tool or manufacturing engineering office; the host computer controls one or more machines with direct numerical control (DNC) or distributed numerical control. Various machining programs are available for writing the operational instructions sent to the controller on the machining center. Figure 1.2*b* shows a detailed view of a typical microprocessor (CNC) control panel used on a machining center. This particular control panel is from an Enshu 550-V machining center, a photograph of which appears in Fig. 1.2*a*.

For a quick approximation of total power requirements of the machine motor, see Fig. 6.34, which gives the maximum metal-removal rates for different horsepower-rated milling machines cutting different materials.

Modern Theory of Milling. The key characteristics of the milling process are

- Simultaneous motion of cutter rotation and feed movement of the workpiece
- Interrupted cut
- Production of tapered chips

It was common practice for many years in the industry to mill *against* the direction of feed. This was due to the type of tool materials then available (HSS) and the absence of antibacklash devices on the machines. This method became known as *conventional* or *up milling* and is illustrated in Fig. 6.28*b*. *Climb milling* or *down milling* is now the preferred method of milling with advanced cutting tool materials such as carbides, cermets, CBN, etc. Climb milling is illustrated in Fig. 6.28*a*. Here, the insert enters the cut with some chip load and proceeds to produce a chip that thins as it progresses toward the end of the cut. This allows the heat generated in the cutting process to dissipate into the chip. Climb-milling forces push the workpiece toward the clamping fixture, in the direction of the feed. Conventional-milling (up-milling) forces are against the direction of feed and produce a lifting force on the workpiece and clamping fixture.

The angle of entry is determined by the position of the cutter centerline in relation to the edge of the workpiece. A negative angle of entry β is preferred and is illustrated in Fig. 6.29*b*, where the centerline of the cutter is below the edge of the workpiece. A negative angle is preferred because it ensures contact with the workpiece at the strongest point of the insert cutter. A positive

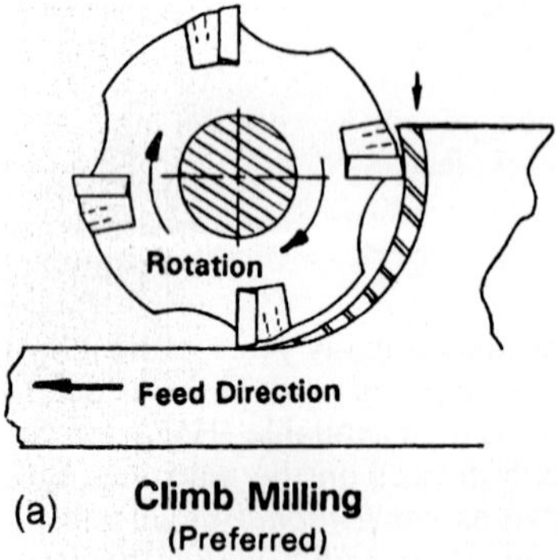

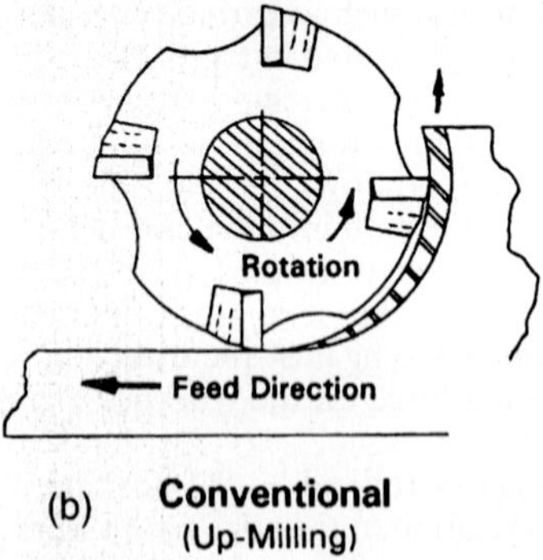

FIGURE 6.28 Climb- and up-milling.

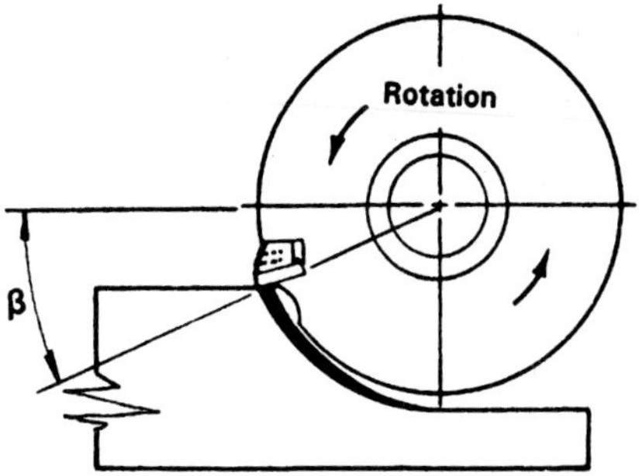

(a) **Positive Entry**

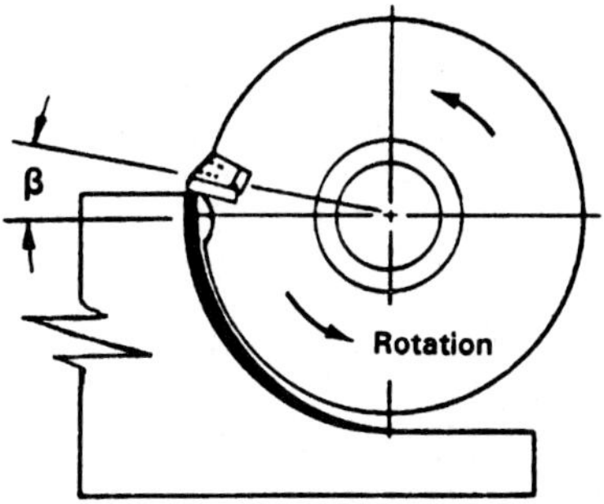

(b) **Negative Entry**

FIGURE 6.29 Milling entry angles.

angle of entry will increase insert chipping. If a positive angle of entry must be employed, use an insert with a honed or negative land. Figure 6.29*a* shows positive entry.

Figure 6.30 shows an eight-tooth cutter climb-milling a workpiece using a negative angle of entry, and the feed, or advance, per revolution is 0.048 in with a chip load per tooth of 0.006 in.

Milling Tables

The following tables give the milling values for *K* factors, Fig. 6.31; feed per tooth, Fig. 6.32; milling cutter speeds, Fig. 6.33; and horsepower ratings, Fig. 6.34.

Cutter Speed (rpm) from Surface Speed (sfpm). A time-saving table of surface speed versus cutter speed is shown in Fig. 6.35 for cutter diameters from 0.25 through 5 in. For cutter speed (rpm) values when the surface speed is greater than 200 sfpm, use the simple equation:

$$\text{rpm} = \frac{12(\text{sfpm})}{\pi D}$$

where D is the effective cutter diameter in inches.

6.3 DRILLS AND DRILLING—FLUTED, SPADE, AND MICRO DRILLS

Drilling is a machining operation for producing round holes in metallic and nonmetallic materials. A *drill* is a rotary-end cutting tool with one or more cutting edges

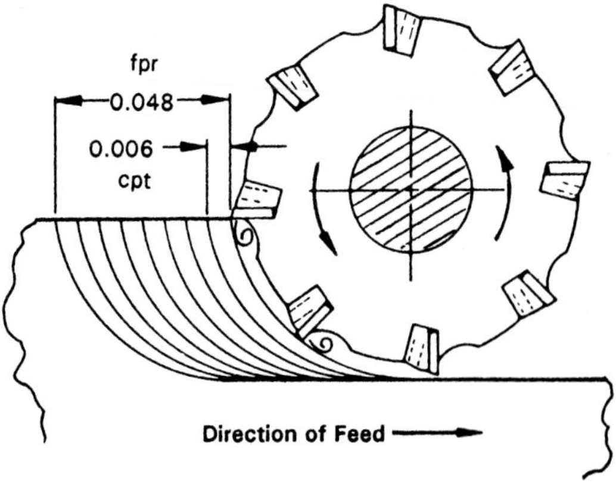

FIGURE 6.30 Milling principle.

or lips and one or more straight or helical grooves or flutes for the passage of chips and cutting fluids and coolants. When the depth of the drilled hole reaches three or four times the drill diameter, a reduction must be made in the drilling feed and speed. A coolant-hole drill can produce drilled depths to eight or more times the diameter of the drill. The gundrill can produce an accurate hole to depths of more than 100 times the diameter of the drill with great precision.

Enlarging a drilled hole for a portion of its depth is called *counterboring,* while a counterbore for cleaning the surface a small amount around the hole is called *spotfacing.* Cutting an angular bevel at the perimeter of a drilled hole is also termed *countersinking.* Countersinking tools are available to produce 82°, 90°, and 100° countersinks and other special angles.

Material	K(in^3/min/(hp$_c$))
Cold drawn steel, SAE 1112, 1120, 1315	1.0
Forged and alloy steel, SAE 3120, 1020, 2320, 2345, 150-300 BHN	0.63 - 0.87
Alloy steel, 300 - 400 BHN	0.5
Malleable iron and cold drawn steel, SAE 6140	0.9
Cast irons:	
Soft	1.5
Medium	0.8 - 1.0
Hard	0.6 - 0.8
Stainless steel, AISI 416, free-machining	1.1
Stainless steel, austenitic, AISI 303, free-machining	0.83
Stainless steel, austenitic, AISI 304	0.72
Tool steel	0.51
Bronze and brass:	
Soft	1.7- 2.5
Medium	1.0 - 1.4
Hard	0.6 - 1.0
Aluminum and magnesium	2.5 - 4.0
Monel metal	0.55
Copper, annealed	0.84
Nickel	0.53
Titanium & alloys	0.75 - 1.1

NOTE: "K" values are in cubic inches per minute per cutter horsepower (in^3/min/hp$_c$)
"K" values for carbide cutters are approximately 25% higher

FIGURE 6.31 *K* factor table for various materials.

Material	Face mills	Helical mills	Slot/side mills	End mills	Form-relieved cutters
Magnesium & alloys	0.022	0.018	0.013	0.011	0.007
Aluminum & Alloys	0.022	0.018	0.013	0.011	0.007
Free-cutting brasses & bronzes	0.022	0.018	0.013	0.011	0.007
Medium brasses & bronzes	0.014	0.011	0.008	0.007	0.004
Hard brasses & bronzes	0.009	0.007	0.006	0.005	0.003
Copper	0.012	0.010	0.007	0.006	0.004
Cast iron, soft (150-180 Bhn)	0.016	0.013	0.009	0.008	0.005
Cast iron, medium (180-220 Bhn)	0.013	0.010	0.007	0.007	0.004
Cast iron, hard (220-300 Bhn)	0.011	0.008	0.006	0.006	0.003
Malleable iron	0.012	0.010	0.007	0.006	0.004
Cast steel	0.012	0.010	0.007	0.006	0.004
Low-carbon steel, free-machining	0.012	0.010	0.007	0.006	0.004
Low-carbon steels	0.010	0.008	0.006	0.005	0.003
Medium-carbon steels	0.010	0.008	0.006	0.005	0.003
Alloy steel, ann'ld (180-220 Bhn)	0.008	0.007	0.005	0.004	0.003
Alloy steel, tough (220-300 Bhn)	0.006	0.005	0.004	0.003	0.002
Alloy steel, hard (300-400 Bhn)	0.004	0.003	0.003	0.002	0.002
Stainless steels, free-machining	0.010	0.008	0.006	0.005	0.003
Stainless steels	0.006	0.005	0.004	0.003	0.002
Monel metal	0.008	0.007	0.005	0.004	0.003
Titanium & alloys	0.008	0.007	0.005	0.004	0.003
Machinable plastics	0.013	0.010	0.008	0.007	0.004

NOTE: Tabular data in inches. For feed per tooth in millimeters, multiply tabular data by 25.4. For carbon-steel cutters, multiply tabular data by 0.50 or divide by 2. Source: Cincinnati Milicron, Inc.

FIGURE 6.32 Suggested feed per tooth for milling; high-speed steel cutters (tabulated data in inches).

Material	High-speed steel tools * * Rough	Finish
Cast iron	50 - 60	80 - 110
Semisteel	40 - 50	65 - 90
Malleable iron	80 - 100	110 - 130
Cast steel	45 - 60	70 - 90
Copper	100 - 150	150 - 200
Brass	200 - 300	200 - 300
Bronze	100 - 150	150 - 180
Aluminum	400 - 450	700 - 750
* Magnesium	600 - 800	1,000 - 1,500
SAE steels:		
1020 (coarse feed), low-carbon	60 - 80	60 - 80
1020 (fine feed), low-carbon	100 - 120	100 - 120
1035, medium-carbon	75 - 90	90 - 120
1330, alloy steel	90 - 110	90 - 110
1050, Med-high-carbon	60 - 80	100 - 125
2315, nickel steel	90 - 110	90 - 110
3150, nickel-chromium	50 - 60	70 - 90
4150, chrome-molybdenum	40 - 50	70 - 90
4340, nickel-chrome-molybdenum	40 - 50	60 - 70
Stainless steel	60 - 80	100 - 120
Titanium, hard alloy	80 - 100	110 - 130

NOTE: Tabular data ranges are in sfpm (surface feet per minute for HSS cutters only).
* A fire hazard is present when machining magnesium at high-speeds.
* * For carbide cutters, increase sfpm × 25% (min.)

FIGURE 6.33 Milling cutting speeds, HSS for various materials (sfpm).

Drills are classified by material, length, shape, number, and type of helix or flute, shank, point characteristics, and size series. Most drills are made for right-hand rotation. Right-hand drills, as viewed from their point, with the shank facing away from your view, are rotated in a counterclockwise direction in order to cut. Left-hand drills cut when rotated clockwise in a similar manner.

Workpiece Material	Rated hp of Machine									
	3	5	7.5	10	15	20	25	30	40	50
	Max. Metal Removal (in^3/min)									
Aluminum	2.7	5.5	8.7	12	18	27	37	48	69	91
Brass, soft	2.4	4.7	7.5	10	16	24	32	41	60	79
Bronze, hard	1.7	3.3	5.3	7.3	11	17	23	30	43	56
Bronze, very hard	0.78	1.6	2.5	3.4	5.3	7.8	11	15	20	26
Cast iron, soft	1.6	3.2	5.2	7.1	11	16	22	28	41	54
Cast iron, hard	1	2	3.3	4.6	7	10	14	18	26	35
Cast iron, chilled	0.78	1.6	2.5	3.4	5.3	7.8	10	13	19	26
Malleable iron	1	2.1	3.4	4.7	7.3	11	14	18	26	36
Steel, soft	1	2	3.3	4.6	7	10	14	18	26	35
Steel, medium	0.78	1.6	2.5	3.4	5.3	7.8	10	13	19	26
Steel, hard	0.56	1.1	1.8	2.5	3.9	5.7	7.7	10	14	19

NOTE: Data source - Kearney & Trecker Corp.

FIGURE 6.34 Milling machine horsepower ratings for maximum metal removal rates for high-speed steel cutters.

Diameter of cutter (in.)	Surface speed (ft. per min.)																
	25	30	35	40	50	55	60	70	75	80	90	100	120	140	160	180	200
	Cutter revolutions per minute																
1/4	382	458	535	611	764	851	917	1,070	1,147	1,222	1,376	1,528	1,834	2,139	2,445	2,750	3,056
5/16	306	367	428	489	611	672	733	856	917	978	1,100	1,222	1,466	1,711	1,955	2,200	2,444
3/8	255	306	357	408	509	560	611	713	764	815	916	1,018	1,222	1,425	1,629	1,832	2,036
7/16	218	262	306	349	437	481	524	611	656	699	786	874	1,049	1,224	1,398	1,573	1,748
1/2	191	229	268	306	382	420	459	535	573	611	688	764	917	1,070	1,222	1,375	1,528
5/8	153	184	214	245	306	337	367	428	459	489	552	612	736	857	979	1,102	1,224
3/4	127	153	178	203	254	279	306	357	381	408	458	508	610	711	813	914	1,016
7/8	109	131	153	175	219	241	262	306	329	349	392	438	526	613	701	788	876
1	95.5	115	134	153	191	210	229	267	287	306	344	382	458	535	611	688	764
1-1/4	76.3	91.8	107	123	153	168	183	214	230	245	274	306	367	428	490	551	612
1-1/2	63.7	76.3	89.2	102	127	140	153	178	191	204	230	254	305	356	406	457	508
1-3/4	54.5	65.5	76.4	87.3	109	120	131	153	164	175	196	218	262	305	349	392	436
2	47.8	57.3	66.9	76.4	95.5	105	115	134	143	153	172	191	229	267	306	344	382
2-1/2	38.2	45.8	53.5	61.2	76.3	84.2	91.7	107	114	122	138	153	184	213	245	275	306
3	31.8	38.2	44.6	51	63.7	69.9	76.4	89.1	95.3	102	114	127	152	178	208	228	254
3-1/2	27.3	32.7	38.2	44.6	54.5	60	65.5	76.4	81.8	87.4	98.1	109	131	153	174	196	218
4	23.9	28.7	33.4	38.2	47.8	52.6	57.3	66.9	71.7	76.4	86	95.6	115	134	153	172	191
5	19.1	22.9	26.7	30.6	38.2	42	45.9	53.5	57.3	61.1	68.8	76.4	91.7	107	122	138	153

FIGURE 6.35 Cutter revolutions per minute from surface speed.

Drill Terminology

Figure 6.36*a* to *c* shows common twist drills made of high-speed steel and coated with titanium nitride. Figure 6.36*d* to *h* show a group of end mills, also coated with titanium nitride. The line just above the flutes shows the limit of the titanium nitride coating. Figure 6.37 shows the standard twist drill form with the appropriate terminology describing its characteristic features.

Drill Types or Styles

- HSS jobber drills
- Solid-carbide jobber drills

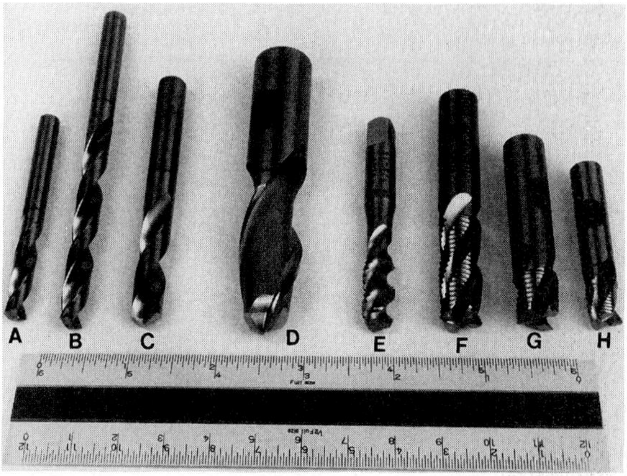

FIGURE 6.36 Common twist drills with titanium nitride coatings A, B, and C.

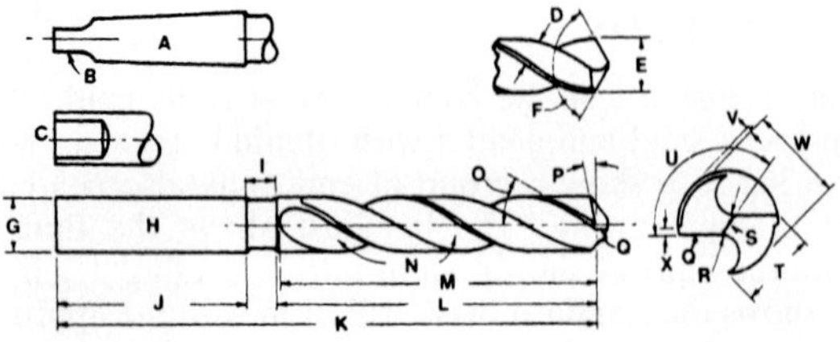

Twist Drill Features

A	Taper shank	M	Flute length
B	Tang	N	Flutes
C	Tang drive	O	Helix angle
D	Land width	P	Lip relief angle
E	Drill diameter	Q	Lip
F	Point angle	R	Web
G	Shank diameter	S	Chisel edge
H	Straight shank	T	Land
I	Neck	U	Chisel edge angle
J	Shank length	V	Body diameter clearance
K	Overall length	W	Clearance diameter
L	Body	X	Margin

FIGURE 6.37 Twist-drill features.

- Carbide-tipped screw-machine drills
- HSS screw-machine drills
- Carbide-tipped glass drills
- HSS extralong straight-shank drills (24 in)
- Taper-shank drills (0 through number 7 ANSI taper)
- Core drills
- Coolant-hole drills
- HSS taper-shank extralong drills (24 in)
- Aircraft extension drills (6 and 12 in)
- Gun drills

- HSS half-round jobber drills
- Spotting and centering drills
- Parabolic drills
- S-point drills
- Square solid-carbide die drills
- Spade drills
- Miniature drills
- Microdrills and microtools

American National Standard Tapers. Figure 6.38 shows the American national standard taper geometry and dimensions for ANSI tapers 1 through 7. Taper number 0 is not listed in the national standards. Table 6.1 follows Fig. 6.38 and lists the detail dimensions.

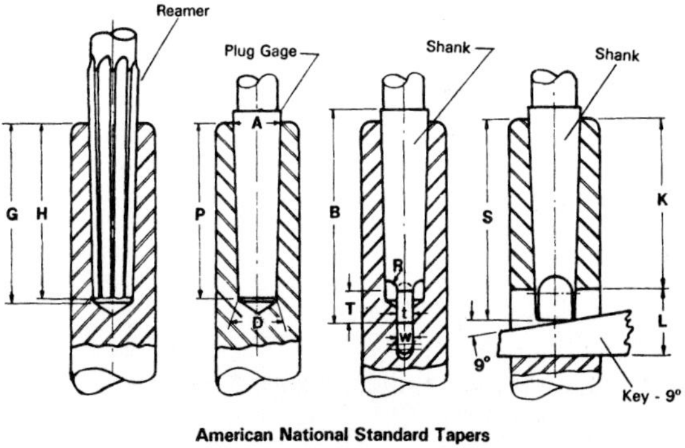

FIGURE 6.38 American national standard tapers.

TABLE 6.1 Detail Dimensions* for American National Standard Tapers

American national standard taper number	Shank							Tang				Tang Slot					
	Diam. of plug at small end	Diam. at end of socket	Whole length	Depth	Depth of drilled hole	Depth of reamed hole	Standard plug depth	Thickness	Length	Radius	Radius	Width	Length	End of socket to tang slot	Taper per inch	Taper per foot	American national standard taper number
	D	A	B	S	G	H	P	t	T	R	a	W	L	K			
0†	0.25200	0.35610	2 11/32	2 7/32	2 1/16	2 1/32	2	5/32	1/4	5/32	3/64	11/64	9/16	1 15/16	0.052050	0.62460	0‡
1	0.36900	0.47500	2 9/16	2 7/16	2 3/16	2 5/32	2 1/8	13/64	3/8	3/16	3/64	7/32	3/4	2 1/16	0.049882	0.59858	1
2	0.57200	0.70000	3 1/8	2 15/16	2 21/32	2 39/64	2 9/16	1/4	7/16	1/4	1/16	17/64	7/8	2 1/2	0.049951	0.59941	2
3	0.77800	0.93800	3 7/8	3 11/16	3 5/16	3 1/4	3 3/16	5/16	9/16	9/32	5/64	21/64	1 3/16	3 1/16	0.050196	0.60235	3
4	1.02000	1.23100	4 7/8	4 5/8	4 3/16	4 1/8	4 1/16	15/32	5/8	5/16	3/32	31/64	1 1/4	3 7/8	0.051938	0.62326	4
4 1/2	1.26600	1.50000	5 3/8	5 1/8	4 5/8	4 9/16	4 1/2	9/16	11/16	3/8	1/8	37/64	1 3/8	4 5/16	0.052000	0.62400	4 1/2
5	1.47500	1.74800	6 1/8	5 7/8	5 5/16	5 1/4	5 3/16	5/8	3/4	3/8	1/8	21/32	1 1/2	4 15/16	0.052626	0.63151	5
6	2.11600	2.49400	8 9/16	8 1/4	7 13/32	7 21/64	7 1/4	3/4	1 1/8	1/2	5/32	25/32	1 3/4	7	0.052138	0.62565	6
7	2.75000	3.27000	11 5/8	11 1/4	10 5/32	10 5/64	10	1 1/8	1 3/8	3/4	3/16	1 5/32	2 5/8	9 1/2	0.052000	0.62400	7

* Table agrees with American nuclear standards for taper shanks except for angle and undercut of tang.
† Size 0 taper shank not listed in American national standards.

Drill Point Styles and Angles

Over a period of many years, the metalworking industry has developed many different drill point styles for a wide variety of applications. All the standard point styles and special points are shown in Fig. 6.39, including the important point angles which differentiate these different points. New drill styles are being introduced periodically, but the styles shown in Fig. 6.39 include some of the newer types as well as the commonly used older configurations.

The old practice of grinding drill points by hand and eye is, at the least, ineffective with today's modern drills and materials. For a drill to perform accurately and efficiently, modern drill-grinding machines such as the models produced by the Darex Corporation are required. Models are also produced which are also capable of sharpening taps, reamers, end mills, and countersinks. Metal- cutting tool sharpening and grinding practices are not detailed in this *Handbook.* The tool suppliers and tool sharpening vendors are equipped to sharpen all the cutting tools they distribute in a more accurate and efficient manner than can be done by hand in modern machine shops. High-speed-steel and cobalt-based turning bits are an exception to this modern grinding and sharpening practice, in that the machinist or machine operator occasionally dresses or hones such a cutting tool.

Recommended general uses for drill point angles shown in Fig. 6.39 are given below. Figure 6.39*k* illustrates web thinning of a standard twist drill.

Typical Uses

A Copper and medium to soft copper alloys

B Molded plastics, Bakelite, etc.

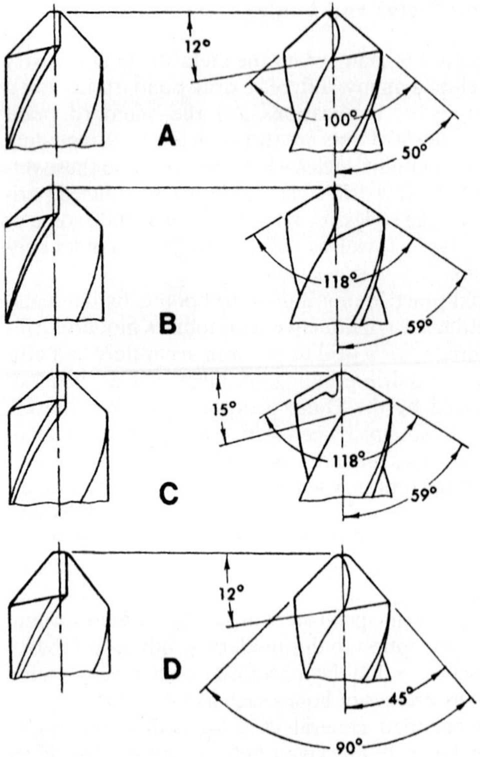

FIGURE 6.39 Drill-point styles and angles.

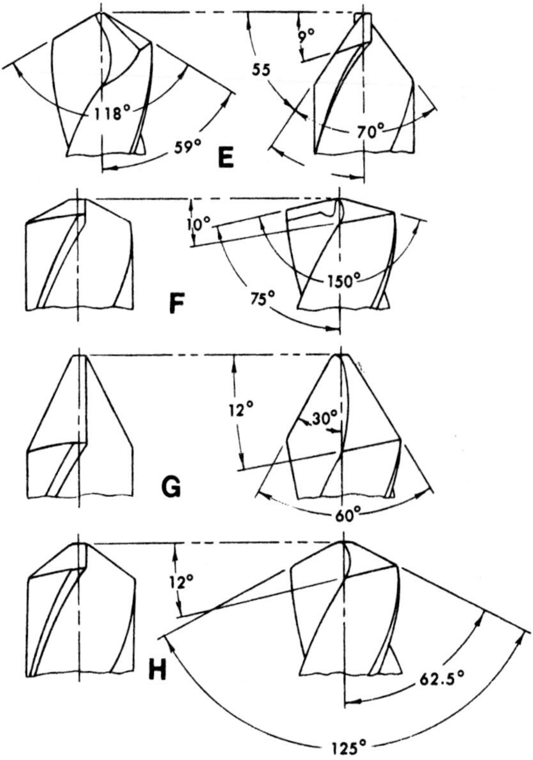

FIGURE 6.39 (*Cont.*)

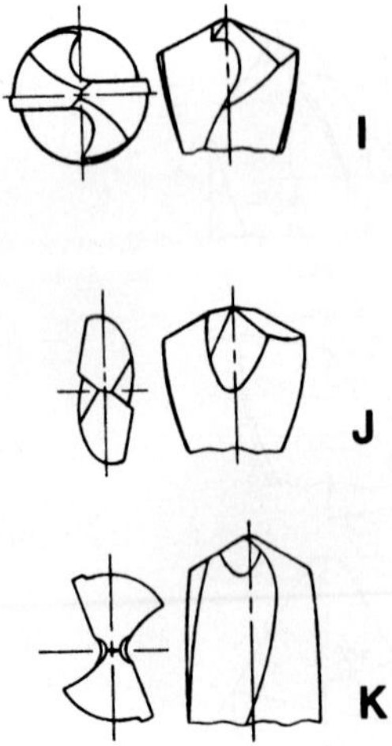

FIGURE 6.39 (*Cont.*)

C Brasses and soft bronzes

D Alternate for G, cast irons, die castings, and aluminum

E Crankshafts and deep holes

F Manganese steel and hard alloys (point angle 125 to 135°)

G Wood, fiber, hard rubber, and aluminum

H Heat-treated steels and drop forgings

I Split point, 118° or 135° point, self-centering (CNC applications)

J Parabolic flute for accurate, deep holes and rapid cutting

K Web thinning (thin the web as the drill wears from resharpening; this restores the chisel point to its proper length)

Other drill styles which are used today include the helical or S-point, which is self-centering and permits higher feed rates, and the chamfered point, which is effective in reducing burr generation in many materials.

Drills are produced from high-speed steel (HSS), solid carbide, or with carbide brazed inserts. Drill systems are made by many of the leading tool manufacturers which allow the use of removable inserts of carbide, cermet, ceramics, and cubic boron nitride (CBN). Many of the HSS twist drills used today have coatings such as titanium nitride, titanium carbide, aluminum oxide, and other tremendously hard and wear-resistant coatings. These coatings can increase drill life by as much as three to five times over premium HSS and plain-carbide drills.

Classification of High-Speed Steels

Figure 6.40 shows the chemical composition and characteristics of the M and T series of high-speed steels. The classification shown applies to drills, turning tools, mills, and other tools made of high-speed steels and is of great value in selecting the proper type of HSS tool for the intended application in machining.

Conversion of Surface Speed to Revolutions per Minute for Drills

Fractional drill sizes. Figure 6.41 shows the standard fractional drill sizes and the revolutions per minute of each fractional drill size for various surface speeds. The drilling speed tables that follow give the allowable drilling speed (sfpm) of the various materials. From these values, the correct rpm setting for drilling can be ascertained using the speed/rpm tables given here.

Wire Drill Sizes (1 through 80). See Fig. 6.42a and b.

Letter Drill Sizes. See Fig. 6.43.

Decimal Sizes of Number and Letter Drills. See Fig. 6.44.

Decimal Sizes of Metric Drills. See Fig. 6.45.

Tap Drill Sizes for Producing Unified Inch Screw Threads, Metric and Pipe Threads (*continued on page 6-81*)

Chemical Composition						Comparative Physical Properties *			
AISI - SAE Grade	Tungsten	Chromium	Vanadium	Molybdenum	Cobalt	Red Hardness	Abrasion Resistance	Edge Strength	Grinding Ability
M - 1	1.5	4.0	1.0	8.5	-	10	10	18	18
M - 2	6.0	4.0	2.0	5.0	-	10	11	17	17
M - 3-1	6.0	4.0	2.4	6.0	-	10	14	12	8
M - 3-2	6.0	4.0	3.0	6.0	-	10	17	11	6
M - 4	5.5	4.5	4.0	4.5	-	10	18	8	5
M - 7	1.7	4.0	2.0	8.75	-	10	12	17	14
M - 10	-	4.0	2.0	8.0	-	10	10	17	17
M - 15	6.5	4.5	5.0	3.0	5.0	13	19	9	2
M - 30	2.0	4.0	1.0	8.0	5.0	13	10	14	14
M - 34	2.0	4.0	2.0	8.5	8.0	15	11	14	14
M - 36	6.0	4.0	2.0	6.0	9.0	16	11	14	14
M - 42	1.5	3.75	1.15	9.5	8.0	17	19	15	14
T - 1	18.0	4.0	1.0	-	-	10	10	16	18
T - 2	18.0	4.0	2.0	-	-	10	11	15	17
T - 3	18.0	4.0	3.0	-	-	10	14	8	7
T - 4	18.0	4.0	1.0	-	5.0	13	10	11	14
T - 5	18.0	4.0	2.0	-	8.0	15	11	11	14
T - 6	22.0	4.5	1.5	-	12.0	20	10	11	14
T - 15	13.0	4.5	5.0	-	5.0	17	20	9	2

* Comparative Range of Physical Properties:
0 - 4 = Poor; 5 - 8 = Fair; 9 - 12 = Good; 13 - 16 = Very Good; 17 - 20 = Excellent

FIGURE 6.40 Classification of high-speed steels (HSS).

FRACTIONAL SIZE DRILLS

Surface Feet per Minute

Diam. Inches	10'	12'	15'	20'	25'	30'	35'	40'	45'	50'	60'	70'	80'	90'	100'
	Revolutions per Minute														
1/64	2445	2934	3667	4889	6112	7334	8556	9778	11001	12223	14668	17112	19557	22001	24446
1/32	1222	1467	1833	2445	3056	3667	4278	4889	5500	6112	7334	8556	9778	11001	12223
3/64	815	978	1222	1630	2037	2445	2852	3259	3667	4074	4889	5704	6519	7334	8149
1/16	611	733	917	1222	1528	1833	2139	2445	2750	3056	3667	4278	4889	5500	6112
5/64	489	587	733	978	1222	1467	1711	1956	2200	2445	2934	3422	3911	4400	4889
3/32	407	489	611	815	1019	1222	1426	1630	1833	2037	2445	2852	3259	3667	4074
7/64	349	419	524	698	873	1048	1222	1397	1572	1746	2095	2445	2794	3143	3492
1/8	306	367	458	611	764	917	1070	1222	1375	1528	1833	2139	2445	2750	3056
9/64	272	326	407	543	679	815	951	1086	1222	1358	1630	1901	2173	2445	2716
5/32	244	293	367	489	611	733	856	978	1100	1222	1467	1711	1956	2200	2445
11/64	222	267	333	444	556	667	778	889	1000	1111	1333	1556	1778	2000	2222
3/16	204	244	306	407	509	611	713	815	917	1019	1222	1426	1630	1833	2037
13/64	188	226	282	376	470	564	658	752	846	940	1128	1316	1504	1692	1880
7/32	175	210	262	349	437	524	611	698	786	873	1048	1222	1397	1572	1746
15/64	163	196	244	326	407	489	570	652	733	815	978	1141	1304	1467	1630
1/4	153	183	229	306	382	458	535	611	688	764	917	1070	1222	1375	1528
9/32	136	163	204	272	340	407	475	543	611	679	815	951	1086	1222	1358
5/16	122	147	183	244	306	367	428	489	550	611	733	856	978	1100	1222
11/32	111	133	167	222	278	333	389	444	500	556	667	778	889	1000	1111
3/8	102	122	153	204	255	306	357	407	458	509	611	713	815	917	1019
13/32	94	113	141	188	235	282	329	376	423	470	564	658	752	846	940
7/16	87	105	131	175	218	262	306	349	393	437	524	611	698	786	873
15/32	81	98	122	163	204	244	285	326	367	407	489	570	652	733	815
1/2	76	92	115	153	191	229	267	306	344	382	458	535	611	688	764
9/16	68	81	102	136	170	204	238	272	306	340	407	475	543	611	679
5/8	61	73	92	122	153	183	214	244	275	306	367	428	489	550	611
11/16	56	67	83	111	139	167	194	222	250	278	333	389	444	500	556
3/4	51	61	76	102	127	153	178	204	229	255	306	357	407	458	509
13/16	47	56	71	94	118	141	165	188	212	235	282	329	376	423	470
7/8	44	52	65	87	109	131	153	175	196	218	262	306	349	393	437
15/16	41	49	61	81	102	122	143	163	183	204	244	285	326	367	407
1	38	46	57	76	95	115	134	153	172	191	229	267	306	344	382
1-1/8	34	41	51	68	85	102	119	136	153	170	204	238	272	306	340
1-1/4	31	37	46	61	76	92	107	122	138	153	183	214	244	275	306
1-3/8	28	33	42	56	69	83	97	111	125	139	167	194	222	250	278
1-1/2	25	31	38	51	64	76	89	102	115	127	153	178	204	229	255
1-5/8	24	28	35	47	59	71	82	94	106	118	141	165	188	212	235
1-3/4	22	26	33	44	55	65	76	87	98	109	131	153	175	196	218
1-7/8	20	24	31	41	51	61	71	81	92	102	122	143	163	183	204
2	19	23	29	38	48	57	67	76	86	95	115	134	153	172	191
2-1/4	17	20	25	34	42	51	59	68	76	85	102	119	136	153	170
2-1/2	15	18	23	31	38	46	53	61	69	76	92	107	122	138	153
2-3/4	14	17	21	28	35	42	49	56	63	69	83	97	111	125	139
3	13	15	19	25	32	38	45	51	57	64	76	89	102	115	127
3-1/2	11	13	16	22	27	33	38	44	49	55	65	76	87	98	109

For speeds higher than tabulated, multiply all values by 10 or 100. For speeds lower than tabulated, divide all values by 10.

FIGURE 6.41 Drill rpm/surface speed, fractional drills.

WIRE SIZE DRILLS

Surface Feet per Minute

Diam. No.	10'	12'	15'	20'	25'	30'	35'	40'	45'	50'	60'	70'	80'	90'	100'
	Revolutions per Minute														
1	168	201	251	335	419	503	586	670	754	838	1005	1173	1340	1508	1675
2	173	207	259	346	432	519	605	691	778	864	1037	1210	1382	1555	1728
3	179	215	269	359	448	538	628	717	807	897	1076	1255	1434	1614	1793
4	183	219	274	366	457	548	640	731	822	914	1097	1280	1462	1645	1828
5	186	223	279	372	465	558	651	743	836	930	1115	1301	1487	1673	1859
6	187	225	281	374	468	562	655	749	843	936	1123	1310	1498	1685	1872
7	190	228	285	380	475	570	665	760	855	950	1140	1330	1520	1710	1900
8	192	230	288	384	480	576	672	768	864	960	1151	1343	1535	1727	1919
9	195	234	292	390	487	585	682	780	877	975	1169	1364	1559	1754	1949
10	197	237	296	395	494	592	691	790	888	987	1184	1382	1579	1777	1974
11	200	240	300	400	500	600	700	800	900	1000	1200	1400	1600	1800	2000
12	202	243	303	404	505	606	707	808	909	1010	1213	1415	1617	1819	2021
13	206	248	310	413	516	619	723	826	929	1032	1239	1450	1652	1859	2065
14	210	252	315	420	525	630	735	839	944	1050	1259	1469	1679	1889	2099
15	212	255	318	424	531	637	743	849	955	1064	1276	1489	1702	1914	2127
16	216	259	324	432	540	647	755	863	971	1079	1295	1511	1726	1942	2158
17	221	265	331	442	552	662	773	883	994	1104	1325	1546	1766	1987	2208
18	225	270	338	451	563	676	789	901	1014	1130	1356	1582	1808	2034	2260
19	230	276	345	460	575	690	805	920	1035	1151	1381	1611	1841	2071	2301
20	237	285	356	474	593	712	830	949	1068	1186	1423	1660	1898	2135	2372
21	240	288	360	480	601	721	841	961	1081	1201	1441	1681	1922	2162	2402
22	243	292	365	487	608	730	852	973	1095	1217	1460	1703	1946	2190	2433
23	248	298	372	496	620	744	868	992	1116	1240	1488	1736	1984	2232	2480
24	251	302	377	503	628	754	880	1005	1131	1257	1508	1759	2010	2262	2513
25	255	307	383	511	639	766	894	1022	1150	1276	1533	1789	2044	2300	2555
26	260	312	390	520	650	780	909	1039	1169	1299	1559	1819	2078	2338	2598
27	265	318	398	531	663	796	928	1061	1194	1327	1592	1857	2122	2388	2653
28	272	326	408	544	680	816	952	1087	1223	1360	1631	1903	2175	2447	2719
29	281	337	421	562	702	843	983	1123	1264	1405	1685	1966	2247	2528	2809
30	297	357	446	595	743	892	1040	1189	1338	1487	1784	2081	2378	2676	2973
31	318	382	477	637	796	955	1114	1273	1432	1592	1910	2228	2546	2865	3183
32	329	395	494	659	823	988	1152	1317	1482	1647	1976	2305	2634	2964	3293
33	338	406	507	676	845	1014	1183	1352	1521	1690	2028	2366	2704	3042	3380
34	344	413	516	688	860	1032	1204	1376	1549	1721	2065	2409	2753	3097	3442
35	347	417	521	694	868	1042	1215	1389	1563	1736	2083	2430	2778	3125	3472
36	359	430	538	717	897	1076	1255	1435	1614	1794	2152	2511	2870	3228	3587
37	367	441	551	735	918	1102	1285	1469	1653	1837	2204	2571	2938	3306	3673
38	376	452	564	753	941	1129	1317	1505	1693	1882	2258	2634	3010	3387	3763
39	384	461	576	768	960	1152	1344	1536	1728	1920	2303	2687	3071	3455	3839
40	390	468	585	780	974	1169	1364	1559	1754	1949	2339	2729	3118	3508	3898

For speeds higher than tabulated, multiply all values by 10 or 100. For speeds lower than tabulated, divide all values by 10.

FIGURE 6.42 *(a)* Wire drill sizes and rpm versus surface speed.

WIRE SIZE DRILLS

Surface Feet per Minute

Diam. No.	10'	12'	15'	20'	25'	30'	35'	40'	45'	50'	60'	70'	80'	90'	100'
	Revolutions per Minute														
41	398	477	597	796	995	1194	1393	1592	1790	1990	2387	2785	3183	3581	3979
42	409	490	613	817	1021	1226	1430	1634	1838	2043	2451	2860	3268	3677	4085
43	429	515	644	858	1073	1288	1502	1717	1931	2146	2575	3004	3434	3863	4292
44	444	533	666	888	1110	1332	1555	1777	1999	2221	2665	3109	3554	3999	4442
45	466	559	699	932	1165	1397	1630	1863	2096	2329	2795	3261	3726	4192	4658
46	472	566	707	943	1179	1415	1650	1886	2122	2358	2830	3301	3773	4244	4716
47	487	584	730	973	1216	1460	1703	1946	2190	2433	2920	3406	3893	4379	4866
48	503	603	754	1005	1256	1508	1759	2010	2262	2513	3016	3518	4021	4523	5026
49	523	628	785	1046	1308	1570	1831	2093	2355	2617	3140	3663	4186	4710	5233
50	546	655	819	1091	1364	1637	1910	2183	2456	2729	3274	3820	4366	4911	5457
51	570	684	855	1140	1425	1710	1995	2280	2565	2851	3421	3991	4561	5131	5701
52	602	722	902	1203	1504	1805	2105	2406	2707	3008	3609	4211	4812	5414	6015
53	642	770	963	1284	1605	1926	2247	2568	2889	3207	3848	4490	5131	5773	6414
54	694	833	1042	1389	1736	2083	2431	2778	3125	3473	4167	4862	5556	6251	6945
55	735	881	1102	1469	1836	2204	2571	2938	3306	3673	4408	5142	5877	6611	7346
56	821	986	1232	1643	2054	2464	2875	3286	3696	4108	4929	5751	6572	7394	8215
57	888	1066	1332	1777	2221	2665	3109	3553	3997	4452	5342	6232	7122	8013	8903
58	909	1091	1364	1819	2274	2728	3183	3638	4093	4547	5456	6367	7275	8186	9095
59	932	1118	1397	1863	2329	2795	3261	3726	4192	4658	5590	6521	7453	8388	9316
60	955	1146	1432	1910	2387	2865	3342	3820	4297	4775	5729	6684	7639	8594	9549
61	979	1175	1469	1959	2449	2938	3428	3918	4407	4897	5876	6856	7835	8815	9794
62	1005	1206	1508	2010	2513	3016	3518	4021	4523	5025	6030	7035	8040	9045	10050
63	1032	1239	1549	2065	2581	3097	3613	4129	4646	5160	6192	7224	8256	9288	10320
64	1061	1273	1592	2122	2653	3183	3714	4244	4775	5305	6366	7427	8488	9549	10610
65	1091	1310	1637	2183	2728	3274	3820	4365	4911	5455	6546	7637	8728	9819	10910
66	1157	1389	1736	2315	2894	3472	4051	4630	5207	5790	6948	8106	9264	10422	11580
67	1194	1432	1790	2387	2984	3581	4178	4775	5371	5970	7164	8358	9552	10746	11940
68	1232	1479	1848	2464	3080	3696	4313	4929	5545	6160	7392	8624	9856	11088	12320
69	1308	1570	1962	2616	3270	3924	4578	5232	5887	6530	7836	9142	10488	11754	13060
70	1364	1637	2046	2728	3410	4093	4775	5457	6139	6820	8184	9548	10912	12276	13640
71	1469	1763	2204	2938	3673	4407	5142	5876	6611	7365	8838	10311	11784	13257	14730
72	1528	1833	2272	3056	3820	4584	5348	6112	6875	7640	9168	10696	12224	13752	15280
73	1592	1910	2387	3183	3979	4775	5570	6366	7162	7960	9552	11144	12736	14328	15920
74	1698	2037	2546	3395	4244	5093	5942	6791	7639	8510	10212	11914	13616	15318	17020
75	1819	2183	2728	3638	4547	5457	6366	7276	8185	9095	10914	12733	14552	16371	18190
76	1910	2292	2865	3820	4775	5730	6684	7639	8594	9550	11460	13370	15280	17190	19100
77	2122	2546	3183	4244	5305	6366	7427	8488	9549	10610	12732	14854	16976	19098	21220
78	2387	2865	3581	4775	5968	7162	8356	9549	10743	11935	14322	16709	19096	21483	23870
79	2634	3161	3951	5269	6586	7903	9220	10537	11854	13170	15804	18438	21072	23706	26340
80	2829	3395	4244	5659	7074	8488	9903	11318	12732	14150	16980	19810	22640	25470	28300

For speeds higher than tabulated, multiply all values by 10 or 100. For speeds lower than tabulated, divide all values by 10.

FIGURE 6.42 (*b*) Drill rpm/surface speed, wire-size drills.

LETTER SIZE DRILLS

Surface Feet per Minute

Letter Size	10'	12'	15'	20'	25'	30'	35'	40'	45'	50'	60'	70'	80'	90'	100'
	Revolutions per Minute														
A	163	196	245	326	408	490	571	653	735	818	982	1145	1309	1472	1636
B	160	193	241	321	401	481	562	642	722	803	963	1124	1284	1445	1605
C	158	189	237	316	395	473	552	631	710	789	947	1105	1262	1420	1578
D	155	186	233	311	388	466	543	621	699	778	934	1089	1245	1400	1556
E	153	183	229	306	382	458	535	611	687	764	917	1070	1222	1375	1528
F	149	178	223	297	372	446	520	595	669	743	892	1040	1189	1337	1486
G	146	176	220	293	366	439	512	585	659	732	878	1024	1170	1317	1463
H	144	172	215	287	359	431	503	574	646	718	862	1005	1149	1294	1436
I	140	169	211	281	351	421	492	562	632	702	842	983	1123	1264	1404
J	138	165	207	276	345	414	483	552	621	690	827	965	1103	1241	1379
K	136	163	204	272	340	408	476	544	612	680	815	951	1087	1223	1359
L	132	158	198	263	329	395	461	527	593	659	790	922	1054	1185	1317
M	129	155	194	259	324	388	453	518	583	648	777	907	1036	1166	1295
N	126	152	190	253	316	379	442	506	569	633	759	886	1012	1139	1265
O	121	145	181	242	302	363	423	484	544	605	725	846	967	1088	1209
P	118	142	177	237	296	355	414	473	532	592	710	828	946	1065	1183
Q	115	138	173	230	288	345	403	460	518	575	690	805	920	1035	1150
R	113	135	169	225	282	338	394	451	507	564	676	789	902	1014	1127
S	110	132	165	220	274	329	384	439	494	549	659	769	878	988	1098
T	107	128	160	213	267	320	373	427	480	533	640	746	853	959	1066
U	104	125	156	208	259	311	363	415	467	519	623	727	830	934	1038
V	101	122	152	203	253	304	355	405	456	507	608	709	810	912	1013
W	99	119	148	198	247	297	346	496	445	495	594	693	792	891	989
X	96	115	144	192	240	289	337	385	433	481	576	672	769	865	962
Y	95	113	142	189	236	284	331	378	425	473	567	662	756	851	945
Z	92	111	139	185	231	277	324	370	416	462	555	647	740	832	925

For speeds higher than tabulated, multiply all values by 10 or 100. For speeds lower than tabulated divide all values by 10.

FIGURE 6.43 Letter drill sizes and rpm versus surface speed.

Tap Drills for Unified Inch Screw Threads. See Fig. 6.46.

Tap-Drill Sizes for Producing Metric Screw Threads. See Fig. 6.47.

Tap-Drill Sizes for Pipe Threads (Taper and Straight Pipe). See Fig. 6.48.

Speeds and Feeds, Drill Geometry, and Cutting Recommendations for Drills

The drilling table recommendation shown in Fig. 6.49 has been derived from data originated by the Society of

DRILL NO.	DECIMAL	DRILL NO.	DECIMAL	DRILL NO.	DECIMAL
97	0.0059	56	0.0465	15	0.180
96	0.0063	55	0.052	14	0.182
95	0.0067	54	0.055	13	0.185
94	0.0071	53	0.0595	12	0.189
93	0.0075	52	0.0635	11	0.191
92	0.0079	51	0.067	10	0.1935
91	0.0083	50	0.070	9	0.196
90	0.0087	49	0.073	8	0.199
89	0.0091	48	0.076	7	0.201
88	0.0095	47	0.0785	6	0.204
87	0.010	46	0.076	5	0.2055
86	0.0105	45	0.082	4	0.209
85	0.011	44	0.086	3	0.213
84	0.0115	43	0.089	2	0.221
83	0.012	42	0.0935	1	0.228
82	0.0125	41	0.096	A	0.234
81	0.013	40	0.098	B	0.238
80	0.0135	39	0.0995	C	0.242
79	0.0145	38	0.1015	D	0.246
78	0.016	37	0.104	E	0.250
77	0.018	36	0.1065	F	0.257
76	0.020	35	0.110	G	0.261
75	0.021	34	0.111	H	0.266
74	0.0225	33	0.113	I	0.272
73	0.024	32	0.116	J	0.277
72	0.025	31	0.120	K	0.281
71	0.026	30	0.1285	L	0.290
70	0.028	29	0.136	M	0.295
69	0.0292	28	0.1405	N	0.302
68	0.031	27	0.144	O	0.316
67	0.032	26	0.147	P	0.323
66	0.033	25	0.1495	Q	0.332
65	0.035	24	0.152	R	0.339
64	0.036	23	0.154	S	0.348
63	0.037	22	0.157	T	0.358
62	0.038	21	0.159	U	0.368
61	0.039	20	0.161	V	0.377
60	0.040	19	0.166	W	0.386
59	0.041	18	0.1695	X	0.397
58	0.042	17	0.173	Y	0.404
57	0.043	16	0.177	Z	0.413

FIGURE 6.44 Drill sizes (American national standard).

Drill	Decimal	Drill	Decimal	Drill	Decimal	Drill	Decimal	Drill	Decimal	Drill	Decimal
.35mm	.0138	1.75mm	.0650	3.70mm	.1457	6.40mm	.2520	9.00mm	.3543	17.00mm	.6693
.38mm	.0150	1.80mm	.0709	3.75mm	.1477	6.50mm	.2559	9.10mm	.3583	17.50mm	.6890
.40mm	.0157	1.85mm	.0728	3.80mm	.1496	6.60mm	.2598	9.20mm	.3622	18.00mm	.7087
.42mm	.0165	1.90mm	.0748	3.90mm	.1535	6.70mm	.2638	9.25mm	.3642	18.50mm	.7283
.45mm	.0177	1.95mm	.0768	4.00mm	.1575	6.75mm	.2658	9.30mm	.3661	19.00mm	.7480
.48mm	.0189	2.00mm	.0787	4.10mm	.1614	6.80mm	.2677	9.40mm	.3701	19.50mm	.7677
.50mm	.0197	2.05mm	.0807	4.20mm	.1654	6.90mm	.2716	9.50mm	.3740	20.00mm	.7874
.55mm	.0217	2.10mm	.0827	4.25mm	.1674	7.00mm	.2756	9.60mm	.3780	20.50mm	.8071
.60mm	.0236	2.15mm	.0846	4.50mm	.1771	7.10mm	.2795	9.70mm	.3819	21.00mm	.8268
.65mm	.0256	2.20mm	.0866	4.60mm	.1811	7.20mm	.2835	9.75mm	.3839	21.50mm	.8465
.70mm	.0276	2.25mm	.0886	4.70mm	.1850	7.25mm	.2855	9.80mm	.3558	22.00mm	.8661
.75mm	.0295	2.30mm	.0905	4.75mm	.1870	7.30mm	.2874	9.90mm	.3898	22.50mm	.8858
.80mm	.0315	2.35mm	.0925	4.80mm	.1890	7.40mm	.2913	10.00mm	.3937	23.00mm	.9055
.85mm	.0335	2.40mm	.0945	4.90mm	.1929	7.50mm	.2953	10.20mm	.4016	23.50mm	.9252
.90mm	.0354	2.45mm	.0965	5.00mm	.1968	7.60mm	.2992	10.50mm	.4134	24.00mm	.9449
.95mm	.0374	2.50mm	.0984	5.10mm	.2008	7.70mm	.3031	10.80mm	.4252	24.50mm	.9646
1.00mm	.0394	2.55mm	.1004	5.20mm	.2047	7.75mm	.3051	11.00mm	.4330	25.00mm	.9843
1.05mm	.0413	2.60mm	.1024	5.25mm	.2067	7.80mm	.3071	11.20mm	.4409		
1.10mm	.0433	2.65mm	.1043	5.30mm	.2087	7.90mm	.3110	11.50mm	.4528		
1.15mm	.0453	2.70mm	.1063	5.40mm	.2126	8.00mm	.3150	11.80mm	.4646	+1.00mm increments	
1.20mm	.0472	2.75mm	.1083	5.50mm	.2165	8.10mm	.3189	12.00mm	.4724	up to 48mm.	
1.25mm	.0492	2.80mm	.1102	5.60mm	.2205	8.20mm	.3228	12.20mm	.4803		
1.30mm	.0512	2.90mm	.1142	5.70mm	.2244	8.25mm	.3248	12.50mm	.4921	+5.00mm increments	
1.35mm	.0531	3.00mm	.1181	5.75mm	.2264	8.30mm	.3268	13.00mm	.5118	from 50mm up to	
1.40mm	.0551	3.10mm	.1220	5.80mm	.2283	8.40mm	.3307	13.50mm	.5315	105mm.	
1.45mm	.0571	3.20mm	.1260	5.90mm	.2323	8.50mm	.3346	14.00mm	.5512		
1.50mm	.0591	3.25mm	.1280	6.00mm	.2362	8.60mm	.3386	14.50mm	.5709		
1.55mm	.0610	3.30mm	.1299	6.10mm	.2401	8.70mm	.3425	15.00mm	.5906		
1.60mm	.0629	3.40mm	.1339	6.20mm	.2441	8.75mm	.3445	15.50mm	.6102		
1.65mm	.0650	3.50mm	.1378	6.25mm	.2461	8.80mm	.3465	16.00mm	.6299		
1.70mm	.0669	3.60mm	.1417	6.30mm	.2480	8.90mm	.3504	16.50mm	6496		

FIGURE 6.45 Drill sizes (metric).

Tap size	Tap drill size	Decimal equiv. of tap drill, in	Theoretical percent of thread, %	Probable mean over-size, in	Probable hole size, in	Probable percent of thread, %
0–80	56	0.0465	83	0.0015	0.0480	74
	3/64	0.0469	81	0.0015	0.0484	71
	1.20 mm	0.0472	79	0.0015	0.0487	69
	1.25 mm	0.0492	67	0.0015	0.0507	57
1–64	54	0.0550	89	0.0015	0.0565	81
	1.45 mm	0.0571	78	0.0015	0.0586	71
	53	0.0595	67	0.0015	0.0610	59
1–72	1.5 mm	0.0591	77	0.0015	0.0606	68
	53	0.0595	75	0.0015	0.0610	67
	1.55 mm	0610	67	0.0015	0.0606	68
2–56	51	0.0670	82	0.0017	0.0687	74
	1.75 mm	0.0689	73	0.0017	0.0706	66
	50	0.0700	69	0.0017	0.0717	62
	1.80 mm	0.0709	65	0.0017	0.0726	58
2–64	50	0.0700	79	0.0017	0.0717	70
	1.80 mm	0.0709	74	0.0017	0.0726	66
	49	0.0730	64	0.0017	0.0747	56
3–48	48	0.0760	85	0.0019	0.0779	78
	5/64	0.0781	77	0.0019	0.0800	70
	47	0.0785	76	0.0019	0.0804	69
	2.00 mm	0.0787	75	0.0019	0.0806	68
	46	0.0810	67	0.0019	0.0829	60
	45	0.0820	63	0.0019	0.0839	56
3–56	46	0.0810	78	0.0019	0.0829	69
	45	0.0820	73	0.0019	0.0839	65
	2.10 mm	0.0827	70	0.0019	0.0846	62
	2.15 mm	0.0846	62	0.0019	0.0865	54
4–40	44	0.0860	80	0.0020	0.0880	74
	2.20 mm	0.0866	78	0.0020	0.0886	72
	43	0.0890	71	0.0020	0.0910	65
	2.30 mm	0.0906	66	0.0020	0.0926	60
4–48	2.35 mm	0.0925	72	0.0020	0.0926	72
	42	0.0935	68	0.0020	0.0955	61
	3/32	0.0938	68	0.0020	0.0958	60
	2.40 mm	0.0945	65	0.0020	0.0965	57
5–40	40	0.0980	83	0.0023	0.1003	76
	39	0.0995	79	0.0023	0.1018	71
	38	0.1015	72	0.0023	0.1038	65
	2.60 mm	0.1024	70	0.0023	0.1047	63

FIGURE 6.46 Tap-drill sizes, unified inch screw threads.

Tap size	Tap drill size	Decimal equiv. of tap drill, in	Theoretical percent of thread, %	Probable mean over-size, in	Probable hole size, in	Probable percent of thread, %
5–44	38	0.1015	79	0.0023	0.1038	72
	2.60 mm	0.1024	77	0.0023	0.1047	69
	37	0.1040	71	0.0023	0.1063	63
6–32	37	0.1040	84	0.0023	0.1063	78
	36	0.1065	78	0.0023	0.1088	72
	7/64	0.1095	70	0.0026	0.1120	64
	35	0.1100	69	0.0026	0.1126	63
	34	0.1100	67	0.0026	0.1136	60
6–40	34	0.1110	83	0.0026	0.1136	75
	33	0.1130	77	0.0026	0.1156	69
	2.90 mm	0.1142	73	0.0026	0.1168	65
	32	0.1160	68	0.0026	0.1186	60
8–32	3.40 mm	0.1339	74	0.0029	0.1368	67
	29	0.1360	69	0.0029	0.1389	62
8–36	29	0.1360	78	0.0029	0.1389	70
	3.5 mm	0.1378	72	0.0029	0.1407	65
10–24	27	0.1440	85	0.0032	0.1472	79
	3.70 mm	0.1457	82	0.0032	0.1489	76
	26	0.1470	79	0.0032	0.1502	74
	25	0.1495	75	0.0032	0.1527	69
	24	0.1520	70	0.0032	0.1552	64
10–32	5/32	0.1563	83	0.0032	0.1595	75
	22	0.1570	81	0.0032	0.1602	73
	21	0.1590	76	0.0032	0.1622	68
12–24	11/64	0.1719	82	0.0035	0.1754	75
	17	0.1730	79	0.0035	0.1765	73
	16	0.1770	72	0.0035	0.1805	66
12–28	16	0.1770	84	0.0035	0.1805	77
	15	0.1800	78	0.0035	0.1835	70
	4.60 mm	0.1811	75	0.0035	0.1846	67
	14	0.1820	73	0.0035	0.1855	66
1/4–20	9	0.1960	83	0.0038	0.1998	77
	8	0.1990	79	0.0038	0.2028	73
	7	0.2010	75	0.0038	0.2048	70
	13/64	0.2031	72	0.0038	0.2069	66
1/4–28	5.40 mm	0.2126	81	0.0038	0.2164	72
	3	0.2130	80	0.0038	0.2168	72
5/16–18	F	0.2570	77	0.0038	0.2608	72
	6.60 mm	0.2598	73	0.0038	0.2636	68
	G	0.2610	71	0.0041	0.2651	66
5/16–24	H	0.2660	86	0.0041	0.2701	78
	6.80 mm	0.2677	83	0.0041	0.2718	75

FIGURE 6.46 *(Cont.)*

Tap size	Tap drill size	Decimal equiv. of tap drill, in	Theoretical percent of thread, %	Probable mean over-size, in	Probable hole size, in	Probable percent of thread, %
	I	0.2720	75	0.0041	0.2761	67
3/8–16	7.80 mm	0.3071	84	0.0044	0.3115	78
	7.90 mm	0.3110	79	0.0044	0.3154	73
	5/16	0.3125	77	0.0044	0.3169	72
	O	0.3160	73	0.0044	0.3204	68
8–24	21/64	0.3281	87	0.0044	0.3325	79
	8.40 mm	0.3307	82	0.0044	0.3351	74
	Q	0.3320	79	0.0044	0.3364	71
	8.50 mm	0.3346	75	0.0044	0.3390	67
7/16–14	T	0.3580	86	0.0046	0.3626	81
	23/64	0.3594	84	0.0046	0.3640	79
	9.20 mm	0.3622	81	0.0046	0.3668	76
	9.30 mm	0.3661	77	0.0046	0.3707	72
	U	0.3680	75	0.0046	0.3726	70
	9.40 mm	0.3701	73	0.0046	0.3747	68
7/16–20	W	0.3860	79	0.0046	0.3906	72
	25/64	0.3906	72	0.0046	0.3952	65
1/2–13	10.50 mm	0.4134	87	0.0047	0.4181	82
	27/64	0.4219	78	0.0047	0.4266	73
1/2–20	29/64	0.4531	72	0.0047	0.4578	65
9/16–12	15/32	0.4688	87	0.0048	0.4736	82
	31/64	0.4844	72	0.0048	0.4892	68
9/16–18	1/2	0.5000	87	0.0048	0.5048	80
5/8–11	17/32	0.5313	79	0.0049	0.5362	75
5/8–18	9/16	0.5625	87	0.0049	0.5674	80
3/4–10	41/64	0.6406	84	0.0050	0.6456	80
	21/32	0.6563	72	0.0050	0.6613	68
1/4–16	11/16	0.6875	77	0.0050	0.6925	71
	17.50 mm	0.6890	75	0.0050	0.6940	69
7/8–9	49/64	0.7656	76	0.0052	0.7708	72
7/8–14	51/64	0.7969	84	0.0052	0.8021	79
1–8	55/64	0.8594	87	0.0059	0.8653	83
	7/8	0.8750	77	0.0059	0.8809	73
1–12	29/32	0.9063	87	0.0059	0.9122	81
	59/64	0.9219	72	0.0060	0.9279	67
1–14	59/64	0.9219	84	0.0060	0.9279	78
1 1/8–7	31/32	0.9688	84	0.0062	0.9750	81
	63/64	0.9844	76	0.0067	0.9911	72
1 1/8–12	1 1/32	1.0313	87	0.0071	1.0384	80

FIGURE 6.46 (*Cont.*) Tap-drill sizes, unified inch screw threads.

Metric Tap size	Tap drill size	Decimal equiv. of tap drill, in	Theoretical percent of thread, %	Probable mean over-size, in	Probable hole size, in	Probable percent of thread, %
M1.6 × 0.35	1.20 mm	0.0472	88	0.0014	0.0486	80
	1.25 mm	0.0492	77	0.0014	0.0506	69
M2 × 0.4	1/16	0.0625	79	0.0015	0.0640	72
	1.60 mm	0.0630	77	0.0017	0.0647	69
	52	0.0635	74	0.0017	0.0652	66
M2.5 × 0.45	2.05 mm	0.0807	77	0.0019	0.0826	69
	46	0.0810	76	0.0019	0.0829	67
	45	0.0820	71	0.0019	0.0839	63
M3 × 0.5	40	0.0980	79	0.0023	0.1003	70
	2.5 mm	0.0984	77	0.0023	0.1007	68
	39	0.0995	73	0.0023	0.1018	64
M3.5 × 0.6	33	0.1130	81	0.0026	0.1156	72
	2.9 mm	0.1142	77	0.0026	0.1163	68
	32	0.1160	71	0.0026	0.1186	63
M4 × 0.7	3.2 mm	0.1260	88	0.0029	0.1289	80
	30	0.1285	81	0.0029	0.1314	73
	3.3 mm	0.1299	77	0.0029	0.1328	69
M4.5 × 0.75	3.7 mm	0.1457	82	0.0032	0.1489	74
	26	0.1470	79	0.0032	0.1502	70
	25	0.1495	72	0.0032	0.1527	64
M5 × 0.8	4.2 mm	0.1654	77	0.0032	0.1686	69
	19	0.1660	75	0.0032	0.1692	68
M6 × 1	10	0.1935	84	0.0038	0.1973	76
	9	0.1960	79	0.0038	0.1998	71
	5 mm	0.1968	77	0.0038	0.2006	70
	8	0.1990	73	0.0038	0.2028	65
M7 × 1	A	0.2340	81	0.0038	0.2378	74
	6 mm	0.2362	77	0.0038	0.2400	70
	B	0.2380	74	0.0038	0.2418	66
M8 × 1.25	6.7 mm	0.2638	80	0.0041	0.2679	74
	17/64	0.2656	77	0.0041	0.2697	71
	H	0.2660	77	0.0041	0.2701	70
	6.8 mm	0.2677	74	0.0041	0.2718	68
M10 × 1.5	8.4 mm	0.3307	82	0.0044	0.3351	76
	Q	0.3320	80	0.0044	0.3364	75
	8.5 mm	0.3346	77	0.0044	0.3390	71
M12 × 1.75	10.25 mm	0.4035	77	0.0047	0.4082	72
	Y	0.4040	76	0.0047	0.4087	71
	13/32	0.4062	74	0.0047	0.4109	69
M14 × 2	15/32	0.4688	81	0.0048	0.4736	76
	12 mm	0.4724	77	0.0048	0.4772	72

FIGURE 6.47 Tap-drill sizes, metric screw threads.

Metric Tap size	Tap drill size	Decimal equiv. of tap drill, in	Theoretical percent of thread, %	Probable mean oversize, in	Probable hole size, in	Probable percent of thread, %
M16 × 2	35/64	0.5469	81	0.0049	0.5518	76
	14 mm	0.5512	77	0.0049	0.5561	72
M20 × 2.5	11/16	0.6875	78	0.0050	0.6925	74
	17.5 mm	0.6890	77	0.0052	0.6942	73
M24 × 3	13/16	0.8125	86	0.0052	0.8177	82
	21 mm	0.8268	76	0.0054	0.8322	73
	53/64	0.8281	76	0.0054	0.8335	73
M30 × 3.5	1 1/32	1.0312	83	0.0071	1.0383	80
	25.1 mm	1.0394	79	0.0071	1.0465	75
	1 3/64	1.0469	75	0.0072	1.0541	70
M36 × 4	1 17/64	1.2656	74	Reaming recommended		

FIGURE 6.47 (*Cont.*) Tap-drill sizes, metric screw threads.

Taper pipe		Straight pipe	
Thread	Drill	Thread	Drill
⅛–27	R	⅛–27	S
¼–18	7/16	¼–18	29/64
⅜–18	37/64	⅜–18	19/32
½–14	23/32	½–14	47/64
¾–14	59/64	¾–14	15/16
1–11½	1 5/32	1–11½	1 3/16
1¼–11½	1½	1¼–11½	1 33/64
1½–1½	1 47/64	1½–11½	1¾
2–11½	2 7/32	2–11½	2 7/32
2½–8	2⅝	2½–8	2 21/32
3–8	3¼	3–8	3 9/32
3½–8	3¾	3½–8	3 25/32
4–8	4¼	4–8	4 9/32

FIGURE 6.48 Pipe taps.

Manufacturing Engineers (SME) and various major drill manufacturers. Figure 6.50 illustrates the standard drill point and split point together with the chisel edge angle range in common use.

Drill selection for various materials is shown in Fig. 6.51.

Spade Drills

Spade drills are used to produce holes ranging from 1 in to over 6 in in diameter. Very deep holes can be produced with spade drills, including core drilling, counterboring, and bottoming to a flat or other shape. The spade drill consists of the spade drill bit and holder. The holder may contain coolant holes through which coolant can be delivered to the cutting edges, under pressure, which cools the spade and flushes the chips from the drilled hole. Typical spade drills, holders, and terminology and spade geometry are shown in Fig. 6.52.

The standard point angle on a spade drill is 130°. The rake angle ranges from 10 to 12° for average-hardness materials. The rake angle should be 5 to 7° for hard steels and 15 to 20° for soft, ductile materials. The back-taper angle should be 0.001 to 0.002 in per inch of blade depth. The outside diameter clearance angle is generally between 7 to 10°.

The cutting speeds for spade drills are normally 10 to 15 percent lower than those for standard twist drills. See the tables of drill speeds and feeds in the preceding section for approximate starting speeds. Heavy feed rates should be used with spade drilling. The table shown in Fig. 6.53 gives recommended feed rates for spade drilling various materials.

Speeds, Feeds, Drill Geometry and Cutting Recommendations for Standard Drill Types *

Material Type	Hardness Bhn	Tool Grade	Drill Type	PA-deg.	LRf deg.	HA-deg.	Point Type	sfpm Speed	ipr Feed x 10^{-3}
Low-alloy steels 4130, 4340, ,4140	to 300	M-1, M-2	A	118-135	7-10	25-30	Split	50-60	3-7
	300-400	M-1, M-2	A	118-135	7-10	25-30	split	40-50	2-6
	400-500	Cobalt	B	118-135	7-10	25-30	split	25-40	1-4
	over 500	C-2	C	118	7-10	0	notched	75-100	0.5-2
Die steels Hot-work	to 300	M-1, M-2	A	118-135	7-10	25-30	split	45-55	3-7
	300-400	M-1, M-2	A	118-135	7-10	25-30	split	35-50	2-6
	400-500	Cobalt	B	118-135	7-10	25-30	split	25-35	1-4
	over 500	C-2	C	118	7-10	0	Notched	70-90	0.5-2
Stainless steels (Austenitic) 300	135-185	M-1, M-2	A	118-135	7-10	25-30	split	70-90	2-6
Stainless steels (Martensitic) 400	150-250	M-1, M-2	A	118-135	7-10	25-30	split	50-70	3-7
	250-450	M-1, M-2	A	118-135	7-10	25-30	split	30-40	2-6
	over 450	Cobalt	B	118-135	7-10	25-30	split	20-30	1-4
Stainless steels Precipitation hardening 17-7PH, etc.	to 200	M-1, M-2	A	118-135	7-10	25-30	split	50-60	3-7
	200-350	M-1, M-2	A	118-135	7-10	25-30	split	35-45	2-6
	over 350	Cobalt	B	118-135	7-10	25-30	split	20-30	1-4
Nickel-cobalt steels High-strength	to 400	M-1, M-2	A	118-135	7-10	25-30	split	55-65	2-6
	400-500	Cobalt	B	118-135	7-10	25-30	split	30-40	1-4
	over 500	C-2	C	118	7-10	0	notched	70-90	0.5-2
High-temperature Cobalt-base alloys	to 300	Cobalt	B	118-135	7-10	25-30	split	15-25	1-4
High-temperature Iron-base alloys	to 250	Cobalt	B	118-135	7-10	25-30	split	20-30	2-6
	over 250	Cobalt	B	118-135	7-10	25-30	split	15-25	2-6
High-temperature Nickel-base alloys	to 265	Cobalt	B	118-135	7-10	25-30	split	20-30	2-6
	265-330	Cobalt	B	118-135	7-10	25-30	split	20-25	2-5
	over 330	Cobalt	B	118-135	7-10	25-30	split	15-20	1-4

Magnesium & alloys	All	M-1, M-2	A	118-135	7-10	25-30	split	150-350	2-7
Aluminum & alloys 2024, 6061, 7075, etc.	All	M-1, M-2	A	118-135	7-10	25-30	split	175-400	2-7
Titanium	to 250	M-34, M-42	B	135	7-12	30-38	split	25-30	5-7
Titanium Alpha alloys	250-300	M-34, M-42	B	135	7-12	30-38	split	20-25	5-7
Titanium Alpha-Beta alloys	to 350 over 350	M-34, M-42 M-34, M-42	B B	135 135	7-12 7-12	30-38 30-38	split split	20-25 15-25	5-7 5-7
Titanium Beta alloys	to 350 over 350	M-34, M-42 M-34, M-42	B B	135 135	7-12 7-12	30-38 30-38	split split	15-20 15-17	1-4 0.5-2
Beryllium copper	250	C-2	D	90-118	10-15	25-30	split	30-45	2-8
Tungsten & alloys	to 350	C-2, C-3	D	90-118	7-10	25-30	notched	200-250	1-4
Brass, free-machining	All	M-1, M-2	A	118	7-10	25-30	standard	100-250	4-10
Bronzes, common	All	M-1, M-2	A	118	7-10	25-30	standard	200-250	3-15
Bronze, phosphur	Hard	M-1, M-2	A	118	7-10	25-30	notched	75-150	2-6
Copper	All	M-1, M-2	A	90-118	7-10	25-30	standard	100-250	1-5
Cast iron Soft to medium	soft-med.	M-1, M-2	A	118	7-10	25-30	std or split	75-150	2-8
Cast iron Hard	Hard	C-2	D	118	7-10	25-30	std or split	40-75	1-5
Zinc	All	M-1, M-2	A	118	7-10	25-30	standard	200-250	3-10
Low-carbon steels	to 300	M-1, M-2	A	118	7-10	25-30	standard	80-100	3-10
Thermoplastics	Medium	M-1, M-2	E	60-90	12-16	17	standard	100-150	2-15

FIGURE 6.49 Drilling recommendation table.

Thermosetting plastics	Soft	M-1, M-2	E	60-90	12-16	17	standard	150	3-8
	Hard	M-1, M-2	E	60-90	12-16	17	standard	100	2-6

NOTE:
Drill Types: A = AIAA type B or C; B = Heavy duty cobalt; C = Carbide tipped; D = Solid carbide; E = Standard with wide, polished flutes.

Coolants and cutting fluids are recommended in the Coolant & Cutting Fluid Section of the handbook.

* Tabular data in the table is for drills of 0.125 through 0.500" diameter and hole depths of 1 to 3 drill diameters.
Adjustments must be made for other conditions, by interpolation or trial drilling. (Smaller drills have a lower ipr feed rate; larger drills have a higher ipr feed rate).

Drill geometry: PA = Point angle, degrees; LRf = Lip relief angle, degrees; HA = helix angle, degrees

Tabular data for ipr - Feed is given in powers of ten notation, i.e. 2 = .002"; 6 = .006"; 0.5 = 0.0005", etc.

FIGURE 6.49 *(Cont.)*

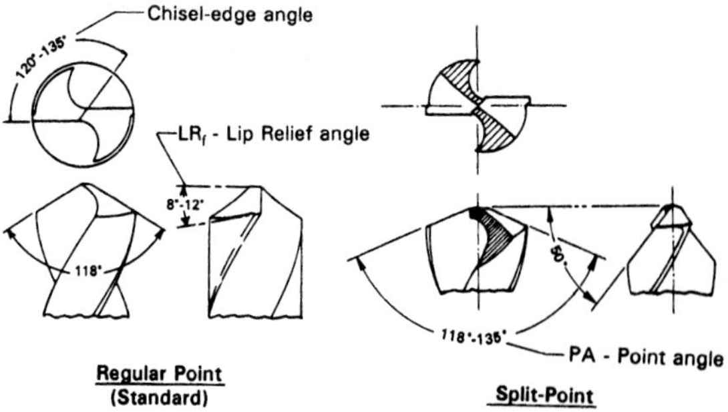

FIGURE 6.50 Standard and split points.

Microdrills

Drills below 0.020 in in diameter are considered to be *microdrills.* Microdrills as small as 0.0016 in in diameter are available as stock items from the drill supply vendors. Microdrills are manufactured from high-speed steel, cobalt-base carbides, and solid-tungsten carbides, and a limited selection is available in diamond and CBN surfaces. Some of the standard commercially available microdrill sizes are listed in Fig. 6.54.

In order to use microdrills, special machines have been developed for controlling these extremely small and fragile tools. Holes of great precision may be made in many materials using the appropriate machine tools and microdrills designed for microdrilling. Needless to say, a microdrilling operation is conducted under a high-

Drill Selection for Various Materials (Metallic and Compositions)

Material to be Drilled	Drill Style or Type
Steel (low, medium and high carbon alloys)	• High-speed steel (HSS) drills, (surface treated)
Brinell hardness up to 225 Bhn	• TiN (Titanium Nitride) coated drills
Steel (high carbon alloy)	• High-speed steel drills (HSS), (surface treated)
Brinell hardness 225 to 325 Bhn	• TiN (Titanium Nitride) coated drills
Steel (high carbon alloy)	• Cobalt drills
Brinell hardness over 325 Bhn	• TiN coated drills
	• Carbide-tipped drills
	• Solid carbide drills
High-temperature alloys (Cobalt, iron and nickel base)	• Cobalt drills • TiN coated drills
Brinell hardness to 180 to 300 Bhn	• Solid carbide, high-performance
Stainless steel (Austenitic, martensitic and precipitation hardened), i.e. 18-8PH, etc.	• High-speed steel drills • TiN coated drills
Brinell hardness below 300 Bhn	• Solid carbide drills
Stainless steel (austenitic, martensitic and precipitation hardened), i.e. 18-8PH, etc.	• Cobalt drills • TiN coated drills
Brinell hardness over 300 Bhn	• Solid carbide drills, high-performance
Titanium alpha and alpha-beta alloys Brinell hardness over 300 Bhn	• All drills in stainless steel list above except TiN coated drills
Titanium alloys (commercially pure)	• High-speed steel drills
Brinell hardness below 250 Bhn	
Cast iron (soft and medium)	• All high-speed steel drills
Brinell hardness up to 250 Bhn	• Carbide-tipped drills
Cast iron (hard)	• Solid carbide drills
Brinell hardness over 250 Bhn	
Aluminim, aluminum alloys, aluminum bronze, zinc.	• High-speed steel drills • TiN coated drills
	• Carbide-tipped drills
	• Solid carbide drills
Brass (free machining) and high tensile bronze	• High-speed steel drills
	• All drills listed for aluminum and zinc, except fast spiral types
Bronze (soft and medium), copper and copper alloys	•High-speed steel and all drills listed for aluminum, zinc, brass and bronze, except fast spiral types
Plastic and other related materials.	• High-speed steel, Carbide-tipped, and solid carbide drills
Composites (carbon, graphite, Kevlar and fiberglass)	• Brad point drills, two and three flute solid carbide drills, and carbide-tipped drills

FIGURE 6.51 Drill selection for various materials (metallic and compositions).

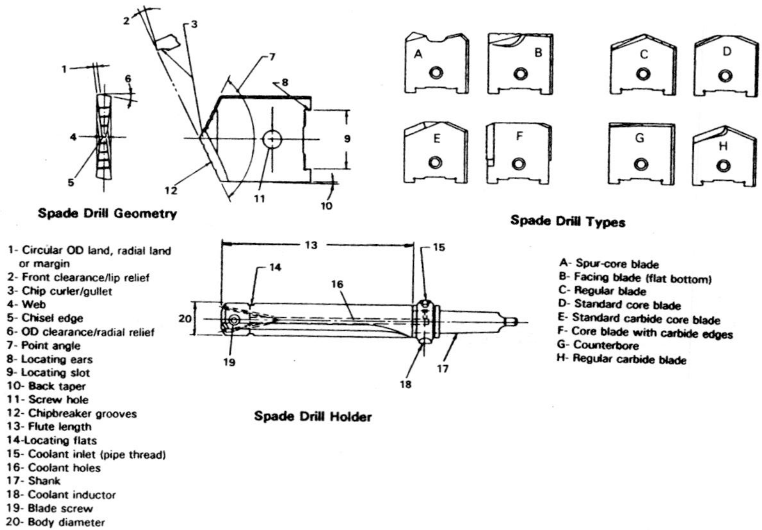

FIGURE 6.52 Spade drill terminology and features.

Recommended Feed Rates for Spade Drilling- (ipr, inches per revolution)

Material	Hardness Bhn	Feed - ipr Spade Drill Diameter, inches					
		1-1.25	1.25-2	2-3	3-4	4-5	5-8
Plain carbon steels	100-225	.012	.015	.018	.022	.025	.030
	225-275	.010	.013	.015	.018	.020	.025
	275-325	.008	.010	.013	.015	.018	.020
Free-machining steels	100-240	.014	.016	.018	.022	.025	.030
	240-325	.010	.014	.016	.020	.022	.025
Free-machining alloy steels	150-250	.014	.016	.018	.022	.025	.030
	250-325	.012	.014	.016	.018	.020	.025
	325-375	.010	.012	.014	.016	.018	.020
Alloy steels	125-180	.012	.015	.018	.022	.025	.030
	180-225	.010	.012	.016	.018	.022	.025
	225-325	.009	.010	.013	.015	.018	.020
	325-400	.006	.006	.010	.012	.014	.016
Grey cast iron	110-160	.020	.022	.026	.028	.030	.034
	160-240	.012	.014	.016	.018	.020	.022
	240-325	.010	.012	.016	.018	.018	.018
Ductile & nodular iron	140-190	.014	.016	.018	.020	.022	.024
	190-250	.012	.014	.016	.018	.018	.020
	250-325	.010	.012	.016	.018	.018	.018

Malleable iron- ferritic	110-160	.014	.016	.018	.020	.022	.024
Malleable iron- pearlitic	160-280	.011	.013	.015	.018	.018	.018
Stainless steels- free-machining							
Ferritic & austenitic (screw stock)	-------	.016	.018	.020	.022	.024	.026
Martensitic (440F, etc.)	-------	.012	.014	.016	.016	.018	.020
Stainless steels							
Ferritic & austenitic (200 & 300 series)	-------	.012	.014	.016	.018	.020	.020
Martensitic (400 series)	-------	.010	.012	.012	.014	.016	.018
Copper alloys							
Soft (ETP-110, etc)	-------	.016	.018	.020	.026	.028	.030
Hard (bronzes)	-------	.010	.012	.014	.016	.018	.018
Aluminum alloys (free-machining)	-------	.020	.022	.024	.028	.030	.040
Magnesium alloys (general) **	-------	.024	.026	.030	.034	.040	.050
High-temperature alloys (general)	-------	.008	.010	.012	.012	.014	.014
Titanium alloys (general)	-------	.008	.010	.012	.014	.014	.016
Tool steels							
Water-hardening & shock resisting	150-250	.012	.-14	.016	.018	.020	.022
Cold work	200-250	.007	.008	.009	.010	.011	.012
Hot work	150-250	.012	.013	.015	.016	.018	.020
High-speed steels	200-250	.010	.012	.013	.015	.017	.018

NOTE: Hardness ranges are Brinell hardness numbers. Tabular data is in inches of feed per revolution, ipr.
** A fire hazard exists when machining or drilling magnesium & alloys.

FIGURE 6.53 Feed rates for spade drills.

Standard Size Microdrills - High-Speed Steel and Cobalt

High-Speed Steel Microdrills

Diameter inches	mm	Flute lth.
0.0059	0.15	0.0469
0.0063	0.16	0.0469
0.0067	0.17	0.0625
0.0071	0.18	0.0625
0.0075	0.19	0.0625
0.0079	0.20	0.0625
0.0083	0.21	0.0625
0.0087	0.22	0.0625
0.0091	0.23	0.0625
0.0094	0.24	0.0781
0.0098	0.25	0.0781
0.0102	0.26	0.0781
0.0106	0.27	0.0781
0.0110	0.28	0.0938
0.0114	0.29	0.0938
0.0118	0.30	0.0938
0.0126	0.32	0.0938
0.0134	0.34	0.0938
0.0142	0.36	0.1875
0.0150	0.38	0.1875
0.0157	0.40	0.1875
0.0165	0.42	0.1875
0.0173	0.44	0.1875
0.0181	0.46	0.1875
0.0189	0.48	0.1875
0.0197	0.50	0.1875

Note: Drill points are 118°. Shank size is same as drill diameter. Drills 0.38mm and less are 0.75" long, all others 0.875".

Cobalt Microdrills

Diameter inches	mm	Flute lth.
0.0019	0.05	0.0158
0.0024	0.06	0.0158
0.0028	0.07	0.0158
0.0032	0.08	0.0197
0.0035	0.09	0.0197
0.0039	0.10	0.0276
0.0043	0.11	0.0276
0.0047	0.12	0.0276
0.0051	0.13	0.0394
0.0055	0.14	0.0394
0.0059	0.15	0.0394
0.0063	0.16	0.0552
0.0067	0.17	0.0552
0.0071	0.18	0.0552
0.0075	0.19	0.0552
0.0079	0.20	0.0709
0.0083	0.21	0.0709
0.0087	0.22	0.0709
0.0091	0.23	0.0709
0.0094	0.24	0.0709
0.0098	0.25	0.0867
0.0102	0.26	0.0867
0.0106	0.27	0.0867
0.0110	0.28	0.0867
0.0114	0.29	0.0867
0.0118	0.30	0.0867
0.0122	0.31	0.1103
0.0126	0.32	0.1103
0.0130	0.33	0.1103
0.0134	0.34	0.1103
0.0138	0.35	0.1103
0.0142	0.36	0.1103
0.0146	0.37	0.1103
0.0150	0.38	0.1103
0.0154	0.39	0.1142
0.0157	0.40	0.1142
0.0161	0.41	0.1142
0.0165	0.42	0.1142
0.0169	0.43	0.1142
0.0173	0.44	0.1142
0.0177	0.45	0.1142
0.0181	0.46	0.1142
0.0185	0.47	0.1142
0.0189	0.48	0.1142
0.0193	0.49	0.1576
0.0197	0.50	0.1576

Note: Cobalt drills are for extremely accurate drilling jobs. Right hand spiral, 118° points and 25° helix angle. All shank diameters are 1mm and total length 25mm.

FIGURE 6.54 Microdrills.

magnification viewing device, with the hand motions of the operator greatly reduced in magnitude by linkage translators on the microdrilling machine.

Specialty flat-style microdrill geometry and terminology are shown in Fig. 6.55, together with the various types of microdrilling tools. The microdrilled hole usually requires additional operations such as countersinking, reaming, counterboring, etc. Figure 6.56 gives the preferred dimensions for the special flat-type microdrills that are usually custom made for various applications. Materials for these special microdrills may be of HSS, cobalt-base carbides, or tungsten carbides.

Solid-Carbide Square Die Drills. For cold drilling extremely accurate holes in hard steels from Rockwell R_c40 to R_c70, the carbide square die drill is standardized in sizes from 0.09375 through 0.5000 in diameter. These are available from national drill distributors and are ideally suited for mold and die repair and maintenance.

6.4 REAMING

A *reamer* is a rotary cutting tool, either cyclindrical or conical in shape, used for enlarging drilled holes to accurate dimensions, normally on the order of ±0.0001 in and closer. Reamers usually have two or more flutes which may be straight or spiral in either left-hand or right-hand spiral. Reamers are made for manual or machine operation. Figure 6.57 shows reamer geometry and terminology.

Reamers are made in various forms, including

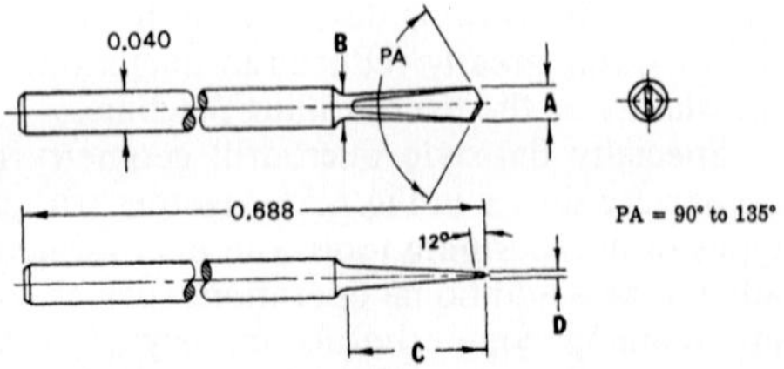

Flat Microdrill Geometry
(See Figure 6.56 for dimensions)

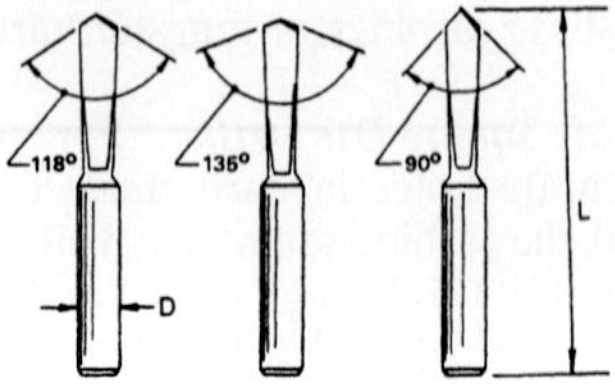

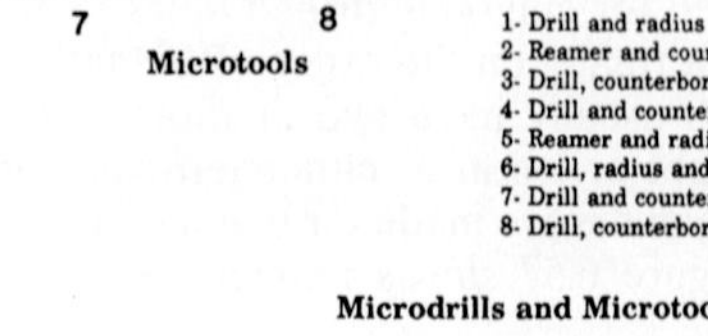

Microtools

Microtool Types

1- Drill and radius
2- Reamer and countersink
3- Drill, counterbore and countersink
4- Drill and countersink
5- Reamer and radius
6- Drill, radius and countersink
7- Drill and counterbore
8- Drill, counterbore and radius

Microdrills and Microtools

FIGURE 6.55 Geometry of microdrills and microtools.

Drilling problems and solutions

Problem	Causes	Remedy
Outer corners breakdown	Revolutions per minute too high Poor lip relief	Reduce feed and speed Check lip relief
Cutting lips chip	Feed too high High lip relief	Reduce feed Check lip relief
Cracks in cutting lips	Running too hot	Repoint drill Check feed and speed
Drill breaks	Improper point Flutes clogging	Use proper point Check feed
Drill splits up center	Feed too high Lip relief wrong	Reduce feed Correct relief
Drill will not penetrate	Reverse chisel	Check chisel angle
Rough hole	Dull point No lubricant	Repoint drill Use lubricant
Oversize hole	Unequal length on cutting lips	Repoint drill
Loose spindle	Inspect spindle	
Chips change form while drilling	Improper point	Repoint drill
Margins chip	Oversize bushing	Change bushing

- Hand reamers
- Machine reamers
- Left-hand flute
- Right-hand flute
- Expansion reamers
- Chucking reamers
- Stub screw-machine reamers
- End-cutting reamers

Flat Microdrill Dimensions - Refer to Figure 6.55 **for drill geometry.**

Diameter A	Back Taper B	Length C	Web Thickness D
0.0010	0.0001	0.0070	0.0005-0.00075
0.0020	0.0001	0.0140	0.0011-0.0015
0.0030	0.0002	0.0210	0.0016-0.0020
0.0040	0.0002	0.0280	0.0018-0.0022
0.0050	0.0003	0.0350	0.0020-0.0024
0.0060	0.0003	0.0420	0.0022-0.0026
0.0070	0.0004	0.0490	0.0024-0.0028
0.0080	0.0004	0.0560	0.0026-0.0030
0.0090	0.0005	0.0630	0.0028-0.0032
0.0100	0.0005	0.0700	0.0030-0.0034
0.0110	0.0006	0.0770	0.0032-0.0036
0.0120	0.0006	0.0840	0.0038-0.0042
0.0130	0.0007	0.0910	0.0042-0.0046
0.0140	0.0007	0.0980	0.0044-0.0048
0.0150	0.0008	0.1050	0.0046-0.0050
0.0160	0.0008	0.1120	0.0048-0.0052
0.0170	0.0009	0.1190	0.0050-0.0054
0.0180	0.0009	0.1260	0.0053-0.0058
0.0190	0.0010	0.1330	0.0056-0.0060
0.0200	0.0010	0.1400	0.0060-0.0064

Note: These microdrill sizes and dimensions are for custom made flat microdrills. See Figure 6.54 **for modern standard size microdrills of HSS and cobalt alloy.**

FIGURE 6.56 Flat microdrill dimensions.

- Jobbers' reamers
- Shell reamers
- Combined drill and reamer

Most reamers are produced from premium-grade HSS. Reamers are also produced in cobalt alloys, and these may be run at speeds 25 percent faster than HSS reamers. Reamer feeds depend on the type of reamer, the material and amount to be removed, and the final finish required. Material-removal rates depend on the size of the reamer and material, but general figures may be used on a trial basis and are summarized below:

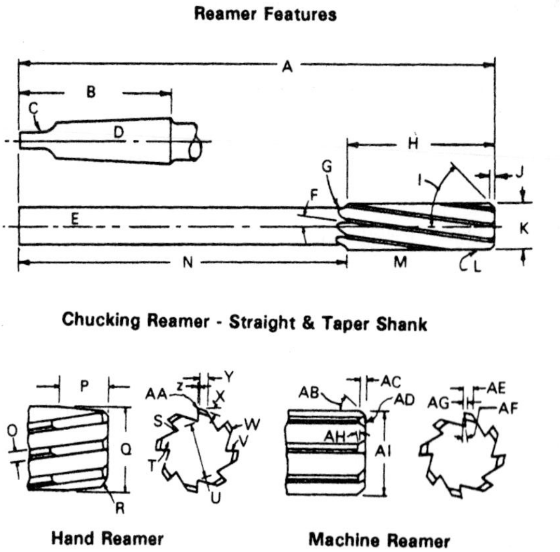

A Overall length
B Shank length
C Tang
D Taper shank
E Straight shank
F Helix angle
G Cutter sweep
H Flute length
I Chamfer angle
J Chamfer length
K Actual size
L Body
M Helix flutes, Right hand shown
N Shank length
O Land width
P Starting taper
Q Actual size diameter
R Bevel

S Flute
T Land
U Core diameter
V Cutting face
W Heel
X Relief angle
Y Relieved land
Z Margin
AA Cutting edge
AB Chamfer angle
AC Chamfer length
AD Chamfer relief
AE Land width
AF Radial rake angle
AG Margin
AH Chamfer relief angle
AI Actual size diameter

FIGURE 6.57 Reamer features.

Hole diameter	Material to be removed
Up to 0.500 in diameter	0.005 in for finishing
More than 0.500 in diameter	0.015 in for finishing
Up to 0.500 in diameter	0.015 in for semifinished holes
More than 0.500 in diameter	0.030 in for semifinished holes

This is an important consideration when using the expansion reamer owing to the maximum amount of expansion allowed by the adjustment on this type of reamer.

Machine Speeds and Feeds for HSS Reamers (see Fig. 6.58)

Note: Cobalt-alloy and carbide reamers may be run at speeds 25 percent faster than those shown in Fig. 6.58.

Carbide-tipped and solid-carbide chucking reamers are also available and afford greater effective life than HHS and cobalt reamers without losing their nominal size dimensions. Speeds and feeds for carbide reamers are generally similar to those for the cobalt-alloy types.

Sharpening Reamers

It would be difficult, if not impossible, to hand sharpen any type of reamer. Reamer sharpening machines are produced by various manufacturers (Darex), and sharpening facilities are available nationwide for this purpose.

Standard reamer sizes are produced and may be purchased separately or in sets for various applications.

Material	Speed (sfpm)	Feed Code (ipr)
Steel - 150 Bhn	80	1
Steel - 200 Bhn	55	2
Steel - 250 Bhn	35	3
Steel - 300 Bhn	30	3
Steel - 350 Bhn	17	4
Steel - 400 Bhn	10	4
Steel, cast	25	3
Steel, forged alloys	30	3
Steel, low carbon	75	2
Steel, high carbon	45	4
Steel, stainless	15	3
Steel, tool	35	4
Titanium	40	1
Zinc alloy	150	1
Aluminum & alloys	150	1
Brass, leaded	175	1
Brass, red & yellow	150	1
Bronzes	160	1
Copper	45	3
Cast iron, chilled	10	4
Cast iron, hard	50	3
Cast iron, pearlitic	60	1
Cast iron, soft	95	1
Malleable iron	65	2
Monels	30	3
Nickels	40	3
Plastic, hard	50	1
Plastic, soft	65	3

Feed Code, ipr (inches per revolution)

Reamer Diameter	Code 1	Code 2	Code 3	Code 4
0.125"	0.006	0.005	0.004	0.003
0.500"	0.012	0.010	0.007	0.005
1.00"	0.020	0.015	0.012	0.008
2.00"	0.032	0.025	0.020	0.012
2.25 - 2.50"	0.043	0.035	0.028	0.018
2.75 - 3.00"	0.055	0.045	0.035	0.024

Note: Reamer feeds may be interpolated for intermediate sizes than those shown in the table. Cobalt reamers may be run at speeds 25% faster than those shown in the table for HSS.

FIGURE 6.58 Machine speeds and feeds for HSS reamers.

Some of the more common types of reamers are shown in Fig. 6.59.

Forms of Reamers

Other forms of reamers include the following:

Morse Taper Reamers. These reamers are used to produce and maintain holes for American standard Morse taper shanks. They usually come in a set of two, one for roughing and the other for finishing the tapered hole.

Taper-Pin Reamers. Taper-pin reamers are produced in HSS with straight, spiral, and helical flutes. They range in size from pin size 7/0 through 14 and

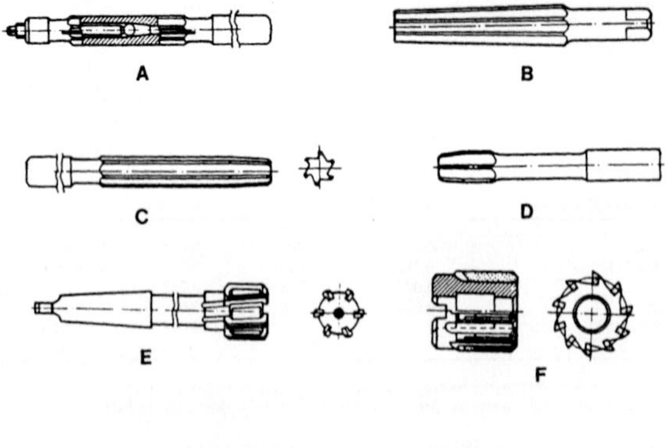

A Expansion Reamer
B Taper Reamer
C Hand Reamer
D Machine Reamer
E Inserted Blade Reamer
F Shell Reamer

FIGURE 6.59 Reamer types.

include 21 different sizes to accommodate all standard taper pins.

Dowel-pin reamers. Dowel-pin reamers are produced in HSS for standard length and jobbers' lengths in 14 different sizes from 0.125 through 0.500 in. The nominal reamer size is slightly smaller than the pin diameter to afford a force fit.

Helical-flute die-makers' Reamers. These reamers are used as milling cutters to join closely drilled holes. They are produced from HSS and are available in 16 sizes ranging from size AAA through O.

Reamer blanks. Reamer blanks are available for use as gages, guide pins, or punches. They are made of HSS in jobbers' lengths from 0.015 through 0.500 in diameters. Fractional sizes through 1.00 in diameter and wire-gage sizes are also available.

Shell reamers (see Fig. 6.59F). These reamers are designed for mounting on arbors and are best suited for sizing and finishing operations. Most shell reamers are produced from HSS. The inside hole in the shell reamer is tapered ⅛ in per foot and fits the taper on the reamer arbor.

Expansion reamers (see Fig. 6.59A). The hand expansion reamer has an adjusting screw at the cutting end which allows the reamer flutes to expand within certain limits. The recommended expansion limits are listed below for sizes through 1.00 in diameter:

Reamer size: 0.25 to 0.625 in diameter
Expansion limit = 0.010 in

Reamer size: 0.75 to 1.000 in diameter
Expansion limit = 0.013 in

Note: Expansion reamer stock sizes up to 3.00 in diameter are available.

6.5 BROACHING

Broaching is a precision machining operation wherein a broach tool is either pulled or pushed through a hole in a workpiece or over the surface of a workpiece to produce a very accurate shape such as round, square, hexagonal, spline, keyway, and so on. Keyways in gear and sprocket hubs are broached to an exact dimension so that the key will fit with very little clearance between the hub of the gear or sprocket and the shaft. The cutting teeth on broaches are increased in size along the axis of the broach so that as the broach is pushed or pulled through the workpiece, a progressive series of cuts is made to the finished size in a single pass.

Broaches are driven or pulled by manual arbor presses and horizontal or vertical broaching machines. A single stroke of the broaching tool completes the machining operation. Broaches are commonly made from premium-quality HSS and are supplied either in single tools or as sets in graduated sizes and different shapes. Figure 6.60 shows a number of different types of broaches such as keyway, square, and hexagonal.

Broaches may be used to cut internal or external shapes on workpieces. Blind holes also can be broached with specially designed broaching tools. The broaching tool teeth along the length of the broach are normally divided into three separate sections. The teeth of a broach include roughing teeth, semifinishing teeth, and

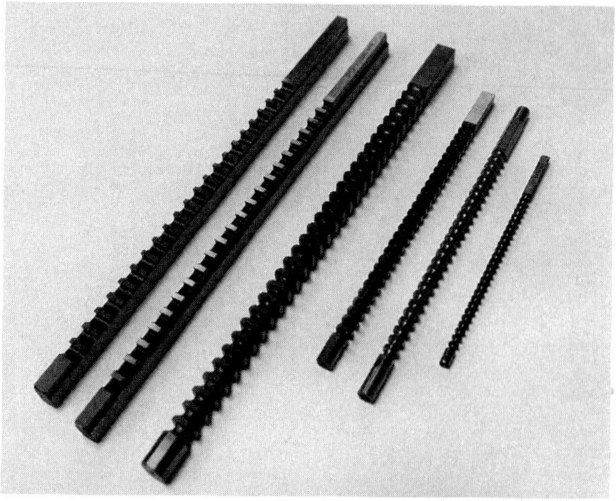

FIGURE 6.60 Keyway, square, and hexagonal broaches.

finishing teeth. All finishing teeth of a broach are the same size, while the semifinishing and roughing teeth are progressive in size up to the finishing teeth.

Figures 6.61 and 6.62 show the terminology and geometry of broaching teeth. A broaching tool must have sufficient strength and stock-removal and chip-carrying capacity for its intended operation. An interval-pull broach must have sufficient tensile strength to withstand the maximum pulling forces that occur during the pulling operation. An internal-push broach must have sufficient compressive strength as well as the ability to withstand buckling or breaking under the pushing forces that occur during the pushing operation.

Material	Rake Angle (Degrees)	Relief Angle (Degrees)
Aluminum	6 - 10	------
Babbitt	8 - 10	------
Brass	-5 to 5	2 - 3
Bronze	0	1 - 2
Cast iron	6 - 10	2 - 5
Copper	12 - 15	2 - 3
Aluminum bronze	12 - 15	2 - 3
Steels: SAE		
1035	15	1 - 2
1112	15	2 - 3
4140	8 - 15	1 - 3
5140	18 - 20 (finishing)	1 - 2
Stainless steels: AISI		
303	15	0.5 - 2
304, 304L	15	0.5 - 2
316	12 - 15	0.5 - 2
410	12 - 15	0.5 - 2
430	15 - 20	0.5 - 2
440	12- 15	0.5 - 2
Titanium:		
Soft	5 - 15	2 - 4
Hard	12 - 15	2.5 - 3
Zinc	6	------

FIGURE 6.61 Typical HSS broach rake angles and relief angles.

Broaches are produced in sizes ranging from 0.050 to as large as 20 in or more. The term *button broach* was used for broaching tools which produced the spiral lands that form the "rifling" in gun barrels from small to large caliber. Broaches may be rotated to produce a predetermined spiral angle during the pull or push operations.

6.6 VERTICAL BORING AND JIG BORING

The increased demand for accuracy in producing large parts initiated the refined development of modern verti-

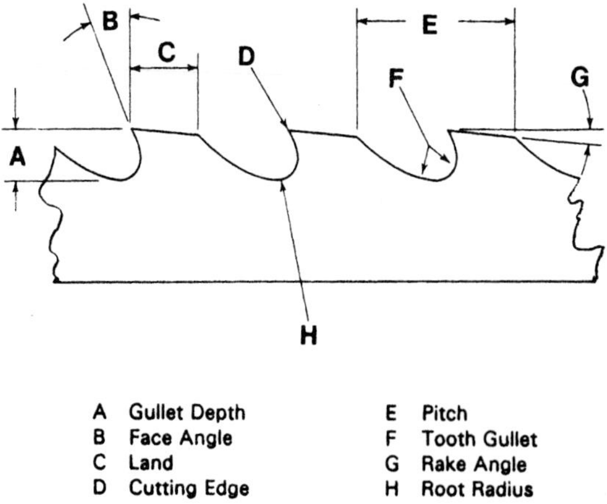

FIGURE 6.62 Broach terminology and geometry.

cal and jig boring machines. Although the modern CNC machining centers can handle small to medium-sized jig boring operations, very large and heavy work of high precision is done on modern CNC jig boring machines or vertical boring machines. Also, any size work which requires extreme accuracy is usually jig bored. The modern jig boring machines are equipped with high-precision spindles and *x*/*y* coordinate table movements of high precision and may be CNC controlled with digital read-out panels. For a modern jig boring operation that is CNC/DNC controlled, the circle diameter and number of equally spaced holes or other geometric pattern is entered into the DNC program and the computer

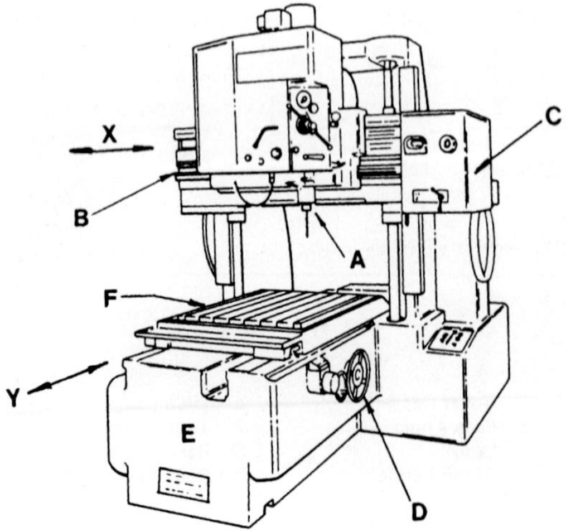

A High precision spindle
B X axis ways
C Controller
D Manual table positioning control
E Front
F Horizontal clamping table
X X axis movement
Y Y axis movement

FIGURE 6.63 Typical jig-boring machine, NC or CNC.

calculates all the coordinates and orientation of the holes from a reference point. This information is either sent to the CNC jig boring machine's controller or the machine operator can load this information into the controller, which controls the machine movements to complete the machining operation. A typical jig-boring machine is illustrated in Fig. 6.63.

Extensive tables of jig boring coordinates are not necessary with the modern CNC jig boring or vertical boring machines. Figures 6.64 and 6.65 are for manually controlled machines, where the machine operator makes the movements and coordinate settings manually.

Vertical boring machines with tables up to 192 in in diameter are produced for machining very large and heavy workpieces.

For manually controlled machines with vernier or digital read-outs, a table of jig boring dimensional coordinates is shown in Fig. 6.64 for dividing a 1-in circle into a number of equal divisions. Since the dimensions or coordinates given in the table are for *xy* table movements, the machine operator may use these directly to make the appropriate machine settings after converting the coordinates for the required circle diameter to be divided.

Figure 6.65 is a coordinate diagram of a jig bore layout for 11 equally spaced holes on a 1-in-diameter circle. The coordinates are taken from the table in Fig. 6.64. If a different-diameter circle is to be divided, simply multiply the coordinate values in the table by the diameter of the required circle; i.e., for an 11-hole circle of 5-in diameter, multiply the coordinates for the 11-hole circle by 5. Thus the first hole x dimension would be $5 \times 0.50000 = 2.50000$ in, and so on.

Hole No.	Horizontal X	Vertical Y	Hole No.	Horizontal X	Vertical Y
Three holes:			Thirteen holes:		
1	0.50000		1	0.50000	
2	0.75000	0.43301	2	0.05727	0.23236
3	----------	0.86602	3, 13	0.15870	0.17913
Five holes:			4, 12	0.22376	0.08486
1	0.50000		5, 11	0.23757	0.02885
2	0.34549	0.47553	6, 10	0.19695	0.13594
3, 5	0.55902	0.18164	7, 9	0.11121	0.21190
4	----------	0.58778	8	----------	0.23932
Six holes:			Fourteen holes:		
1, 3, 6	0.50000		1	0.50000	
2, 4, 5	0.25000	0.43301	2, 8, 9	0.04951	0.21694
Seven holes:			3, 7, 10, 14	0.13875	0.17397
1	0.50000		4, 6, 11, 13	0.20048	0.09655
2	0.18826	0.39091	5, 12	0.22252	
3, 7	0.42300	0.09655	Fifteen holes:		
4, 6	0.33923	0.27052	1	0.50000	
5	----------	0.43388	2	0.04323	0.20337
Eight holes:			3, 15	0.12221	0.16820
1	0.50000		4,14	0.18005	0.10396
2, 5, 6	0.14645	0.35355	5, 13	0.20677	0.02173
3, 4, 7, 8	0.35355	0.14645	6, 12	0.19774	0.06425
Nine holes:			7, 11	0.15451	0.13912
1	0.50000		8, 10	0.08456	0.18994
2	0.11698	0.32139	9	----------	0.20790
3, 9	0.29620	0.17101	Sixteen holes:		
4, 8	0.33682	0.05939	1	0.50000	
5, 7	0.21084	0.26200	2, 9, 10	0.03806	0.19134
6	----------	0.34202	3, 8, 11, 16	0.10839	0.16221
Ten holes:			4, 7, 12, 15	0.16221	0.10839
1	0.50000		5, 6, 13, 14	0.19134	0.03806
2, 6, 7	0.09549	0.29389	Seventeen holes:		
3, 5, 8, 10	0.25000	0.18164	1	0.50000	
4,9	0.30902		2	0.03377	0.18062
Eleven holes:			3, 17	0.09672	0.15623
1	0.50000		4, 16	0.14664	0.11073
2	0.07937	0.27032	5, 15	0.17674	0.05028
3, 11	0.21292	0.18450	6, 14	0.18296	0.01695
4, 10	0.27887	0.04009	7, 13	0.16449	0.08190
5, 9	0.25626	0.11704	8, 12	0.12379	0.13580
6, 8	0.15233	0.23701	9, 11	0.06637	0.17134
7	----------	0.28172	10	----------	0.18374
Twelve holes:			Eighteen holes:		
1	0.50000		1	0.50000	
2, 7, 8	0.06699	0.25000	2, 10, 11	0.03016	0.17101
3, 6, 9, 12	0.18301	0.18301	3, 9, 12, 18	0.08682	0.15038
4, 5, 10, 11	0.25000	0.06699	4, 8, 13, 17	0.13302	0.11162
			5, 7, 14, 16	0.16318	0.05939
			6, 15	0.17364	

End of left side table - Fig.

FIGURE 6.64(a) Jig-boring coordinates for dividing the circle.

Hole No.	Horizontal X	Vertical Y	Hole No.	Horizontal X	Vertical Y
Nineteen holes:			11, 15	0.07076	0.11634
1	0.50000		12, 14	0.03673	0.13112
2	0.02709	0.16235	13	----------	0.13616
3, 10	0.07834	0.14475	Twenty-four holes:		
4, 18	0.12110	0.11148	1	0.50000	
5, 17	0.15073	0.06612	2, 13, 14	0.01704	0.12941
6, 16	0.16403	0.01358	3, 12, 15, 24	0.04995	0.12059
7, 15	0.15956	0.04039	4, 11, 16, 23	0.07946	0.10355
8, 14	0.13779	0.09003	5, 10, 17, 22	0.10355	0.07946
9, 13	0.10110	0.12989	6, 9, 18, 21	0.12059	0.04995
10, 12	0.05344	0.15567	7, 8, 19, 20	0.12941	0.01704
11	----------	0.16460	Twenty-five holes:		
Twenty holes:			1	0.50000	
1	0.50000		2	0.01508	0.12434
2, 11, 12	0.02447	0.15451	3, 25	0.04677	0.11653
3, 10, 13, 20	0.07102	0.13938	4, 24	0.07367	0.10140
4, 9, 14, 19	0.11062	0.11062	5, 23	0.09657	0.07989
5, 8, 15, 18	0.13938	0.07102	6, 22	0.11340	0.05337
6, 7, 16, 17	0.15451	0.02447	7, 21	0.12312	0.02348
Twenty-one holes:			8, 20	0.12508	0.00787
1	0.50000		9, 19	0.11920	0.03873
2	0.02221	0.14738	10, 15	0.10582	0.06716
3, 21	0.06467	0.13428	11, 17	0.08580	0.09136
4, 20	0.10138	0.10925	12, 16	0.06038	0.10983
5, 19	0.12908	0.07452	13, 15	0.03116	0.12140
6, 18	0.14530	0.03317	14	----------	0.12532
7, 17	0.14862	0.01114	Twenty-six holes:		
8, 16	0.13874	0.05445	1	0.50000	
9, 15	0.11652	0.09293	2, 14, 15	0.01454	0.11966
10, 14	0.08397	0.12314	3, 13, 16, 26	0.04273	0.11270
11, 13	0.04393	0.14242	4, 12, 17, 25	0.06848	0.09920
12	----------	0.14904	5, 11, 18, 24	0.09022	0.07993
Twenty-two holes:			6, 10, 19, 23	0.10673	0.05601
1	0.50000		7, 9, 20, 22	0.11703	0.02885
2, 12, 13	0.02025	0.14086	8, 21	0.12054	
3, 11, 14, 22	0.05912	0.12946	Twenty-seven holes		
4, 10, 15, 21	0.09321	0.10755	1	0.50000	
5, 9, 16, 20	0.11971	0.07695	2	0.01348	0.11530
6, 8, 17, 19	0.13655	0.04009	3, 27	0.03971	0.10910
7, 18	0.14232		4, 26	0.06379	0.09699
Twenty-three hole			5, 25	0.08444	0.07967
1	0.50000		6, 24	0.10054	0.05805
2	0.01854	0.13490	7, 23	0.11121	0.03329
3, 23	0.05425	0.12480	8, 22	0.11580	0.00675
4, 22	0.08593	0.10562	9, 21	0.11433	0.02016
5, 21	0.11125	0.07853	10, 20	0.10660	0.04598
6, 20	0.12830	0.04560	11, 19	0.09312	0.06933
7, 19	0.13585	0.00930	12, 18	0.07462	0.08893
8, 18	0.13331	0.02771	13, 17	0.05210	0.10374
9, 17	0.12091	0.06264	14, 16	0.02678	0.11297
10, 16	0.09951	0.09295	15	----------	0.11608

FIGURE 6.64(b)

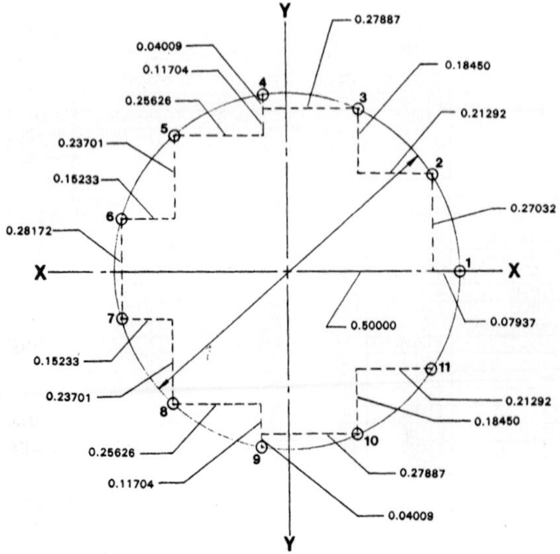

FIGURE 6.65 Coordinate diagram.

6.7 GRINDING, LAPPING, HONING, AND SUPERFINISHING (SURFACE FINISHES)

Grinding

The *grinding* process is an abrasive machining operation where material is removed from a workpiece in small chips or particles by the mechanical action of abrasive particles of irregular shape, size, and hardness. Grinding can be a rough or a precision operation for producing smooth surfaces, either flat, cylindrical, or irregularly shaped. The medium of the grinding operation is the grinding wheel, which is used for both external and internal grinding procedures. Grinding wheel shapes and other specifications are defined by the following ANSI standards:

ANSI B74.2-1982, *Shapes and sizes of grinding wheels*

ANSI B74.3-1986, *Shapes and sizes of diamond and cubic boron nitride abrasive products*

ANSI B74.13-1990, *Markings for identifying grinding wheels and other bonded abrasives*

Other ANSI standards define chemical analysis, bulk density, size of abrasive grains, and other specifications for grinding products and testing procedures.

An ideal grinding abrasive has the ability to fracture before serious dulling occurs and offers maximum resistance to point wear. Each abrasive has a special crystal structure and fracture characteristics, making it suitable for grinding operations on specific materials.

Grinding wheels are composed of abrasive grains of preselected size bonded together with different bonding media. Five important considerations must be given to the selection of a grinding wheel to suit a particular application:

1. Abrasive type
2. Grain size
3. Bonding media
4. Grade or hardness of the wheel
5. Structure

Wheel Abrasives

Natural abrasives: Corundum, emery, and diamond (corundum is natural aluminum oxide containing varying amounts of impurities).

Manufactured abrasives: Synthesized diamond, silicon carbide, aluminum oxide, and cubic boron nitride (CBN). These abrasives all have well-defined physical and chemical characteristics.

Grain size: Abrasive grains vary from 6 to 8 coarse grit to 1000 to 2000 grit for polishing and lapping.

Designation of Grinding Wheels. A standard marking system defined by ANSI 74.13-1990 for the identification of grinding wheels (excluding diamond and cubic boron nitride) is shown in Fig. 6.66. From this marking system, you may determine the characteristics of the grinding wheel from its markings, as defined by Fig. 6.66.

A standard marking system defined by ANSI B74.3-1986 for the identification of diamond and CBN is shown in Fig. 6.67. From this marking system, you may determine the characteristics of diamond and CBN grinding wheels.

Grinding Wheel Speeds. The most efficient operating speeds in surface feet per minute (sfpm) for general use are summarized in Fig. 6.68. The manufacturer of your particular wheel may recommend a different surface speed based on its experience with its product. Too low a speed will result in wasted abrasive and lower efficiency, and too high a speed may result in too hard grinding action and breakage of the grinding wheel. Do not exceed the maximum speed in revolutions per

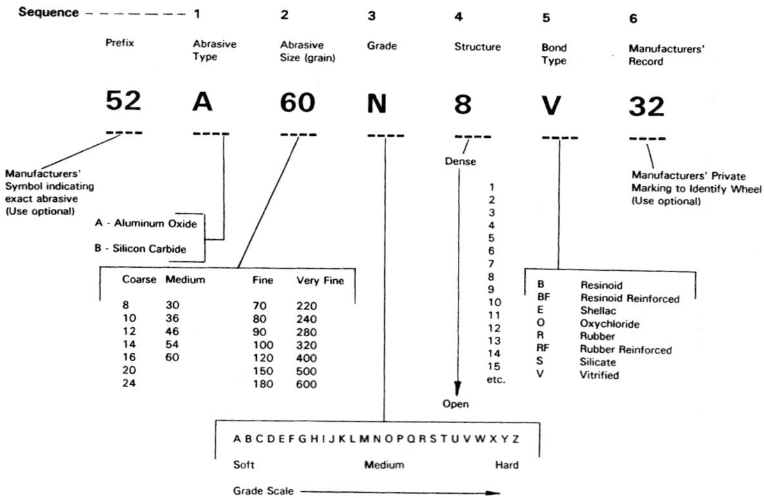

FIGURE 6.66 Standard grinding wheel marking system (ANSI B74.13-1990).

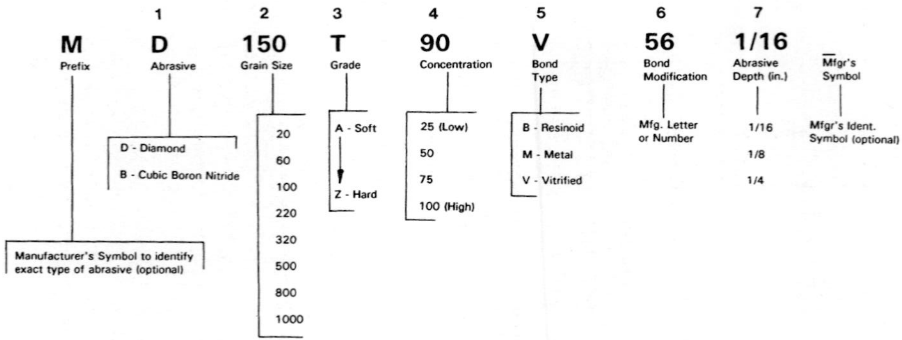

FIGURE 6.67 Standard marking system for diamond and CBN (ANSI B74.3-1986).

Type of Operation	sfpm
Cutlery (large-offhand)	4,000-5,000
Cut-off (Rubber, shellac, resinoid)	9,000-16,000
Cylinders (Including hemming)	2,500-5,000
Cylindrical grinding	5,000-12,000
Disc grinding	4,000-5,500
Internal grinding	4,000-12,000
Knife grinding (machine knives)	3,500-4,500
Portable grinding	6,500-12,500
Snagging (vitrified small hole)	5,000-6,000
Snagging (resinoid and rubber)	7,000-9,500 - 12,500
Surface grinding	4,000-6,500
Tool grinding	5,000-6,000
Weld grinding	9,500-14,200

Note: The higher speeds are permitted only where bearings, protection devices and machine rigidity are adequate for the intended operation.
To convert sfpm to rpm use the following equation: rpm = (12 x sfpm)/(π x D)
Where: sfpm = surface feet per minute; π = 3.1416 and D = wheel diameter, inches.
CAUTION: Do not operate the grinding wheel above its marked maximum rpm limit.

FIGURE 6.68 Efficient grinding wheel speeds.

minute that is marked on each wheel. Severe injury can result from the flying fragments of a broken grinding wheel.

It is difficult in a modern machining and metalworking handbook to give the exact type or number of grinding wheel to use for any specific application because of the many variables that arise in actual production situations. The data given in this section are for reference and approximate applicational uses. The manufacturer of the grinding wheels and abrasives employed should be contacted for precise applications on any grinding operation.

Modern machine grinding equipment has been developed to a high degree over the past 40 years. Figure 6.69 shows a modern surface grinder used in tool and die making and other precision applications. In this machine, the grinding wheel is stationary, and the table traverses from left to right and reverses itself in a continuous

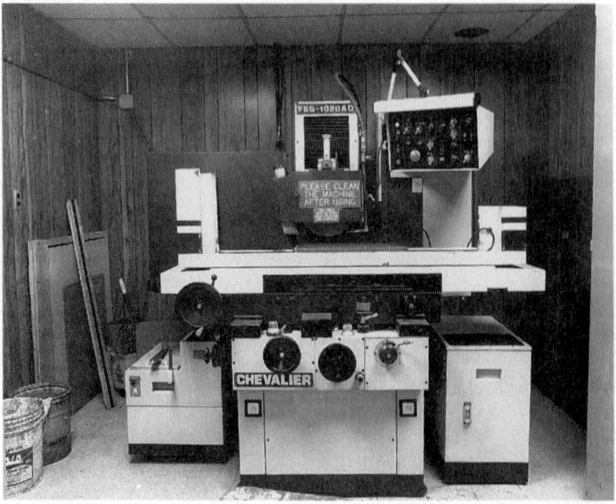

FIGURE 6.69 A modern surface grinder machine with CNC controls..

travel until the final finish is produced on the workpiece. The grinding wheel moves vertically in exact increments to produce the required cut and surface finish. This type of machine is used for flat-surface grinding only.

The final quality or surface texture of machined and ground parts is usually given on the engineering drawing in root mean square (RMS) numbers. These finishes range from 500 to 2 RMS, with the numbers representing microinches (μin) average. 500 RMS would represent a roughing operation on a milling machine or shaper to a superfinishing operation of 2 μm, average or RMS (root mean square). Finishes finer than 2 RMS can be made using polishing compounds such as cerium oxide or rouge. Cerium oxide and rouge are used extensively in the optical industries for polishing lenses and reflecting mirrors. For finishes of higher RMS values (ranging from 5 to 3 μin), compounds such as aluminum oxide powder may be used.

A convenient table of surface feet per minute converted to revolutions per minute of the grinding wheel is shown in Fig. 6.70. The figures in the table may be calculated using the following equation for converting surface feet per minute (sfpm) to revolutions per minute (rpm):

$$\text{sfpm} = \frac{\pi D(\text{rpm})}{12} \quad \text{and} \quad \text{rpm} = \frac{12(\text{sfpm})}{\pi D}$$

Characteristics of Grinding

Wheel speed: As wheel speed is increased, less work is required of each individual abrasive grain, and this promotes slower wheel wear.

Work speed: This is the speed at which the workpiece traverses the wheel or rotates about a center.

In-feed rates: The rate at which the wheel enters the workpiece during the grinding action. High in-feed rates increase wheel wear and produce a rougher finish than low in-feed rates.

Traverse or cross-feed: This is the rate at which the workpiece is moved across the face of the wheel. It is not the same as work speed.

Material to be Ground. Materials to be ground are either metallic or nonmetallic, and the metallics are divided into low- or high-tensile types. Aluminum oxide wheels are generally used for grinding metallic materials, while diamond and CBN are used to grind the extremely

Revolutions per Minute for Various Grinding Speeds in sfm - (Surface feet per minute)

Wheel Dia. Inches	4000	4500	5000	5500	6000	Periferal 6,500 Rev.	surface 7,000 per	speed 7,500 Minute	"sfm" 8,000 "rpm"	8500	9000	9500	10000	12000	14000	16000	Wheel Dia. Inches
1	15,279	17,189	19,099	21,008	22,918	24,828	26,738	28,648	30,558	32,468	34,377	36,287	38,197	45,837	53,476	61,115	1
2	7,639	8,594	9,549	10,504	11,459	12,414	13,369	14,324	15,279	16,234	17,189	18,144	19,099	22,918	26,738	30,558	2
3	5,093	5,730	6,366	7,003	7,639	8,276	8,913	9,549	10,186	10,823	11,459	12,096	12,732	15,279	17,825	20,372	3
4	3,820	4,297	4,775	5,252	5,730	6,207	6,685	7,162	7,639	8,117	8,594	9,072	9,549	11,459	13,369	15,279	4
5	3,056	3,438	3,820	4,202	4,584	4,966	5,348	5,730	6,112	6,494	6,875	7,257	7,639	9,167	10,695	12,223	5
6	2,546	2,865	3,183	3,501	3,820	4,138	4,456	4,775	5,093	5,411	5,730	6,048	6,366	7,639	8,913	10,186	6
7	2,183	2,456	2,728	3,001	3,274	3,547	3,820	4,093	4,365	4,638	4,911	5,184	5,457	6,548	7,639	8,731	7
8	1,910	2,149	2,387	2,626	2,865	3,104	3,342	3,581	3,820	4,058	4,297	4,536	4,775	5,730	6,685	7,639	8
9	1,698	1,910	2,122	2,334	2,546	2,759	2,971	3,183	3,395	3,608	3,820	4,032	4,244	5,093	5,942	6,791	9
10	1,528	1,719	1,910	2,101	2.292	2,483	2,674	2,865	3,056	3,247	3,438	3,629	3,820	4,584	5,348	6,112	10
12	1,273	1,432	1,592	1,751	1,910	2,069	2,228	2,387	2,546	2,706	2,865	3,024	3,183	3,820	4,456	5,093	12
14	1,091	1,228	1,364	1,501	1,637	1,773	1,910	2,046	2,183	2,319	2,456	2,592	2,728	3,274	3,820	4,365	14
16	955	1,074	1,194	1,313	1,432	1,552	1,671	1,790	1,910	2,029	2,149	2,268	2,387	2,865	3,342	3,820	16
18	849	955	1,061	1,167	1,273	1,379	1,485	1,592	1,698	1,804	1,910	2,016	2,122	2,546	2,971	3,395	18
20	764	859	955	1,050	1,146	1,241	1,337	1,432	1,528	1,6623	1,719	1,814	1,910	2,292	2,674	3,056	20
22	694	781	868	955	1,042	1,129	1,215	1,302	1,389	1,476	1,563	1,649	1,736	2,083	2,431	2,778	22
24	637	716	796	875	955	1,035	1,114	1,194	1,273	1,353	1,432	1,512	1,592	1.910	2,228	2,546	24
26	588	661	735	808	881	955	1,028	1.102	1,175	1,249	1,322	1,396	1,469	1,763	2,057	2,351	26
28	546	614	682	750	819	887	955	1,023	1,091	1,160	1,228	1,296	1,364	1,637	1,910	2,183	28
30	509	573	637	700	764	828	891	955	1,019	1,082	1,146	1,210	1,273	1,528	1,783	2,037	30
32	477	537	597	657	716	776	836	895	955	1,015	1,074	1,134	1,194	1,432	1,671	1,910	32
34	449	506	562	618	674	730	786	843	899	955	1,011	1,067	1,123	1,348	1,573	1,798	34
36	424	477	531	584	637	690	743	796	849	902	955	1,008	1,061	1,273	1,485	1,698	36
38	402	452	503	553	603	653	704	754	804	854	905	955	1,005	1,206	1,407	1,608	38

FIGURE 6.70 Standard grinding wheels.

hard metallics as well as ceramics and other hard nonmetallics. Silicon carbide wheels are generally used to grind the softer-grade nonmetallics. Specific and specialized grinding applications should be referred to the grinding wheel manufacturer. Arbitrarily selecting a grinding wheel or relying on handbook listings of specific grinding wheels for exact operations may not be the best solution to your grinding requirements, especially as far as productivity and wheel wear are concerned.

Grinding Wheel Dressing. There are various forms of grinding wheel dressing tools available. Some of the many types of grinding wheel dressers include

- Helical hooded dressers
- Abrasive wheel dressers
- Single-stone diamond dressers
- Multiple-point diamond dressers
- T-type diamond hand dressers
- Diamond stick dressers (silicon carbide)
- Dressing sticks for diamond and CBN grinding wheels
- Ball bearing dressers (for extremely accurate dressing)
- Diamond surface grinder wheel dressers (diamond nib)

Lapping

Lapping is a final finishing operation which results in four major improvements to a workpiece:

1. Extreme accuracy of dimension
2. Correction of minor imperfections
3. Better surface finish
4. Close fit between mating or faying surfaces.

In normal lapping operations, less heat is generated than in most other finishing operations. In manual and semiautomatic machine lapping, the end results depend on the following major factors:

- Type of lap material
- Type of lapping medium or compound
- Speed of the lapping motion
- The material to be lapped

Lap Materials. Cast iron is the most efficient machine lapping material. Other materials used for hand lapping include soft steel, bronze, brass, lead, leather, and various cloths. Leather and cloth are used for polishing. The material of the lap should be softer than the material that is being lapped.

Lapping Media

- Silicon carbide—for rapid stock removal
- Fused aluminum oxide—for soft steel and nonferrous alloys
- Unfused aluminum oxide—for excellent polishing action
- Diamond—for precious stones and tungsten carbides

The manufacturers who produce the lapping media also

provide the proper mixtures and viscosities of the lapping solutions.

Lapping Speeds. Efficient lapping speeds range from 300 to 8000 sfpm. Higher speeds will improve the surface finish. Lapping pressures range between 1 and 3 psi for soft materials and 10 psi for hard materials. In manual lapping, the final surface finish depends on the skill of the operator and the lapping media. New materials which can be used for lapping include cerium oxide and microfine aluminum oxides of optical grade. There are no known materials other than optical rouge and cerium oxide for producing ultrafine finishes on glass and metallic materials.

Honing

Honing is a low-speed abrading process using bonded abrasive sticks for removing stock from metallic and nonmetallic materials. Honing corrects surface errors produced by other machining or grinding operations. Honing has its most important function in the final finishing of internal cylindrical surfaces.

Honing Speeds. Figure 6.71 gives the approximate honing speeds for cast irons and steels, which are the most commonly honed materials. The combined rotation and reciprocation of a hone produces a cross-hatched surface finish on an internal cylindrical part.

Honing Abrasives. Honing sticks are produced with the following abrasives: silicon carbide, aluminum oxide, CBN, and diamond. Silicon carbide is used to hone cast

Material	Hardness R_C	Bore Character	sfpm
Steel	15-35	Interrupted	150
	15-35	Plain	80
	35-50	Interrupted	80
	35-50	Plain	60
	50-65	Interrupted	80
	50-65	Plain	50
Cast Irons	15-50	Interrupted	200
	15-50	Plain	110
	50-65	Interrupted	110
	50-65	Plain	60

Note: Tabular values are approximate and should be used as a guide.
Observe the honing action and take appropriate steps to refine the operation.

FIGURE 6.71 Honing speeds.

irons and nonferrous materials, aluminum oxide is used to hone steels, and diamond and CBN are used on surfaces that have been chromium plated and other extremely hard materials.

Surfaces as fine as 3 to 4 μin are obtainable using 500 grit silicon carbide on steel parts. Grain sizes for manual and power stroke honing range from 150 to 1200 grit, according to the honing media and the application.

Superfinishing

Utilizing a bonded stick for cylindrical parts or a cup wheel for flat and spherical workpieces with an abrasive action, superfinishing may be performed. *Superfinishing* produces a highly wear-resistant finish on parts that are applicable for the superfinishing process. The objective in superfinishing is the removal of fragmentation or smear metal irregularities to restore surface geometry and the surface of the workpiece by eliminating surface stresses and burns. Stock removal may range from 0.0002 to 0.001 in. Scratch patterns of 30 μin RMS or more to a mirror finish may be produced.

General-purpose and high-production superfinishing machines are available to produce a superfinish on almost any symmetrical type of workpiece. The superfinishing stone is ground to the contour of the part to be superfinished, such as a cylindrical outside surface.

Superfinishing is possible on the hardest of steels and other metallic alloys. Very fine mesh aluminum oxide is employed for many superfinishing applications.

Superfinishing Speeds. The recommended speed for most superfinishing operations is from 50 to 60 sfpm at a pressure on the stone contact area of 10 to 35 psi. Superfinishing times are suprisingly fast, with steel of hardness R_c35 being finished from 20 to 1 μin in approximately 2 min under average pressure and with 500 mesh aluminum oxide as the abrasive media. The use of optical-grade cerium oxide will also improve the finish after the aluminum oxide operation is completed.

6.8 JIC AND ISO CARBIDE CODES

The Joint Industrial Council (JIC) carbide classification code system is shown in Fig. 6.72. Since this system evolved around the early cemented and sintered carbide grades, no provision was made for the newer, more advanced cutting tool materials. Also, the wide variety of cemented carbide compositions prevented the universal acceptance of a single classification system. Nevertheless, two grade-classification systems are presently

Code	Application	Carbide Characteristics
C-1	Roughing	Medium-high shock resistance Medium-low wear resistance
C-2	General purpose	Medium shock resistance Medium wear resistance
C-3	Finishing	Medium-low shock resistance Medium-high wear resistance
C-4	Precision finishing	Low shock resistance High wear resistance
C-5	Roughing	High resistance to cutting temperatures Shock and cutting load Medium wear resistance
C-6	General purpose	Medium-high shock resistance Medium wear resistance Medium shock resistance
C-7	Finishing	Medium shock resistance Medium wear resistance
C-8	Precision finishing	Very high wear resistance Low shock resistance

Note: Hardness increases from top to bottom; toughness increases from bottom to top

FIGURE 6.72 JIC carbide classification code.

accepted: the JIC system and the International Standards Organization (ISO) system. The ISO carbide grade-classification system is shown in Fig. 6.73.

6.9 CUTTING FLUIDS AND COOLANTS FOR MACHINING OPERATIONS

Cutting fluids and machining coolants have been developed over the past 20 to 25 years which are very different than those used previously in industry. Many of the

Main Machining Group	Color Marking	Application Group	Operations and Working conditions
P Group: steel cast steel, long chip malleables	Blue	P01	High precision turning and boring. High cutting speeds. Good surface finish and vibration-free machining
		P10	Turning, thread cutting and milling. High cutting speeds. Small/medium chip cross-section
		P20	Turning, milling, medium cutting speeds and medium chip cross-section. Planing - small chip.
		P30	Turning, milling, planing. Medium to low cutting speeds. Medium/large chip cross-section. Unfavorable conditions
		P40	Planing, turning, milling, shaping. Low cutting speeds. Large chip cross-section. High rake angles. Unfavorable conditions
		P50	Where highest demands are made on carbide toughness. Turning, planing and shaping. Low cutting speeds. Large chips and high rakes. Unfavorable conditions.
M Group: steel, cast steel austenitic manganese steels, cast iron alloys, austenitic steels, malleable cast iron, free-cutting steels.	Yellow	M10	Turning. Medium-high cutting speeds. Small to medium chip cross-section.
		M20	Turning, milling. Medium cutting speeds and medium chip cross-section.
		M30	Turning, milling, planing. Medium cutting speeds. Medium/large chip cross-section.
		M40	Turning, form turning, parting off and recessing - for automatics.
K Group: cast iron, chilled cast irons, short chip malleables hardened steels, nonferrous metals and nonmetallic materials.	Red	K01	Turning, precision turning/boring. Finish milling and scraping.
		K10	Turning, milling, boring, countersinking. reaming, scraping and broaching.
		K20	Turning, milling, planing, countersinking. scraping, reaming and broaching under tougher conditions than K10.
		K30	Turning, milling, planing, shaping under unfavorable conditions. High rakes.
		K40	Turning, milling, planing, shaping under unfavorable conditions. High rakes.

Note: Cutting speed and wear resistance increase from bottom to top; feed and carbide toughness increase from top to bottom.

FIGURE 6.73 ISO carbide classification system.

modern cutting fluids and coolants are designed and formulated to be environmentally safe and biodegradable. Some cutting fluids have been developed for use on specific types of materials, while others are suitable for a very wide range of different materials and cutting conditions. Some of the modern cutting fluids are supplied in aerosol cans and are sprayed on the workpiece while the machining operation is being performed. These spray-application fluids adhere strongly to the workpiece and allow easier machinability while preventing corrosion on the workpiece, such as rusting on ferrous materials.

Combination cutting fluids and coolants are being produced which may be used for most of the high-speed machining operations afforded by the advanced cutting tool materials. With the advent of universal-type cutting/coolant combination fluids, the inventory of cutting fluids and coolants can be kept to a minimum within a manufacturing facility or machine shop.

The older types of coolants and cutting fluids are not ineffective, but they are outdated performance-wise, and some are environmentally unsafe for disposal into any sewage system. Lard oils, kerosene, and other older cutting oils are difficult to degrade chemically in the sewage-treatment plants.

A list of typical cutting fluids and coolants used many years ago with various materials in industrial machining applications follows:

Cast iron	Usually worked dry
Mild steel	Oil or soapy water
Hard steels	Mineral lard oil
Monel metal	Dry or mineral lard oil
Bronze	Dry or mineral lard oil
Brass	Dry (kerosene or turpentine was used on the harder compositions)
Copper	Dry or mixture of lard oil and turpentine
Babbitt	Dry or mixture of lard oil and kerosene

Aluminum	Dry or mixture of lard oil and kerosene or plain kerosene
Threading	Mineral lard oil

Coolants

- Lard oil—One of the oldest and best lubricants/coolants
- Mineral lard oil mixtures—More fluid and cheaper than lard oils
- Soluble oils—Specially prepared mineral oils mixed with water
- Soda water mixtures—Made by mixing 1 lb sodium carbonate, 1 qt lard oil, 1 qt soft soap, and 8 to 10 gal water; boil the mixture for 30 min and allow to cool before using.

Many of these outdated cutting fluids and coolants accounted for the smoke that was produced during the cutting operations. Bear in mind that during the period that these fluids were used, the only cutting tool materials were heat-treated high-carbon and alloy steels and early variations of HSS (high-speed steel).

Modern Coolants/Cutting Fluids. Cutting oils are divided into three basic types: petroleum oils, fixed (animal or vegetable) oils, and synthetic oils. Chemical additives give oils additional or enhanced properties such as resistance to oxidation and foaming and the ability to perform under extreme pressures and temperatures.

Some of the modern coolants/cutting fluids and compounds include

Nonhazardous cutting compounds—Biodegradable, contain no 1,1,1-trichloroethane

Plumbers' lard oil—Used for mixing with mineral oils

Wax cut cutting oil—Chlorinated waxes suspended in clear oil

Medium-duty soluble oil—Used where tool life and surface finish are critical

Thred Kut—Heavy-duty brown sulfochlorinated cutting oil, antiweld, antiwear solution.

Sulfur-base cutting oils—Used for all metals and high-alloy steels

Kleen Kut soluble oil—Water-soluble emulsion, economical; mixed with water at 5 to 10% concentration

Thred Kut 99—A dark, heavy sulfochlorinated fatty oil for machining and grinding soft, tough, and stringy metals such as stainless steels, low-carbon steels, jet-engine alloys, and monel metals

Trampol-X cutting fluid—Economical, water-soluble, and efficient cutting fluid/coolant for all types of machining operations; designed to withstand recycling in central coolant systems

Synthetic coolant—Heavy-duty synthetic cutting and grinding concentrate for machining ferrous alloys, aluminum, and brasses

Blasocut—A high-efficiency cutting/coolant fluid for all types of machining operations; used in central coolant systems and fed at high pressures into the tool and workpiece cutting area; water-soluble; made in Switzerland; frequently used on turning center and machining center CNC machines.

6.10 DRILL-ROD TYPES AND SIZES FOR MACHINING AND TOOLING APPLICATIONS

U.S. Customary or Inch-Size Drill Rods

U.S. Customary or inch-system drill rods are generally supplied in 36-in lengths, in the following grades of material:

0-1 Oil-hardening for general toolroom use. Good nondeforming characteristics. Moderate resistance to abrasion. Machinability 90 percent.

S-7 Air-hardening molybdenum, chromium, shock resisting. High-impact strength with moderate wear resistance. Machinability 95 percent.

A-2 Air-hardening. Maximum abrasion resistance. Machinability 65 percent.

D-2 Air-hardening. High carbon, high chrome, high wear-resistance. Machinability 50 percent.

W-1 Water-hardening offers a combination of high abrasion resistance and toughness. Machinability 100 percent.

The five grades listed may be heat-treated and drawn or tempered to various hardness classes by using the heat-treating data shown in Chap. 9.

Available Diameters. U.S. Customary drill rods are available in the following diameters (inches): 0.0625, 0.1250, 0.1875, 0.2500, 0.3125, 0.3750, 0.4375, 0.5000, 0.5625, 0.6250, 0.7500, 0.8750, 1.000, 1.1250, 1.2500, and 1.5000. Other diameters are also available. W-1 is the most economical, and D-2 is the most expensive, with the other grades falling between these two limits.

Tolerances on Drill Rod

Up to 0.125 in diameter ±0.003 in

0.1875 to 0.4375 in diameter ±0.0005 in

0.5000 to 1.5000 in diameter ±0.001 in

Metric Drill Rods

Metric drill rods are made of chrome vanadium alloy steel in 1-meter lengtrhs (approximately 39 in). Accuracy is metric h9. Heating for hardening: 1490–1580 °F quench in oil; 1470–1510°F quench in water. Drawing temperatures: HRC-62, 400°F; HRC-60, 485°F; HRC-56, 570°F.

Available Diameters. Metric drill rods are available in diameters ranging from 1 through 50 mm, in various steps or increments. Accuracy is metric h9.

CHAPTER 7

TOOL STEELS

The majority of tool-steel applications may be categorized under a small number of groups or types of operations: cutting, shearing, forming, drawing, bending, extruding, rolling, and ramming or battering. For each of these groups, certain metallurgic characteristics are of utmost importance. Cutting tools must possess high hardness and extreme strength, high resistance to the effects of elevated temperatures, and high wear resistance. Shearing tools require high wear resistance combined with toughness, and these characteristics must be balanced properly, depending on the tool design, thickness of the stock being sheared, and temperature of the shearing operation. Forming tools must possess high toughness and high strength, and many require high resistance to the softening effect of elevated temperatures. In battering or ramming tools, high toughness is the most important characteristic.

Heat treatment plays an important role in attaining these desired characteristics. It should be kept in mind that tools are actually being processed, not steels. To ensure that proper heat-treatment procedures are employed, Chapter 9 will present tables that list guidelines for forging, annealing, hardening, tempering, and normalizing the commonly used tool steels to which AISI/SAE designations have been assigned. These guidelines, which include temperature ranges, heating rates, cooling rates, and time at temperature, should be adapted to fit the specific application.

7.1 TOOL STEEL CHARACTERISTICS

A. Cold-work tool steels (medium alloy, air hardening). These steels show a minimum of distortion during heat treatment and have greater safety in hardening than the oil-hardening grades. Dies which could crack during heat treatment when made from oil-hardening grades generally can be handled safely when made from the air-hardening group. The low-carbon types A8 and A9 offer greater shock resistance than the other steels in this group but are lower in wear resistance. Type A7 exhibits maximum abrasion resistance but should be restricted to lower-toughness applications.

D. Cold-work tool steels (high carbon, high chromium). These steels are more wear resistant than the medium-alloy air-hardening steels used where long-run dies are required. Of the group, D1, D2, and D5 have the greater toughness but the least abrasion resistance. These steels, being very wear-resistant, are difficult to machine and grind, and a minimum of stock should be allowed for grinding after hardening. These grades are all air-hardening and show little distortion or movement during heat treatment. Wear resistance increases as the carbon and vanadium contents increase.

E. Special-purpose tool steels (carbon-tungsten types). These are often called *finishing steels* and provide relatively high wear resistance in water-hardening steels. These steels are typically used for wire, bar, or tube cold-drawing dies, finishing tools, and fine-cutting-edge tools such as broaches, taps, and reamers. F2 is recommended where high wear resistance and sharp cutting edge capability are required. These steels are somewhat difficult to grind after hardening because of their wear and abrasion resistance.

H. Hot-work tool steels. These steels are chromium-, tungsten-, and molybdenum-based. The chromium-based types are air-hardening, showing very little distortion during heat treatment. The tungsten- and molybdenum-based steels are either air- or oil-hardening and show more distortion during oil quenching. These are high-heat-resisting steels with moderate wear resistance. The tungsten types (H20 to H39) are for those hot-work applications where resistance to the softening effect of elevated temperature is of greatest importance and a lesser degree of toughness is allowed. The molybdenum types (H40 to H59) offer excellent high-heat resistance but should be restricted to those applications where less ductility is acceptable. A wide range of hot-work tools are made from these steels for die-casting dies, extrusion-press parts, forging dies, hot punches, and shears.

I. Special-purpose tool steels (low alloy). These steels have deeper hardening characteristics than the W group and have similar properties to O1 and O2 as general-purpose tool steels but exhibit a greater tendency to distort during heat treatment. Of the group, L6 has greater toughness, at the sacrifice of wear resistance. High-carbon type L3 may be used for short-run and special-purpose tools and dies.

M. High-speed tool steels (molybdenum base). These steels are commonly used for high-speed cutting-tool applications. They decarburize better than the tungsten type and are somewhat more critical in their heat treatment. Vanadium and cobalt are added, as is the case with the tungsten types, with corresponding advantages and disadvantages. The M2 type is considered the molybdenum counterpart of T1 as a general-purpose cutting-tool material. The series beginning with M41 has the characteristic of exceptionally high hardness in heat treatment. These steels are also used for cold-header die inserts, punches, and thread-rolling and blanking dies. The steels in this group are *underhardened* for these applications in order to increase toughness.

O. Cold-work tool steels (oil hardening). All these steels listed under the symbol O are low-alloy types that must be oil quenched in heat treatment. Types O1 and O2 are widely used tool steels for many types of die applications, having greater hardenability and generally less distortion during heat treatment than the water-hardening steels. These steels will harden throughout in sections up to 2.50 in thick with relative freedom from distortion and cracking during heat treatment. Type O7 attains the greatest wear or abrasion resistance but is usually used in special applications.

P. Special-purpose tool steels (mold steels). These steels are generally considered as plastic-mold steels but have found other applications outside the mold-making field. These steels are usually supplied at very low hardness to facilitate machining and other processing prior to carburizing to develop the required surface properties for injection and compression molds for plastics. Types P20 and P21 are usually supplied in the hardened condition so that they may be machined and put into immediate service. These types are suitable for plastic molds, zinc die-casting dies, and holder blocks.

S. Shock-resisting tool steels. These steels contain less carbon but higher alloy content and are therefore tougher than water-hardening grades. They have less wear resistance than the water-hardening steels but are used where punching, shearing, or trimming is being performed. Type S1 has the greatest wear resistance in the group and may be carburized for even greater hardness and wear resistance, but with a loss in toughness. Because of the moderately high hardening temperatures and rapid quenching required, these steels are subject to distortion during heat treatment.

T. High-speed tool steels (tungsten base). These steels are generally used for metal-cutting tools, although type T1 is sometimes used as a die material because of its high strength and wear resistance. Type T1 has been an established general-purpose high-speed steel (HSS) for many years. See Sec. 7.2, "Classification of High-Speed

TABLE 7.1 **Identification and Type Classification of Tool Steels**

Type	C	W	Mo	Cr	V	Co
	Identifying elements, %					
	High Speed: Tungsten Types (1)					
T1	0.75 (2)	18.00	—	4.00	1.00	—
T2	0.80	18.00	—	4.00	2.00	—
T4	0.75	18.00	—	4.00	1.00	5.00
T5	0.80	18.00	—	4.00	2.00	8.00
T6	0.80	20.00	—	4.50	1.50	12.00
T8	0.75	14.00	—	4.00	2.00	5.00
T15	1.50	12.00	—	4.00	5.00	5.00
	High Speed: Molybdenum Types (1)					
M1	0.85 (2)	1.50	8.50	4.00	1.00	—
M2	0.85: 1.00 (2)	6.00	5.00	4.00	2.00	—
M3 class 1	1.05	6.00	5.00	4.00	2.40	—
M3 class 2	1.20	6.00	5.00	4.00	3.00	—
M4	1.30	5.50	4.50	4.00	4.00	—
M6	.80	4.00	5.00	4.00	1.50	12.00
M7	1.00	1.75	8.75	4.00	2.00	—
M10	0.85: 1.00 (2)	—	8.00	4.00	2.00	—
M30	0.80	2.00	8.00	4.00	1.25	5.00
M33	0.90	1.50	9.50	4.00	1.15	8.00
M34	0.90	2.00	8.00	4.00	2.00	8.00
M36	0.80	6.00	5.00	4.00	2.00	8.00
M41	1.10	6.75	3.75	4.25	2.00	5.00
M42	1.10	1.50	9.50	3.75	1.15	8.00
M43	1.20	2.75	8.00	3.75	1.60	8.25
M44	1.15	5.25	6.25	4.25	2.00	12.00
M46	1.25	2.00	8.25	4.00	3.20	8.25
M47	1.10	1.50	9.50	3.75	1.25	5.00

Type	C	Mn or Si	W	Mo	Cr	V	Co	Other
	Identifying elements, %							
	Cold Work: Medium-Alloy Air-Hardening Types (1)							
A2	1.00	—	—	1.00	5.00	—	—	—
A3	1.25	—	—	1.00	5.00	1.00	—	—
A4	1.00	2.00 Mn	—	1.00	1.00	—	—	—
A6	0.70	2.00 Mn	—	1.25	1.00	—	—	—
A7	2.25	—	1.00 (3)	1.00	5.25	4.75	—	—
A8	0.55	—	1.25	1.25	5.00	—	—	—
A9	0.50	—	—	1.40	5.00	1.00	—	1.50 Ni
A10 (4)	1.35	1.80 Mn: 1.25 Si	—	1.50 —	—	—	—	1.80 Ni
	Cold Work: Oil-Hardening Types							
01	0.90	1.00	.50	—	.50	—	—	—
02	0.90	1.60	—	—	—	—	—	—
06 (4)	1.45	0.80 Mn: 1.00 Si	 —	25	—	—	—	—
07	1.20	—	1.75	—	.75	—	—	—
—	—							
W5	1.10	—	—	—	0.50	—	—	—
	Shock-Resisting Types							
S1	0.50	—	2.50	—	1.50	—	—	—
S2	0.50	1.00 Si	—	0.50	—	—	—	—
S5	0.55	0.80 Mn: 2.00 Si	—	0.40	—	—	—	—
S6	0.45	1.40 Mn: 2.25 Si	—	0.40	1.50	—	—	—
S7	0.50	—	—	1.40	3.25	—	—	—

Hot Work: Chromium Types (1)						
H10	0.40	—	2.50	3.25	0.40	—
H11	0.35	—	1.50	5.00	0.40	—
H12	0.35	1.50	1.50	5.00	0.40	—
H13	0.35	—	1.50	5.00	1.00	—
H14	0.40	5.00	—	5.00	—	—
H19	0.40	4.25	—	4.25	2.00	4.25
Hot Work: Tungsten Types						
H21	0.35	9.00	—	3.50	—	—
H22	0.35	11.00	—	2.00	—	—
H23	0.30	12.00	—	12.00	—	—
H24	0.45	15.00	—	3.00	—	—
H25	0.25	15.00	—	4.00	—	—
H26	0.50	18.00	—	4.00	1.00	—
Hot Work: Molybdenum Types						
H42	0.60	6.00	5.00	4.00	2.00	—
Cold Work: High-Carbon High-Chromium Types (1)						
D2	1.50	—	1.00	12.00	1.00	—
D3	2.25	—	—	12.00	—	—
D4	2.25	—	1.00	12.00	—	—
D5	1.50	—	1.00	12.00	—	3.00
D7	2.35	—	1.00	12.00	4.00	—

Special-Purpose Low-Alloy Types								
L2	0.50/1.10 (5)	—	—	—	1.00	0.20	—	—
L6	0.70	—	—	0.25 (3)	0.75	—	—	1.50 Ni
Water-Hardening Types								
W1	0.60/1.40 (5)	—	—	—	—	—	—	—
W2	0.60/1.40 (5)	—	—	—	—	0.25	—	—
W5	1.10	—	—	—	0.50	—	—	—
Mold Types								
P2	0.07	—	—	0.20	2.00	—	—	0.50 Ni
P3	0.10	—	—	—	0.60	—	—	1.25 Ni
P4	0.07	—	—	0.75	5.00	—	—	—
P5	0.10	—	—	—	2.25	—	—	—
P6	0.10	—	—	—	1.50	—	—	3.50 Ni
P20	0.35	—	—	0.40	1.70	—	—	—
P21	0.20	—	—	—	—	—	—	4.00 Ni: 1.20 A1

(1) Some of the types can be produced with a sulfur addition to improve machinability.
(2) Other carbon contents may be available.
(3) Optional.
(4) Contains free graphite in the microstructure to improve machinability.
(5) Various carbon contents are available.

Steels," for compositions of both the T and M types of HSS and characteristics.

W. Water-hardening tool steels. Being low in alloy content, these steels are shallow-hardening, developing a tough case with a hard core. They are quenched in water or brine, are subject to distortion, and do not retain their properties above 350°F. Water-hardening tool steels are used for short production runs in all types of tools and dies. Their shallow hardenability adapts them well for cold-heading dies. Although used for many applications, they are limited in their properties and capabilities.

7.2 IDENTIFICATION AND CLASSIFICATION OF TOOL STEELS

Table 7.1 shows the identification and type classification of tool steels. Figure 7.1 provides a list of the main tool-steel groups.

7.3 SELECTION OF TOOL STEELS

The tool steels listed in the following tables are generally available. The relative availability within a group, however, will vary depending on whether the steel is designed for specific applications or finds use in general applications. Experience indicates that in the majority of cases the choice is not limited to a single type of tool steel or even to a particular family of tool steels for a

Type of Tool-Steel	Designation
High-speed tool steels	M - molybdenum types T - tungsten types
Hot-work tool-steels	H H1 to H19 - chromium types H20 to H39 - tungsten types H40 to H59 - molybdenum types
Cold-work tool-steels	D - high-carbon high-chromium types A - medium-alloy air-hardening types O - oil-hardening types
Shock-resisting tool-steels	S
Mold tool-steels	P
Special-purpose tool-steels	L - low-alloy types F - carbon-tungsten types
Water-hardening tool-steels	W

FIGURE 7.1 Main tool-steel groups.

workable solution to an individual tooling problem. It is desirable to select a tool steel that will give the most economical overall performance when weighed against expected productivity, ease of fabrication, and cost.

The majority of tool-steel applications can be divided into a small number of groups or types of operations, such as cutting, shearing, forming, drawing, extruding, rolling, and battering. The previous characteristic descriptions of the tool steels (Sec. 7.1) are intended as a guide, and it must be understood that the proper selection of a tool steel for a specific application cannot always be made with 100 percent assurance. Because of this, consultation with the steel producer is recommended.

TABLE 7.2 **Comparison of Tool Steels on the Basis of Properties Affecting Selection**

AISI/SAE steel designation	Nondeforming properties	Safety in hardening	Depth of hardening*	Toughness	Resistance to softening effect of heat	Wear resistance
Water-hardening tool steels						
W108	Poor	Fair	Shallow	Good	Poor	Fair
W109	Poor	Fair	Shallow	Good	Poor	Fair
W110	Poor	Fair	Shallow	Good	Poor	Good
W112	Poor	Fair	Shallow	Good	Poor	Good
W209	Poor	Fair	Shallow	Good	Poor	Fair
W210	Poor	Fair	Shallow	Good	Poor	Good
W310	Poor	Fair	Shallow	Good	Poor	Good
Shock-resisting tool steels						
S1—Chromium-tungsten	Fair	Good	Medium	Good	Fair	Fair
S2—Silicon-molybdenum	W Poor O Fair	W Poor O Good	Medium	Best	Fair	Fair
S5—Silicon-manganese	W Poor O Fair	W Poor O Good	Medium	Best	Fair	Fair
Cold-work tool steels						
Oil-hardening types						
O1—Low manganese	Good	Good	Medium	Fair	Poor	Good
O2—High manganese	Good	Good	Medium	Fair	Poor	Good
O6—Molybdenum graphitic	Fair	Good	Medium	Fair	Poor	Good
Medium-alloy air-hardening types						
A2—5% Chromium air hard	Best	Best	Deep	Fair	Fair	Good
High-carbon–high-chromium types						
D2—High-carbon–high-chromium (air)	Best	Best	Deep	Fair	Fair	Best
D3—High-carbon–high-chromium (oil)	Good	Good	Deep	Poor	Fair	Best
D5—High-carbon–high-chromium-cobalt	Best	Best	Deep	Fair	Fair	Best
D7—High-carbon–high-chromium–high-vanadium	Best	Best	Deep	Poor	Fair	Best

Hot-work tool steels						
Chromium base types						
H11—Chromium-molybdenum-V	Good	Good	Deep	Good	Good	Fair
H12—Chromium-molybdenum-tungsten	Good	Good	Deep	Good	Good	Fair
H13—Chromium-molybdenum-VV	Good	Good	Deep	Good	Good	Fair
Tungsten base types						
H21—Tungsten	Good	Good	Deep	Good	Good	Fair
High-speed tool steels						
Tungsten base types						
T1—Tungsten 18-4-1	Good	Good	Deep	Poor	Good	Good
T2—Tungsten 18-4-2	Good	Good	Deep	Poor	Good	Good
T4—Cobalt-tungsten 18-4-1-5	Good	Fair	Deep	Poor	Best	Good
T5—Cobalt-tungsten 18-4-2-8	Good	Fair	Deep	Poor	Best	Good
T8—Cobalt-tungsten 14-4-2-5	Good	Fair	Deep	Poor	Best	Good
Molybdenum base types						
M1—Molybdenum 8-2-1	Good	Fair	Deep	Poor	Good	Good
M2—Molybdenum-tungsten 6-6-2	Good	Fair	Deep	Poor	Good	Good
M3—Molybdenum-tungsten 6-6-3	Good	Fair	Deep	Poor	Good	Best
M4—Molybdenum-tungsten 6-6-4	Good	Fair	Deep	Poor	Good	Best
Special-purpose tool steels						
Low-alloy types						
L6—Nickel-chromium	Fair	Good	Medium	Fair	Poor	Fair
L7—Chromium	Fair	Good	Medium	Fair	Poor	Good

* These are intended to emphasize major differences between the groups of steels and do not account for the minor differences in depths of hardening that exist between steels of the same group. This is particularly true of the water-hardening W steels which are frequently furnished with varying degrees of hardenability.

TABLE 7.3 Approximate Comparison of Tool and Die Steels on the Basis of Some Heat-Treating Characteristics

AISI/SAE steel designation	Quench medium	Preheat temperature, °F	Hardening temperature range, °F*	Hardness after quenching, Rockwell C	Tempering temperature range, °F*	Hardness after tempering, Rockwell C
Water-hardening tool steel						
W108	Water	—	1420–1450	65–67	350–525	65–56
W109	Water	—	1420–1450	65–67	350–525	65–56
W110	Water	—	1420–1450	65–67	350–525	65–56
W112	Water	—	1420–1500	65–67	350–525	65–56
W209	Water	—	1420–1500	65–67	350–525	65–56
W210	Water	—	1420–1500	65–67	350–525	65–56
W310	Water	—	1420–1500	65–67	350–525	65–56
Shock-resisting tool steels						
S1—Chromium-tungsten	Oil	1200–1300	1650–1800	57–59	300–1000	57–45
S2—Silicon-molybdenum	Water	—	1550–1575	60–62	300–500	60–54
	Oil	—	1600–1625	58–60	300–500	58–54
S5—Silicon-manganese	Water	—	1550–1600	60–62	300–650	60–54
	Oil	—	1600–1675	58–60	300–650	58–54
Cold-work tool steels						
Oil-hardening types						
O1—Low manganese	Oil	—	1450–1500	63–65	300–800	62–50
O2—High manganese	Oil	—	1420–1450	63–65	375–500	62–57
O6—Molybdenum graphitic	Oil	—	1450–1500	63–65	300–800	63–50
Medium-alloy air-hardening types						
A2—5% Chromium air hard	Air	1200–1300	1725–1775	61–63	400–700	60–57
High-carbon–high-chromium types						
D2—High-carbon–high-chromium	Air	1200–1300	1800–1875	61–63	400–700	60–58
D3—High-carbon–high-chromium	Oil	1200–1300	1750–1800	62–64	400–700	62–58
D5—High-carbon–high-chromium-cobalt	Air	1200–1300	1800–1875	60–62	400–700	59–57
D7—High-carbon–high-chromium–high-vanadium	Air	1200–1300	1850–1950	63–65	300–500 850–1000	65–63 62–58

Hot-work tool steels						
Chromium base types						
H11—Chromium-molybdenum-V	Air	1450–1500	1825–1875	53–55	1000–1100	51–43
H12—Chromium-molybdenum-tungsten	Oil, Air	1450–1500	1800–1900	53–55	1000–1100	51–43
H13—Chromium-molybdenum-VV	Air	1400–1450	1825–1875	53–55	1000–1100	51–43
Tungsten base types						
H21—Tungsten	Oil, Air	1500–1550	2100–2150	50–52	950–1150	50–47
High-speed tool steels						
Tungsten base types						
T1—Tungsten 18-4-1	Oil, air, salt	1500–1550	2300–2375	63–65	1025–1100	65–63
T2—Tungsten 18-4-2	Oil, air, salt	1500–1550	2300–2375	63–65	1025–1100	65–63
T4—Cobalt-tungsten 18-4-1-5	Oil, air, salt	1500–1550	2300–2375	63–65	1025–1100	65–63
T5—Cobalt-tungsten 18-4-2-8	Oil, air, salt	1500–1550	2300–2400	63–65	1050–1100	65–63
T8—Cobalt-tungsten 14-4-2-5	Oil, air, salt	1500–1550	2300–2375	63–65	1025–1100	65–63
Molybdenum base types						
M1—Molybdenum 8-2-1	Oil, air, salt	1400–1500	2150–2250	63–65	1025–1050	65–63
M2—Molybdenum-tungsten 6-6-2	Oil, air, salt	1450–1500	2175–2250	63–65	1025–1075	65–63
M3—Molybdenum-tungsten 6-6-3	Oil, air, salt	1450–1500	2150–2225	63–65	1025–1075	65–63
M4—Molybdenum-tungsten 6-6-4	Oil, air, salt	1450–1500	2150–2225	63–65	1025–1075	65–63
Special-purpose tool steels						
Low-alloy types						
L6—Nickel-chromium	Oil	—	1500–1600	62–64	400–800	62–48
L7—Chromium	Oil	—	1525–1550	63–65	350–500	62–60

* The purpose of these columns is to show the usual ranges of temperature employed in hardening and tempering and is not to be used as a specification.

Table 7.2 shows comparison of tool steels on the basis of properties affecting selection, and Table 7.3 lists approximate comparison of tool and die steels on the basis of some heat-treating characteristics.

NOTE: HEAT TREATMENT OF TOOL STEELS (SEE CHAPTER 9)

CHAPTER 8

SHEET METAL PRACTICES

The branch of metalworking known as *sheet metal* comprises a large and important element. Sheet metal parts are used in countless commercial and military products. Sheet metal parts are found on almost every product produced by the metalworking industries throughout the world. Sheet metals are available in various steels, copper alloys, aluminum alloys, magnesium alloys, and titanium alloys.

Sheet metal gages run from under 0.001 in to 0.500 in. Hot-rolled steel products can run from ½ in thick to no. 18 gage (0.0478 in) and still be considered as "sheet." Cold-rolled steel sheets are generally available from stock in sizes from no. 10 gage (0.1345 in) down to no. 28 gage (0.0148 in). Other sheet thicknesses are available as special-order "mill-run" products when the order is large enough. Large manufacturers who use vast tonnages of steel products such as the automobile makers, switch-gear producers, and other sheet metal fabricators may order their steel to their own specifications (composition, gages, and physical properties).

The steel sheets are supplied in flat form or rolled into coils. Flat-form sheets are made to specific standard sizes unless ordered to special nonstandard dimensions.

8.1 CARBON AND LOW-ALLOY STEEL SHEETS

Carbon steel sheets and coils are produced in the following grades or classes:

Hot Rolled

- Low carbon (commercial quality)
- Pickled and oiled
- 0.40/0.50 carbon
- Abrasion resisting
- Hi-Form (A715), high strength/low alloy (HSLA) (grade 50, grade 80)
- A607 specification, high strength/low alloy (HSLA) (INX 45, INX 50, Ex-Ten 50)
- A606 specification, high strength/low alloy (HSLA) (Cor-Ten)

Note: The code numbers indicate the yield strength of the high-strength/low-alloy steels; i.e., INX 45 = 45,000 psi yield; grade 80 = 80,000 psi yield.

Cold Rolled

- Low carbon (commercial quality)
- Special killed (drawing quality)
- Auto prototype (special killed drawing quality)

- Vitreous enameling
- Plating quality
- Stretcher leveled

The applications for the previously listed sheet steels are as follows.

Hot-Rolled Applications

Low carbon (commercial quality): Conforms to ASTM A569 and is used for tanks, barrels, farm implements, and other applications where surface quality is not critical or important.

Pickeled and oiled: Conforms to ASTM A569 and is used for automotive parts, switch gear, appliances, toys, and other applications where a better surface quality is required and paint and enamel adhere well. Carbon content is 0.10 percent maximum, and this material may be easily formed and welded, the same as low-carbon sheet (commercial quality).

0.40/0.50 carbon: Has 50 percent more yield strength and abrasion resistance than low-carbon sheets. May be heat-treated for more strength and hardness. Used for scrapers, blades, tools, and other applications requiring a strong, moderate-cost steel sheet.

Abrasion resistant: Medium carbon content and higher manganese greatly improve resistance to abrasion. Brinell hardness = 210 minimum. Uses include scrapers, liners, chutes, conveyors, and other applications requiring a strong, abrasion-resistant steel sheet. Formability is moderate.

A607 specification: Lowest-cost low-alloy steel sheet. Low carbon content ensures good formability. Excellent weldability. Typical uses include utility poles, transmission towers, automotive parts, truck trailers, and other applications requiring a low-cost, high-strength alloy steel sheet.

A606 specification: Five times more resistant to atmospheric corrosion than low-carbon steel. Excellent weldability and formability.

A715 specification: Fine-grained columbium bearing series of high-strength steel. Enhanced bending and forming properties. Tough and fatigue resistant, with excellent weldability using all welding processes. Yield-point levels range from 40,000 to 80,000 psi.

Cold-Rolled Applications

Low carbon (commercial quality): Produced with a high degree of gage accuracy, with uniform physical characteristics. Excellent surface for painting (enamel or lacquer). Good for stamping and moderate drawing applications. Improved welding and forming characteristics with uses such as household appliances, truck bodies, signs, panels, and many other applications.

Special killed (drawing quality): Used for severe forming and drawing applications. Freedom from age hardening and fluting. Conforms to ASTM A365 specification.

Auto prototype (special killed drawing quality): Used for proto-type work and other deep drawing applications. Closely controlled gage thickness with better tolerances compared with commercial-quality grades.

Vitreous enameling: Cold rolled from commercially pure iron ingots for porcelain-enameled products. Textured surface and suitable for forming and moderate drawing applications and flatwork. Conforms to ASTM A424, grade A.

Plating quality: Two finishes are provided which are suitable for most plating applications: commercial bright and extra light matte.

Stretcher leveled: Uniform, high-quality matte sheets, further processed by stretching to provide superior flatness. Furnished resquared or not resquared. Resquared sheets have the stretching gripper marks removed. Used in the manufacture of table tops, cabinets, truck body panels, partitions, templates, and many other applications. Conforms to ASTM A336 specification.

Galvanized Sheet and Coil. The galvanic coating is zinc, and it is applied to standard steel sheets or coils in two basic methods: hot-dipped galvanized and electrogalvanized. The hot-dipped galvanized processes are known as Ti-Co galvanized, galvanized bonderized, galvannealed, galvannealed A, and hot-dipped galvanized. Galvanizing specifications are found in ASTM A526 and A527. Some of the hot-dipped sheets may have 1.25 oz zinc per square foot of surface area and others a lighter deposit.

Electrogalvanized sheets are cold-rolled steel sheets zinc coated by electrolytic deposition and conform to ASTM A591. These sheets should be painted if they are to be exposed to outdoor conditions. These sheets can be formed, rolled, or stamped without flaking, peeling, or cracking of the zinc coating. These galvanized sheets have the same gage thickness as cold-rolled sheets.

Applications include cabinets, signs, light fixtures, and others where an excellent finish is required. Coating weight is typically 0.1 oz/ft^2, or each side is 0.00008 in thick. Trade names include Paint-Lok, Bethzin, Gripcoat, Lifecote 1, Weirzin Bonderized, and others.

Aluminized and Long-Terne Sheets. Sheet steel is also aluminized and produced in long-terne sheets. Aluminized steel sheet is hot-dip coated on both sides with aluminum-silicon alloy by the continuous method. Strong and corrosion resistant, aluminized sheet is also inexpensive. The aluminum coating is typically 0.001 in thick on both sides or 0.40 oz/ft^2. Aluminized sheet conforms to ASTM A463 specification. Applications include dry kiln fan walls, dryers, incinerators, mufflers, and oven and space-heater components. Long-terne sheet is a soft steel coated with an 85 percent lead and 15 percent tin alloy for maximum ease in soldering. These long-terne sheets conform to ASTM A308 and are used for soldered tanks, automotive accessories, hood and radiator work, and many other stamped and formed products.

As can be seen from the preceding description of sheet steels that are commercially available, the selection of a particular steel for a particular sheet metal application is relatively easy. Not only are there a great number of different sheet metal stocks available, but special sheet steels may be ordered to your specifications when quantities are large enough to justify their production by the American steel makers.

Nonferrous Sheet Metal

The nonferrous sheet metals include aluminum and aluminum alloys, copper and copper alloys, magnesium alloys, titanium alloys, and other special alloys. Supplier catalogs are also available from companies such as Ryerson, Vincent, Atlantic, Alcoa, Reynolds, Anaconda, Chase, and others, which may be selected from an industrial supplier master index such as the *Thomas Register of American Manufacturers.*

8.2 MACHINERY FOR SHEET METAL FABRICATION AND GAGING SYSTEMS

Some of the typical machinery found in a large manufacturing plant for processing and producing sheet metal parts includes:

- Shears, hydraulic and squaring
- Press brakes
- Leaf brakes
- Roll-forming machines
- Automatic (CNC), multistation punch presses
- Single-die punch presses, strippit and unipunch setups
- Slitting machines
- Stretcher-bending machines
- Hydropresses (Marforming presses, Martin-Marietta Corp.)
- Pin routers
- Yoder hammers
- Spin forming machines
- Tumbling and deburring machines

- Sand-blasting equipment
- Explosive forming facilities
- Ironworkers (for structural shapes)

The designer and tool engineer should be familiar with all machinery used to manufacture parts in a factory. These specialists must know the limitations of the machinery that will produce the parts as designed and tooled. Coordination of design with the tooling and manufacturing departments within a company is essential to the quality and economics of the products that are manufactured. Our modern machinery has been designed and is constantly being improved to allow us to manufacture a quality product at an affordable price to the consumer. Medium- to large-sized companies can no longer afford to manufacture products whose quality standards do not meet the demands and requirements of the end user.

Modern Sheet Metal Manufacturing Machinery

The processing of sheet metal begins with the hydraulic shear, where the material is squared and cut to size for the next operation. Figure 8.1 shows a typical hydraulic shear with the capacity to do a 120-in cut in up to no. 7 gage steel sheet. These types of machines are the "workhorses" of the typical sheet metal department, since all operations on sheet metal parts start at the shear.

Figure 8.2 shows a Wiedemann Optishear, which shears and squares the sheet metal to a high degree of accuracy. Blanks which are used in blanking, punching,

FIGURE 8.1 The hydraulic shear used to cut sheet metal up to 120-in cuts on no.7 gage.

and forming dies are produced on this machine, as are other flat and accurate pieces which proceed to the next stage of manufacture.

The flat, sheared sheet metal parts may then be routed to the punch presses, where holes of various sizes and patterns are produced. Figure 8.3 shows a medium-sized CNC-controlled multistation turret punch press, which is both highly accurate and very high speed. (See Chap. 1, "Modern Metalworking Machinery and Measuring Devices," for other types of machinery used by the metalworking industry.)

After a sheet metal part is sheared and punched, it

FIGURE 8.2 A highly accurate squaring shear which is CNC controlled.

may require a press-brake operation to form flanges or produce hemmed edges. Figure 8.4 shows a small, digital-readout press brake which has a digital readout for the "back-gage" dimension. A larger press brake may be seen in Fig. 8.5. A small press brake such as that shown in Fig. 8.4 may be used for bending small sheet metal parts, flat bar stock, and copper and aluminum bus bars.

To quickly check a sheet metal gage, either ferrous or nonferrous, a tool similar to that shown in Fig. 8.6 is frequently used in the sheet metal department. The tool gage shown in the figure is for nonferrous metals such as stainless steel sheets and nonferrous wires.

FIGURE 8.3 The Amada Pega high-speed, medium-sized CNC-controlled multistation turret punch press.

Gaging Systems

To specify the thickness of different metal products such as steel sheets, wire, strip and tubing, music wire, and others, a host of gauging systems were developed over the course of many years. Shown in Fig. 8.7 are the common gaging systems used for commercial steel sheets, strip and tubing, and brass and steel wire. The steel sheets column in Fig. 8.7 lists the gages and equivalent thicknesses used by American steel sheet manufacturers and steel makers. This gaging system can be recognized immediately by its no. 11 gage equivalent of 0.1196 in, which is standard today for this very common and high-usage gage of sheet steel.

FIGURE 8.4 A small press brake with digital back gage.

Figure 8.8 is a table of gaging systems that were used widely in the past, although some are still in use today, including the American or Brown and Sharpe system. The Brown and Sharpe system is also shown in Fig. 8.7, but there it is indicated in fewer decimal places.

Aluminum Sheet Metal Standard Thicknesses Aluminum is used widely in the aerospace industry, and over the years, the gage thicknesses of aluminum sheets have developed on their own. Aluminum sheet is now generally available in the thicknesses shown in Fig. 8.9. The fact that the final weight of an aerospace vehicle is very critical to its performance has played an important role in the development of the standard aluminum sheet

FIGURE 8.5 A large, high-tonnage press brake.

gages, wherein the strength-to-weight ratio is critical and a few thousandths of an inch extra on an aluminum sheet will mean more final weight of the aerospace vehicle. The gages or thicknesses of other metals and alloys, together with their stock tolerances, are shown in Chap. 3, "Materials—Physical Properties, Characteristics and Uses."

8.3 METHODS OF SHEET METAL FABRICATION

Many methods have been developed for working or fabricating sheet metal parts, including those employed to cut, punch, and form sheet metal parts.

FIGURE 8.6 American standard wire gage-measuring tool. (*L.S. Starrett Co.*)

Sheet Metal Cutting Methods

Shearing: The sheet metal stock sheet is cut on a hydraulic-powered shear to its appropriate blank size or minimum size for punching or profiling the part. See Figs. 8.1 and 8.2 for hydraulic shears.

Slitting: The sheet metal stock sheet is run through a slitting machine, where accurate widths are slit with slitting knives to various lengths. The slit edges are sharp and accurate, with excellent straightness along both edges. Slit sheet metal is usually fed directly into the feed gates of punching or bending dies, or roll-forming equipment. The roll-forming machine produces a specific type of cross-sectional shape to the

Gauge No.	Brass (Brown & Sharpe)	Steel Sheets *	Strip & Tubing	Steel Wire Ga. ♦
6-0	0.5800	--------	--------	0.4615
5-0	0.5165	--------	0.500	0.4305
4-0	0.4600	--------	0.454	0.3938
3-0	0.4096	--------	0.425	0.3625
2-0	0.3648	--------	0.380	0.3310
0	0.3249	--------	0.340	0.3065
1	0.2893	--------	0.300	0.2830
2	0.2576	--------	0.284	0.2625
3	0.2294	0.2391	0.259	0.2437
4	0.2043	0.2242	0.238	0.2253
5	0.1819	0.2092	0.220	0.2070
6	0.1620	0.1943	0.203	0.1920
7	0.1443	0.1793	0.180	0.1770
8	0.1285	0.1644	0.165	0.1620
9	0.1144	0.1495	0.148	0.1483
10	0.1019	0.1345	0.134	0.1350
11	0.0907	0.1196	0.120	0.1205
12	0.0808	0.1046	0.109	0.1055
13	0.0720	0.0897	0.095	0.0915
14	0.0641	0.0747	0.083	0.0800
15	0.0571	0.0673	0.072	0.0720
16	0.0508	0.0598	0.065	0.0625
17	0.0453	0.0538	0.058	0.0540
18	0.0403	0.0478	0.049	0.0475
19	0.0359	0.0418	0.042	0.0410
20	0.0320	0.0359	0.035	0.0348
21	0.0285	0.0329	0.032	0.0317
22	0.0253	0.0299	0.028	0.0286
23	0.0226	0.0269	0.025	0.0258
24	0.0201	0.0239	0.022	0.0230
25	0.0179	0.0209	0.020	0.0204
26	0.0159	0.0179	0.018	0.0181
27	0.0142	0.0164	0.016	0.0173
28	0.0126	0.0149	0.014	0.0162
29	0.0113	0.0135	0.013	0.0150
30	0.0100	0.0120	0.012	0.0140
31	0.0089	0.0105	0.010	0.0132
32	0.0080	0.0097	0.009	0.0128
33	0.0071	0.0090	0.008	0.0118
34	0.0063	0.0082	0.007	0.0104
35	0.0056	0.0075	0.005	0.0095
36	0.0050	0.0067	0.004	0.0090
37	0.0045	0.0064	--------	0.0085
38	0.0040	0.0060	--------	0.0080

*= Common Commercial Standard; ♦ = Reference only

FIGURE 8.7 Common modern gaging decimals system table.

Number of wire gauge	American or Brown & Sharpe	Birmingham or Stubs' Iron wire	Washburn & Moen, Worcester, Mass.	W. & M. steel music wire	American S. & W. Co's. music wire gauge	Stubs' steel wire	U.S. Standard gauge for sheet and plate iron and steel	Number of wire gauge
00000000				0.0083				00000000
0000000				0.0087				0000000
000000				0.0095	0.004		0.46875	000000
00000				0.010	0.005		0.4375	00000
0000	0.460	0.454	0.3938	0.011	0.006		0.40625	0000
000	0.40964	0.425	0.3625	0.012	0.007		0.375	000
00	0.3648	0.380	0.3310	0.0133	0.008		0.34375	00
0	0.32486	0.340	0.3065	0.0144	0.009		0.3125	0
1	0.2893	0.300	0.2830	0.0156	0.010	0.227	0.28125	1
2	0.025763	0.284	0.2625	0.0166	0.011	0.219	0.265625	2
3	0.22942	0.259	0.2437	0.0178	0.012	0.212	0.250	3
4	0.20431	0.238	0.2253	0.0188	0.013	0.207	0.234375	4
5	0.18194	0.220	0.2070	0.0202	0.014	0.204	0.21875	5
6	0.16202	0.203	0.1920	0.0215	0.016	0.201	0.203125	6
7	0.14428	0.180	0.1770	0.023	0.018	0.199	0.1875	7
8	0.12849	0.165	0.1620	0.0243	0.020	0.197	0.171875	8
9	0.11443	0.148	0.1483	0.0256	0.022	0.194	0.15625	9
10	0.10189	0.134	0.1350	0.027	0.024	0.191	0.140625	10
11	0.090742	0.120	0.1205	0.0284	0.026	0.188	0.125	11
12	0.080808	0.109	0.1055	0.0296	0.029	0.185	0.109375	12
13	0.071961	0.095	0.0915	0.0314	0.031	0.182	0.09375	13
14	0.064084	0.083	0.0800	0.0326	0.033	0.180	0.078125	14

15	0.057068	0.072	0.0720	0.0345	0.035	0.178	0.0703125	15
16	0.05082	0.065	0.0625	0.036	0.037	0.175	0.0625	16
17	0.045257	0.058	0.0540	0.0377	0.039	0.172	0.05625	17
18	0.040303	0.049	0.0475	0.0395	0.041	0.168	0.050	18
19	0.03589	0.042	0.0410	0.0414	0.043	0.164	0.04375	19
20	0.031961	0.035	0.0348	0.0434	0.045	0.161	0.0375	20
21	0.028462	0.032	0.03175	0.046	0.047	0.157	0.034375	21
22	0.025347	0.028	0.0286	0.0483	0.049	0.155	0.03125	22
23	0.022571	0.025	0.0258	0.051	0.051	0.153	0.028125	23
24	0.0201	0.022	0.0230	0.055	0.055	0.151	0.025	24
25	0.0179	0.020	0.0204	0.0586	0.059	0.148	0.021875	25
26	0.01594	0.018	0.0181	0.0626	0.063	0.146	0.01875	26
27	0.014195	0.016	0.0173	0.0658	0.067	0.143	0.0171875	27
28	0.012641	0.014	0.0162	0.072	0.071	0.139	0.015625	28
29	0.011257	0.013	0.0150	0.076	0.075	0.134	0.0140625	29
30	0.010025	0.012	0.0140	0.080	0.080	0.127	0.0125	30
31	0.008928	0.010	0.0132		0.085	0.120	0.0109375	31
32	0.00795	0.009	0.0128		0.090	0.115	0.01015625	32
33	0.00708	0.008	0.0118		0.095	0.112	0.009375	33
34	0.006304	0.007	0.0104			0.110	0.00859375	34
35	0.005614	0.005	0.0095			0.108	0.0078125	35
36	0.005	0.004	0.0090			0.106	0.00703125	36
37	0.004453					0.103	0.006640625	37
38	0.003965					0.101	0.00625	38
39	0.003531					0.099		39
40	0.003144					0.097		40

FIGURE 8.8 Gaging systems used in the past, and sometimes employed today.

Standard Thickness, in.	Weight, lbs/sq. ft.
0.010	0.141
0.016	0.226
0.020	0.282
0.025	0.353
0.032	0.452
0.040	0.564
0.050	0.706
0.063	0.889
0.071	1.002
0.080	1.129
0.090	1.270
0.100	1.411
0.125	1.764
0.160	2.258
0.190	2.681
0.250	3.528

Weight based on an average aluminum weight of 0.098 lb/in^3

FIGURE 8.9 Standard aluminum sheet metal thicknesses and weights.

finished part. Samples of roll-formed parts can be seen in Fig. 8.10.

Welding-process cutting and water-jet cutting: The sheet metal stock sheet may be oxyacetylene torch cut with an automatic torch cutting machine, or it may be plasma cut, arc cut, or water-jet cut. All of these methods are used today to cut both metallic and nonmetallic flat sheet parts.

Laser-beam cutting: The technology of laser-beam cutting of sheet metal flat stock has progressed dramatically in the past five years. The older laser-cutting machines produced a ragged edge, which made parts processing difficult. The new laser-cutting machines can cut a variety of sheet metal stock, includ-

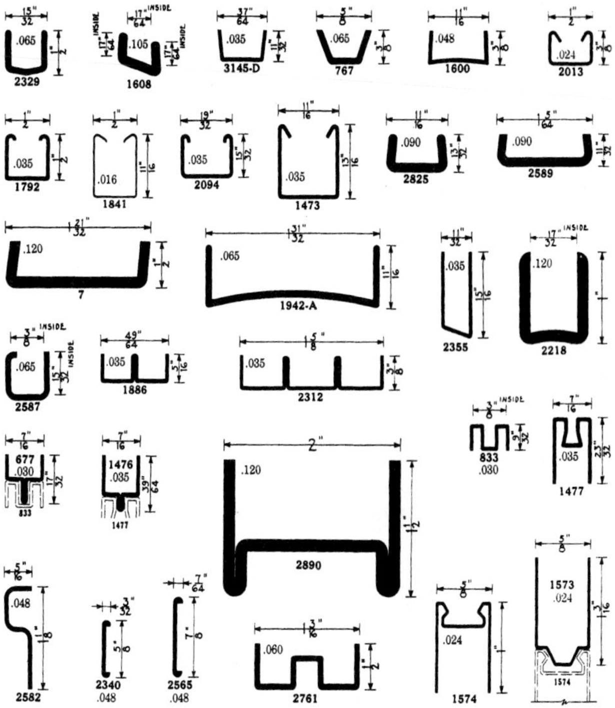

FIGURE 8.10 Samples of roll-formed sections of sheet metal. (*Dahlstrom Manufacturing Corp, Jamestown, N.Y.*)

ing steel, aluminum, stainless steel, and nonmetallic parts, while achieving excellent accuracy and edge qualities. Laser cutting allows sheet-metal parts to be cut to a level of accuracy that would not be possible using conventional punch presses or other equipment.

Stretcher bending: This process for sheet metal forming uses a machine known as a *stretcher bender,* wherein a sheet metal formed section such as an angle, channel, or Z section is pulled against a radiused die block to accurately form structural frame sections to a specified radius of curvature. The frame structures of rockets and aerospace vehicles, which have a single radius of curvature, are formed on the stretcher-bending machine. This is the only cost-effective method known for producing such sheet metal frame parts quickly and accurately on a relatively simple machine. In this operation, the straight frame section which was press-brake bent is held in a set of grippers at each end of the part. The part is stretched and pulled against the radiused die block simultaneously by the gripper arms, thus forming the part to the specified radius. Allowance is made for "spring back" of the formed part by overbending and then allowing the metal to spring back or return to the correct form.

Roll forming: In the roll-forming process, a flat strip of sheet metal is fed into the roll-forming machine which has a series of rolling dies whose shape gradually changes as the metal is being fed past each stage of rolls, until the final roll-formed section is completed. The number of different cross sections of roll-formed sheet metal parts is limitless. The roll-formed part is usually made to a specific length or stock length or may be produced to any special length required. Figure 8.10 is a sample page of roll-formed sections taken from the Dahlstrom catalog of molded and rolled sections, Dahlstrom Manufacturing Corporation, Jamestown, New York.

Figure 8.11 shows the bending capabilities of the press-brake when equipped with the appropriate bending dies.

8.4 SHEET METAL FLAT PATTERNS

The correct determination of the flat-pattern dimensions of a sheet metal part which is formed or bent is of prime importance to sheet metal workers, designers, and design drafters. There are three methods for performing the calculations to determine flat patterns which are considered normal practice. The method chosen also can determine the accuracy of the results. The three common methods employed for doing the work include:

- By bend deduction (B.D.) or setback
- By bend allowance (B.A.)
- By inside dimensions (IML), for sharply bent parts only

Other methods are also used for calculating the flat-pattern length of sheet metal parts. Some take into consideration the ductility of the material, and others are based on extensive experimental data for determining the bend allowances. The methods included in this section are accurate when the bend radius has been

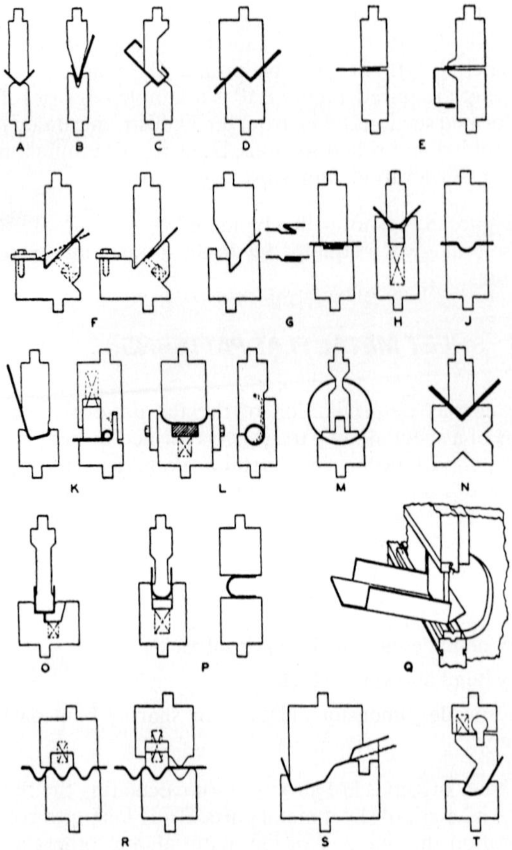

FIGURE 8.11 Bending capabilities of the press brake using the indicated die forms.

selected properly for each particular gage and condition of the material. When the proper bend radius is selected, there is no stretching of the *neutral axis* within the part (the neutral axis is generally accepted as being located 0.445 × material thickness inside the inside mold line (IML) (see Fig. 8.13).

Methods of Determining Flat Patterns. Refer to Fig. 8.12.

Method 1: By bend deduction or setback

$$L = a + b - \text{setback}$$

Method 2: By bend allowance

$$L = a' + b' + c$$

where c = bend allowance or length along neutral axis.

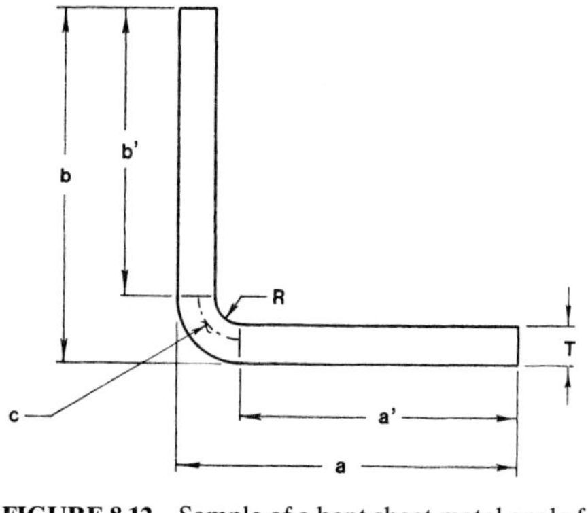

FIGURE 8.12 Sample of a bent sheet metal angle for determining flat pattern procedures.

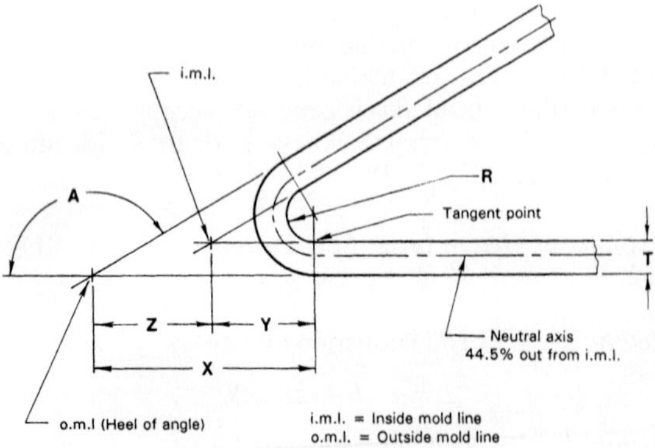

FIGURE 8.13 Geometry of a sheet metal bend angle.

Method 3: By inside dimensions or inside mold line (IML)

$$L = (a - T) + (b - T)$$

The calculation of bend allowance and bend deduction (setback) is keyed to Fig. 8.13 and is as follows:

$$\text{Bend allowance (B.A.)} = A(0.01745R + 0.00778T)$$

$$\text{Bend deduction (B.D.)} = \left(2 \tan \frac{1}{2}A\right)(R + T) - \text{(B.A.)}$$

$$X = \left(\tan\frac{1}{2}A\right)(R + T)$$

$$Z = T\left(\tan\frac{1}{2}A\right)$$

$$Y = X - Z \qquad \text{or} \qquad R\left(\tan\frac{1}{2}A\right)$$

On "open" angles that are bent less than 90° (see Fig. 8.14),

$$X = \left(\tan\frac{1}{2}A\right)(R + T)$$

The method used to calculate the sheet metal flat pattern may be determined by designer option, com-

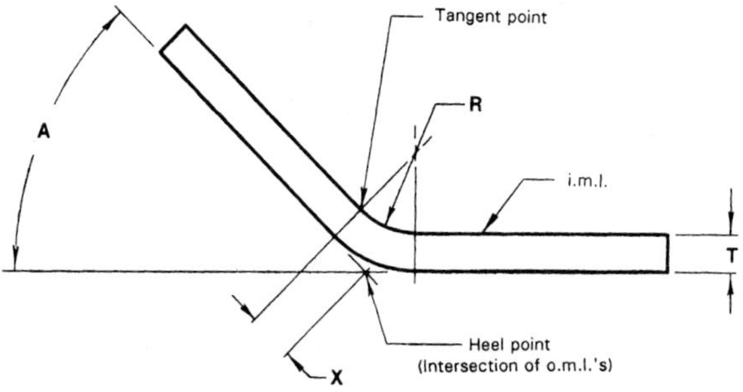

FIGURE 8.14 An open-angle sheet metal bend.

pany standards, order of accuracy, and method of manufacture. For soft steel (1010, etc.), the inside dimension method (method 3 described previously) is used whenever the material may be bent with a sharp or minimal inside bend radius (0.062 in or less). The inside bend method is accurate enough for all gages up to and including 0.375-in-thick stock where the tolerance of bent parts is ±0.032 in. On stock thicknesses from no. 7 gage to 0.375 in, 0.062 in is added to the sum of the inside bend dimensions and divided across the bend, 0.032 in going into each leg or flange inside dimension. The 0.062-in allowance is added for each 90° bend in the part. This method is popular in industries such as the electrical switch-gear industry, general sheet metal industry, and others where great accuracy is not required.

For recommended bend radii on various materials and gages, see Sec. 8.8. Also, your company may have the recommended bend radii listed in the design manuals used in the design and tool engineering departments. All the aerospace companies and automobile makers have this information as part of the company design standards.

For very accurate flat-pattern dimensions intended for aerospace vehicles, automotive work, appliances, and other consumer products, the bend deduction or setback method is used in *lofting* procedures and standard sheet metal tooling drawings. The tooling department is usually responsible for generating the flat-pattern drawings of parts produced with stamping dies, punching dies, and drawing dies. The engineering department is usually responsible for generating the lofting drawings for flat patterns of regular and irregular shapes. In lofting, the part is drawn very accurately in flat pattern on flat metal sheets with specially prepared surfaces or directly onto heavy Mylar drafting film. The loft, or drawing, is then photographed, and the pattern is transferred to another metal sheet in full scale.

The part is then accurately cut out and becomes a master pattern. A stack of sheets can then be pin routed or tracer milled using the master template as a guide or jig. The cutout parts are then sent to the forming dies or the brakes for the final bending or forming operations. With modern CNC equipment, the part outline may be programmed and then automatically cut on the appropriate machine prior to the bending or forming operations.

When the sheet metal part is to be press-brake bent only on radiused bending dies, the bend allowance method can be used to calculate the flat pattern whenever accuracy and a specific inside bend radius are required. Bend deduction or setback also may be used in this case.

Setback or J Chart for Determining Bend Deductions

Figure 8.15 shows a form of bend deduction (B.D.) or setback chart known as a *J chart.* You may use this chart to determine bend deduction or setback when the angle of bend, material thickness, and inside bend radius are known. The chart in the figure shows a sample line running from the top to the bottom and drawn through the 3⁄16-in radius and the material thickness of 0.075 in. For a 90° bend, read across from the right to where the line intersects the closest curved line in the body of the chart. In this case, it can be seen that the line intersects the curve whose value is 0.18. This value is then the required setback

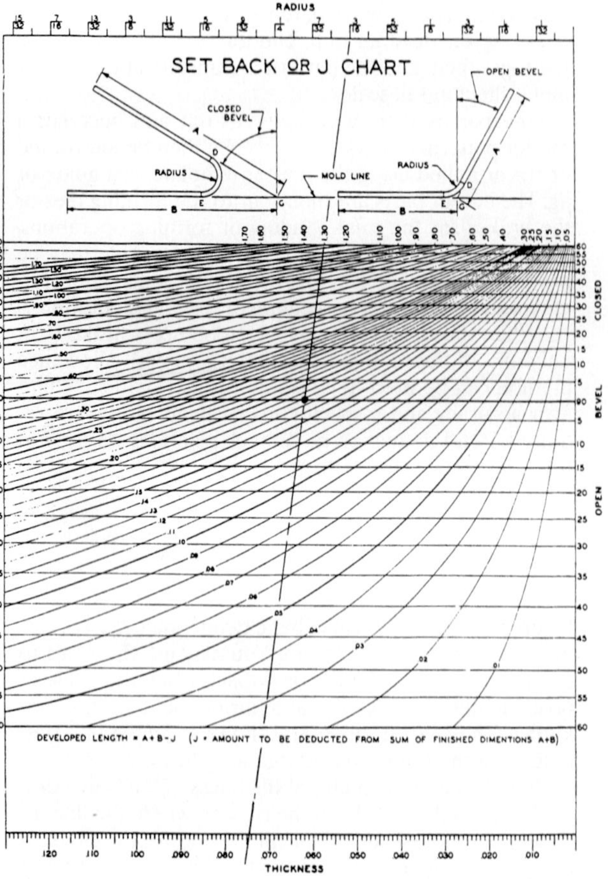

FIGURE 8.15 Setback or J chart.

or bend deduction for a bend of 90° in a part whose thickness is 0.075 in with an inside bend radius of 3⁄16 in. If we check this setback or bend deduction value using the appropriate equations shown previously, we can check the value given by the J chart. (The setback or bend deduction calculations will be very close to the "J chart" values).

Developing Flat Patterns

Developing flat patterns can be done by bend deduction or setback. Figure 8.16 shows a type of sheet metal part that may be bent on a press brake. The flat-pattern part is bent on the brake, with the center of bend line (CBL) held on the bending die centerline. The machine's back gage is set by the operator in order to form the part. If you study the figure closely, you can see how the dimensions progress: The bend deduction is drawn in from the right edge of the part, and the next dimension is taken from the end of the first bend deduction. The next dimension is then measured, the bend deduction is drawn in for that bend, and then the next dimension is taken from the end of the second bend deduction, etc. Notice that the second bend deduction is larger because of the larger radius of the second bend (0.16*R*).

8.5 STIFFENING SHEET METAL PARTS

On many sheet metal parts that have large areas, stiffening can be achieved by creasing the metal in an X configuration by means of brake bending. On certain parts where great stiffness and rigidity are required, a method called *beading* is employed. The beading is carried out at

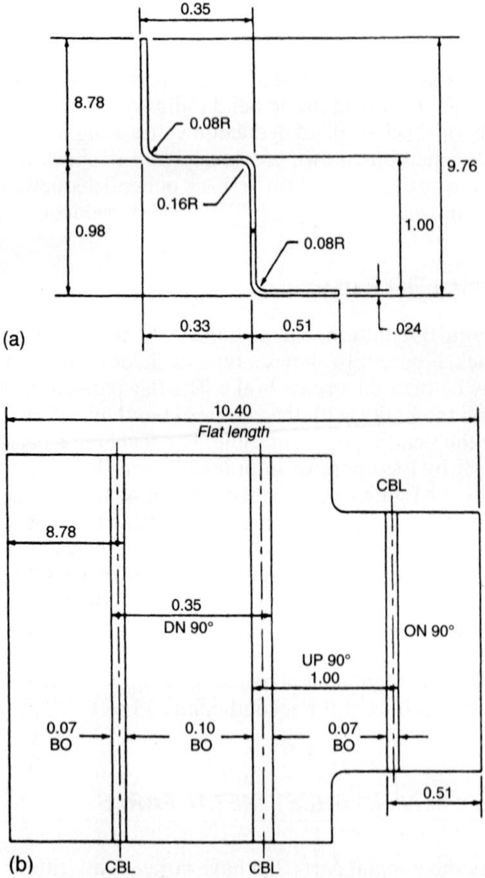

FIGURE 8.16 Layout of bent sheet metal by bend-deduction method. (*a*) Section of a sheet metal part. (*b*) The flat pattern of the part, with bend deductions.

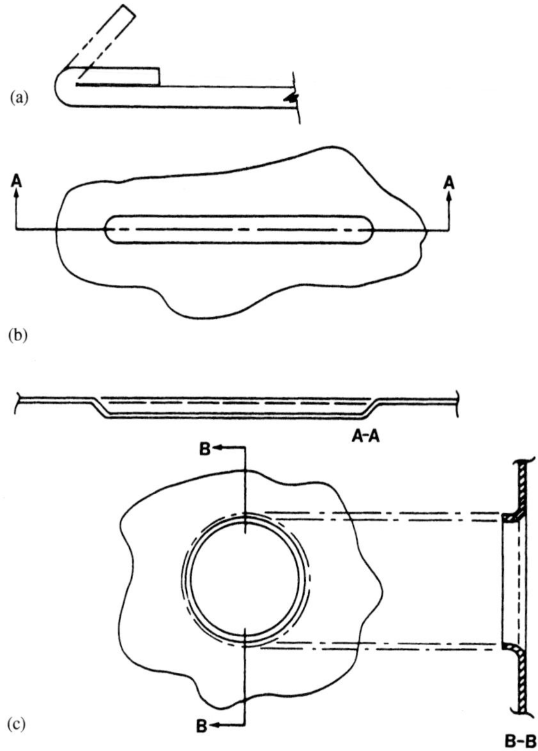

FIGURE 8.17 (*a*) Stiffening bend or "Dutch" bend. (*b*) A stiffening bead in sheet metal. (*c*) A lightening hole.

the same time as the part is being hydropressed, Marformed, or hard-die formed. Figure 8.17*b* shows a typical sheet metal stiffening bead.

Another method for stiffening the edge of a long sheet metal part is to hem or "Dutch bend" the edge as shown in Fig. 8.17*a.* In aerospace and automotive sheet metal parts, flanged *lightening holes* are used as shown in Fig. 8.17*c.* The lightening hole not only makes the part lighter in weight but also more rigid. This method is used commonly in wing ribs, airframes, and gussets or brackets. The lightening hole need not be circular but can take any convenient shape as required by the application.

Sheet Metal Faying Surfaces

Faying surfaces are those where two sheet metal parts come into contact with each other, such as the joining of a flange with a web section or the joining of two flanges. All faying surfaces should be primer painted or finished in some manner to prevent corrosion. A standard finish for many applications is zinc-chromate primer.

Other finishes include zinc plating, nickel plating, cadmium plating, chrome plating, etc. Since cadmium plating and plating solutions are toxic, zinc plating is used in its place where the application allows this. New designs should specify zinc or nickel in place of cadmium unless there is a specific technical reason for using cadmium.

8.6 DESIGN CRITERIA FOR SHEET METAL PARTS

- Do not specify flanges that are too narrow for the type of bending operation. That is, do not design a part with a 0.375-in flange width unless you have the equipment or dies to produce such a short flange on the press brake or other type of machine. Bottoming dies can enable the bending of such a flange in very light gages up to no. 11 gage.

- Do not specify gages that are heavier than necessary for the function of the part. An exception to this rule is code specifications such as for electrical switch-gear equipment, where a minimum of no. 11 gage steel is specified for certain sections of the equipment, whether it is structurally required or not. The heavy gage limit in this application is for the prevention of the spread of fires between adjacent units if there is an electrical fault.
- Use proper edge distances for hardware components, keeping the flanges as narrow as practical with respect to their required rigidity or strength.
- Do not design a part that is impractical to bend on the type of machinery with which your operation is equipped.
- Keep brake-formed parts in a size range where the parts can be handled manually by the brake operators, unless your operation is equipped with automatic machinery or other special-handling equipment.
- When using hot-rolled steel sheets, use tolerances on your parts that are functionally related to the equipment used in your operation. That is, do not expect to hold dimensions to $\pm 1/64$ in when you are producing no. 11 gage hot-rolled sheet metal parts on a press brake. Normal shop tolerance on general sheet metal parts is usually $\pm 1/32$ in for parts under 3 ft in length in both directions. One of the reasons that hot-rolled

steel sheet metal parts between nos. 16 and 7 gages need a generous tolerance is that the steel sheet metal thickness varies from sheet to sheet. This variation can reach ±½ a gage step on a batch of steel of the same gage. The variation of the gage causes the flanges to be bent to different outside dimensions even though the back-gage dimension is the same and is very apparent when more than two bends are made on the same part.

In high-speed, mass-production operations there is no remedy for this condition of variable thickness within the same gage of hot-rolled steel sheet material except a favorable tolerance spread on the dimensions of the parts. Some of the larger companies specify the gage-variation limits when they order steel sheet metal, but the mill order must be large in order to do this. These problems do not occur when using gage-accurate cold-rolled steel sheets.

Sheet Metal Procedures—Design Applications

Typical methods of preventing corner cracks from occurring in bent-up flanged sheet metal parts are shown in Fig. 8.18*a, b* and *c*. Three types of corner arrangements or designs are shown in Fig. 8.19. A method of brake bending a partially closed sheet metal shape that cannot be bent using a "goose-neck" die is shown in Fig. 8.20. In the first step, the outer flanges are bent using a normal V bending die set in the press-brake, and then the center web is bent as shown in step 3, after which the part is closed using a flattening die to form the finished part. For brake bending return flanges that cannot be

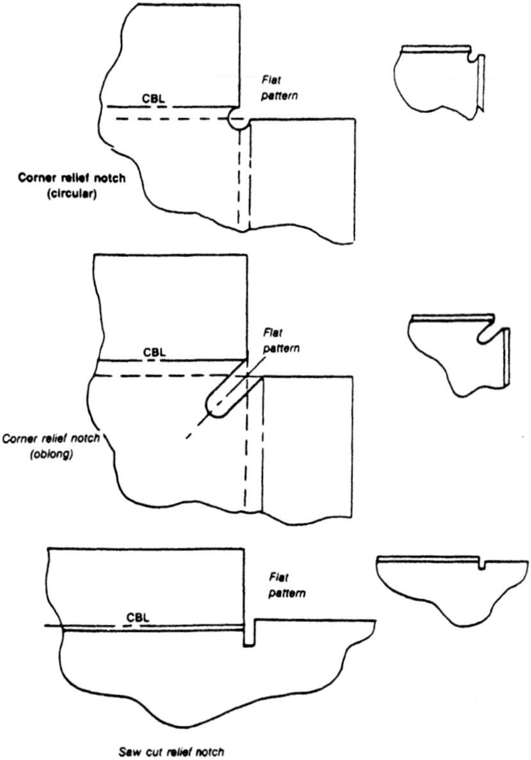

FIGURE 8.18 Three types of corner relief notches for sheet metal parts.

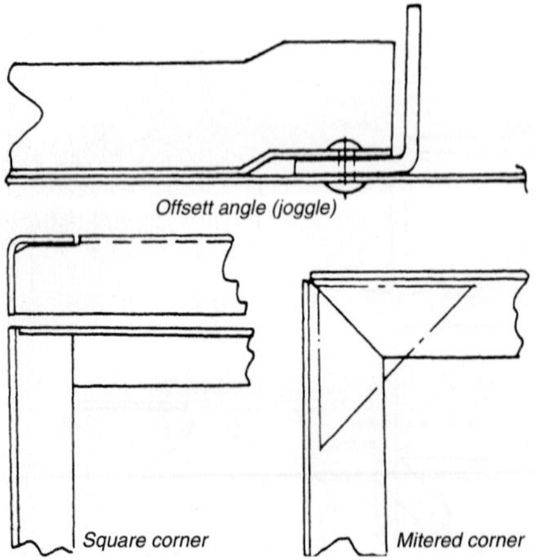

FIGURE 8.19 Three types of corner designs or arrangements for sheet metal assemblies.

bent using a normal V bending die, the goose-neck die is used as shown in Fig. 8.21. Figure 8.22*a* shows a sheet metal "stake" which is often used to anchor the looped ends of extension springs and other applications; while Fig. 8.22*b* shows an integral bent up flange using part of the web of the sheet metal part. Both of the methods shown in Fig. 8.22*a* and *b* require special tooling dies to perform these operations. A widely used method of strengthening a light-gage sheet metal angle is shown in Fig. 8.23*a,* where a corner "break" is placed into the bend radius of the part with a special die. Air-venting louvers are shown in Fig. 8.23*b,* which require a special combination piercing and forming die. The louvers are

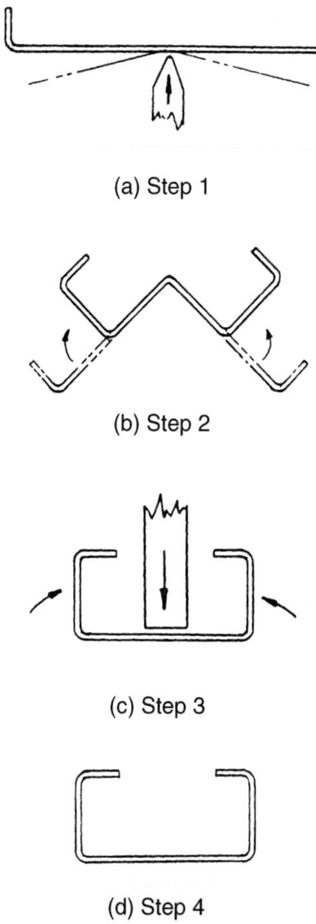

FIGURE 8.20 The goose-neck brake die, used for return flange bending.

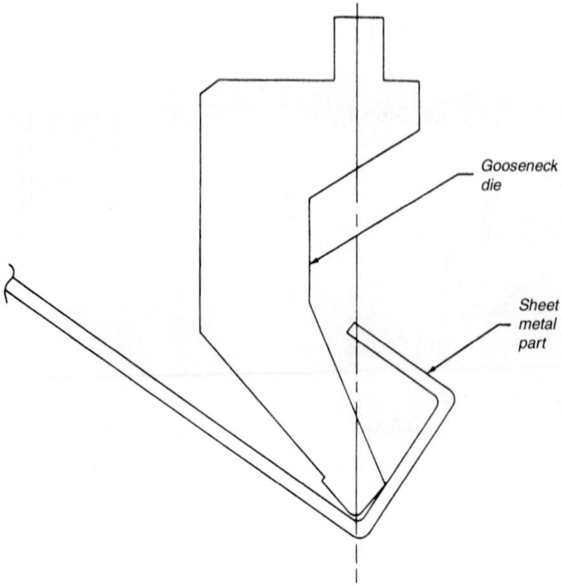

FIGURE 8.21 A method of brake bending a partially closed section using standard AV dies.

produced in one stroke of the press brake. In this case, the press brake is used only to apply sufficient pressure to the die-set to produce the louvers.

8.7 THE EFFECTS OF COLD-WORKING STEEL SHEET METAL

It has long been known that any cold working, such as cold stretching, bending, twisting, etc., affects the me-

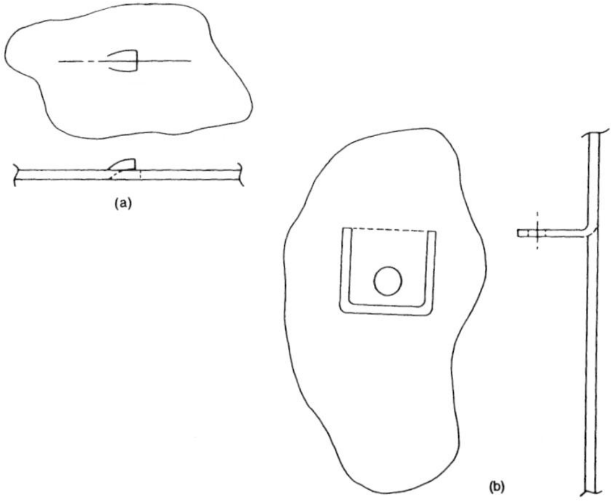

FIGURE 8.22 (*a*) Sheet metal "stake." (*b*) Integral bent-up flange.

chanical properties of steel. Generally, such operations produce strain hardening; that is, they increase the yield strength and, to a lesser degree, the ultimate tensile strength of steels while decreasing the ductility. Cold working of one sort or another occurs in all cold-forming operations, such as roll forming or forming in press brakes. The properties of the cold-worked parts are thus different from the metal prior to the cold-forming operations. The effects of cold forming depend strongly on the details of the particular cold-forming process. The effects of cold forming are also much more evident in the bent corners than in the flat sections. Metallurgically different kinds of structural carbon steels react differently to the same cold-forming process. The actual effects of any cold-forming operation on sheet steels are,

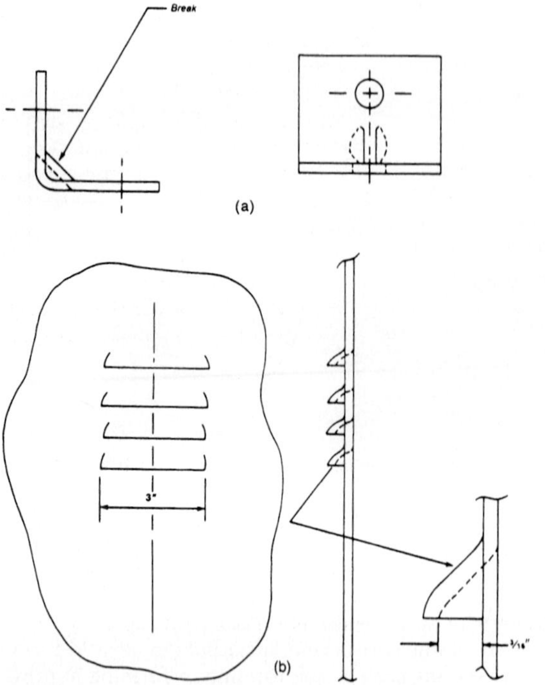

FIGURE 8.23 (*a*) Stiffening corner break used in light gage sheet metal. (*b*) Air-vent louvers.

of course, of an extremely complex nature. Unusual or excessive cold working of structural sheet steel may render the formed section unsuitable for a particular application. In other words, the cold-worked section may be weakened by excessive cold-forming operations.

An interesting aspect of the cold-working applications may be seen in Fig. 8.24. Here we see a copper bus bar which has been axially twisted into a 90° spiral bend. This practice is common in industry, where the direction of the flats on the bar must be changed in a limited amount of space. This copper bus bar was bent on a specially designed machine on which various sized bars may be bent axially to different angles. The machine used to bend this particular bar was designed by the author and has been in operation for 20 years with no maintenance required. The machine referred to will bend bars of cold-drawn ETP no. 110 copper from 0.125 × 1.00 in through 0.625 × 8 in, and a 90° bending operation requires

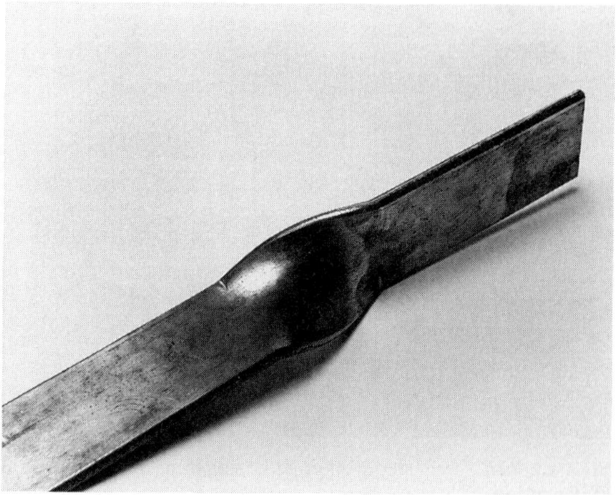

FIGURE 8.24 A copper bus bar with a 90° axial twist.

approximately 15 seconds. The gripper jaws which hold the bar are spaced an appropriate distance (according to bar size) to allow an almost perfect spiral bend, unsupported.

Figure 8.25 shows a machine that is used to cut and notch structural steel shapes, such as angles, channels, and Zs or sheet metal roll-formed shapes efficiently and accurately. In structural welding work, the structural shapes often must be notched, mitered, or cut at an angle prior to the welding operation. This machine allows accurate notches to be made for corners and other structural intersections.

FIGURE 8.25 The Peddiworker (a notching machine) structural shape cutter.

Figure 8.26 shows a lineup of electrical switch gear that is made almost entirely of sheet metal cold-formed parts. The common gauges of sheet steel used on this type of industrial equipment include no. 7 gage, no. 11 gage, no. 13 gage, and no. 16 gage. Most of the sheet steel used in this type of application is hot-rolled, pickeled, and oiled, commercial quality. See also Fig. 8.27, showing a partially assembled electric power distribution substation.

The sheet metal flat-pattern calculation method for producing this type of large equipment consists mainly of the inside bend method, as outlined earlier, where the sheet metal is sharply bent on the press brake.

FIGURE 8.26 Electrical power distribution equipment which consists mainly of formed sheet steel (no. 11 gage).

FIGURE 8.27 A partially assembled electric power control panel fabricated from no. 11-gage steel sheet metal.

8.8 BEND RADII FOR SHEET METAL—(MINIMUM VALUES) (SEE FIG. 8.28)

8.9 SHEAR STRENGTHS FOR STEELS, METAL ALLOYS, PLASTICS, AND COMPOSITES (SEE FIG. 8.29)

Minimum Bend Radii for Metals and Alloys- Given in Multiples of Material Thickness (inches)

Material		Thickness, inches						
		0.015	0.031	0.063	0.093	0.125	0.188	0.250
Carbon steels:								
SAE 1010		S	S	S	S	S	0.5	0.5
SAE 1020-1025		0.5	0.5	1.0	1.0	1.0	1.1	1.25
SAE 1070 & 1095		3.75	3.0	2.6	2.7	2.5	2.7	2.8
Alloy steels:								
SAE 4130 & 8630		0.5	2.0	1.5	1.7	1.5	1.7	1.9
Stainless steels:								
AISI 301, 302, 304 (A)		0.5					0.5	0.75
AISI 316 (A)		0.5					0.5	0.75
AISI 410, 430 (A)		1.0					1.0	1.25
AISI 301, 302, 304 (CR) ¼ H		0.5		0.5	1.0		1.0	1.25
AISI 316 ¼ H		1.0					1.0	1.25
AISI 301, 302, 304 ½ H		1.0					1.0	1.25
AISI 316 ½ H		2.0	2.0	3.0	2.0	2.0	2.0	2.5
AISI 301, 302, 304 H		2.0	2.0	1.5	1.5	1.5	1.5	1.5
Aluminum alloys:								
1100	O	0	0	0	0	0	0	0
	H12	0	0	0	0	0	3.0	6.0
	H14	0	0	0	0	0	3.0	6.0
	H1	0	0	2.0	3.0	4.0	8.0	16.0
	H18	1.0	2.0	4.0	6.0	8.0	16.0	24.0
2014 & Alclad	O	0	0	0	0	0	3.0	6.0
	T6	2.0	4.0	8.0	15.0	20.0	36.0	64.0
2024 & Alclad	O	0	0	0	0	0	3.0	6.0
	T3	2.0	4.0	8.0	15.0	20.0	30.0	48.0
3003, 5005,	O	0	0	0	0	0	0	0
5357, 5457	H12/H32	0	0	0	0	0	3.0	6.0
	H14/H34	0	0	0	1.0	2.0	4.0	8.0

FIGURE 8.28 Bend radii for sheet metal, given in multiples of material thickness (in.).

	H16/H36	0	1.0	3.0	5.0	6.0	12.0	24.0
	H18/H38	1.0	2.0	5.0	9.0	12.0	24.0	40.0
5050, 5052,	O	0	0	0	0	2.0	3.0	4.0
5652	H32	0	0	2.0	3.0	4.0	6.0	12.0
	H34	0	0	2.0	4.0	5.0	9.0	16.0
	H36	1.0	1.0	4.0	5.0	8.0	18.0	24.0
	H38	1.0	2.0	6.0	9.0	12.0	24.0	40.0
6061	O	0	0	0	0	2.0	3.0	4.0
	T6	1.0	2.0	4.0	6.0	9.0	18.0	28.0
7075 & Alclad	O	0	0	2.0	3.0	5.0	9.0	18.0
	T6	2.0	4.0	12.0	18.0	24.0	36.0	64.0
7178	O	0	0	2.0	3.0	5.0	9.0	18.0
	T6	2.0	4.0	12.0	21.0	28.0	42.0	80.0
Copper & alloys:								
ETP #110	soft	S	S	S	S	0.5	0.5	1.0
	hard	S	1.0	1.5	2.0	2.0	2.0	2.0
Alloy 210	¼ H	S	S	S	S	S	0.5	1.0
	½ H	S	S	S	S	S	1.0	1.5
	H	S	S	S	S	S	....	....
	EH	S	0.5	0.5	0.5	0.5	....	....
Alloy 260	¼ H	S	s	S	S	S	0.5	1.0
	½ H	S	S	S	0.3	0.3	....	....
	H	S	0.5	0.5	0.5	1.0	....	....
	EH	2.0	2.0	1.5	2.0	2.0	....	....
Alloy 353	¼ H	S	S	S	S	S	0.5	1.0
	½ H	S	0.5	0.5	0.7	0.3	....	....
	H	2.0	2.0	1.5	2.0	2.0	....	....
	EH	6.0	6.0	4.0	4.0	4.0	....	....
Magnesium sheet @ 70°F:								
	AZ31B-O (S.B.)	3.0						3.0
	AZ31B-O	5.5						5.5

FIGURE 8.28 (*cont.*) Bend radii for sheet metal, given in multiples of material thickness (in.).

Material							
AZ31B-H24	8.0 ———						8.0
HK31A-O	6.0 ———						6.0
HK31A-H24	13.0 ———						13.0
HM21A-T8	9.0 ———						9.0
HM21A-T81	10.0 ———						10.0
LA141A-O	3.0 ———						3.0
ZE10A-O	5.5 ———						5.5
ZE10A-H24	8.0 ———						8.0
Titanium & alloys @ 70°F:							
Pure (A)	3.0	3.0	3.0	3.5	3.5	3.5	3.5
Ti-8Mn (A)	4.0	4.0	4.0	4.0	5.0	5.0	5.0
Ti-5Al-2.5Sn (A)	5.5	5.5	5.5	5.5	6.0	6.0	6.0
Ti-6Al-4V (A)	4.5	4.5	4.5	5.0	5.0	5.0	5.0
Ti-6Al-4V (ST)	7.0						7.0
Ti-6Al-6V-2Sn (A)	4.0						4.0
Ti-13V-11Cr-3Al (A)	3.0	3.0	3.0	3.5	3.5	3.5	3.5
Ti-4Al-3Mo-1V (A)	3.5	3.5	3.5	4.0	4.0	4.0	4.0
Ti-4Al-3Mo-1V (ST)	5.5	5.5	5.5	6.0	6.0	6.0	6.0

Note: S = sharp bend; 0 = sharp bend; S.B. = special bending quality; A = annealed; ST = solution treated; H = hard; EH = extra hard; Magnesium sheet may be bent at temperatures to 800°F; titanium may be bent at temperatures to 1,000°F; on copper and alloys, direction of bending is at 90° to direction of rolling (bend radii must be increased 10-20% @ 45° and 25-35% parallel to direction of rolling). The tabulated values of the minimum bend radii are given in multiples of the material thickness. The values of the bend radii should be tested on a test specimen prior to die-design or production bending finished parts.

FIGURE 8.28 *(Cont.)*

Material	Shear Strength, psi
Carbon Steels:	
SAE 1010 HR	21,500
SAE 1020 HR	32,000
SAE 1045 QT	55,000
SAE 1045 A	44,000
SAE 1095 QT	90,000
SAE 1095 A	63,000
SAE 1117 HR	32,000
Alloy Steels:	
SAE 4130 N	43,500
SAE 4130 T (150,000)	90,000
SAE 4140 N	66,500
SAE 3120 HT-D (800°F)	95,000
SAE 3140 HT-D (800°F)	130,000
SAE 3250 HT-D (800°F)	165,000
Stainless Steels:	
AISI 201	52,000
AISI 301	50,000
AISI 302	41,000
AISI 304	38,500
AISI 310	42,750
AISI 316	38,250
AISI 321	38,250
Cold rolled S/S strip (full hard)	
AISI 300 Series	112,000
Stainless Steels: Annealed	
AISI 410	33,750
AISI 416	33,750
AISI 440C	49,500
AISI 430	33,750
Monel Metal:	
70,000 UTS	42,900
110,000 UTS	65,500
K Monel:	
155,500 UTS	98,500
Nickel:	
68,000 UTS	52,300
121,000 UTS	75,300
Inconel Alloys:	
80,000 UTS	59,000
100,000 UTS	66,000
150,000 UTS	80,000
175,000 UTS	87,000

FIGURE 8.29 Shear strengths for metallic and nonmetallic materials.

Copper and Alloys:	
CA 110 (ETP 110)	22,000-28,000
CA 210 (Guilding)	26,000-37,000
CA 220 (Bronze)	28,000-38,000
CA 230 (Red brass)	31,000-42,000
CA 260 (Cartridge brass)	33,000-44,000
CA 268 (Yellow brass)	33,000-43,000
Beryllium copper: Strip & sheet	
C 17200 (25)	34,200-54,000
C 17000 (165)	34,200-94,500
C 17510 (3)	24,750-67,500
C 17500 (10)	24,750-67,500
C 17410 (174) HT	58,500
Beryllium Nickel:	
UNS-N033 HT	123,750
Aluminum and Alloys:	
1100-O	9,000
1100-H18	13,000
2014-O	18,000
2014-T4, T451	38,000
2014-T6, T651	42,000
2024-O	18,000
2024-T3, T4, T351	41,000
3003-O	11,000
3003-H14	14,000
3003-H18	16,000
5052-O	18,000
5052-H32, H38	60,000-77,000
6061-O	12,000
6061-T4, T451	24,000
6061-T6, T651	30,000
7075-O	22,000
7075-T6, T651	48,000
7178-O	23,000
7178-T6, T651	53,300
Magnesium Alloys:	
Soft (annealed)	19,000
Hardened	28,500 max.
Titanium & Alloys:	
Pure	27,000-49,500
Typical alloys	45,000-77,000
Nonmetallics:	
Polyester-glass (GPO-1, 2 & 3)	12,000-17,000
Polycarbonate (Lexan)	6,000-10,000
Cycolac	4,400-7,400
ABS (Acrylonitrile Butadene Styrene)	1,500-4,000
Acetal (Delrin)	3,000-6,000
Acetate (Cellulose)	2,000-4,000
Epoxy-glass	4,000-10,000
Nylon	3,000-12,000

FIGURE 8.29 *(Cont.)*

Phenolic resins (cloth)	26,000 (Hot-blanked)
Paper	3,500-6,400
Mica	10,000
Teflon, rigid (TFE)	1,500-3,000
Hard rubber	20,000
Polystyrene	10,000 max.
Asbestos board	5,000

Notes: For metallic materials- when the tabulated shear values are given in ranges, the shear values run from the annealed or soft condition to the hardest condition. Interpolate intermediate values between ranges. For nonmetallic materials- the shear value ranges are given from soft to hard grades or glass-filled grades.

A = annealed; HR = hot-rolled; N= normalized; T = tempered; HT-D = heat-treated and drawn (tempered); QT = quenched and tempered; UTS = ultimate tensile strength; HT = heat-treated.

FIGURE 8.29 (*Cont.*) Shear strengths for metallic and nonmetallic materials.

CHAPTER 9

HEAT-TREATING STEELS AND NONFERROUS ALLOYS

The heat treatments of annealing, normalizing, hardening, and tempering of steels and nonferrous alloys are important aspects of metalworking. Carbon and alloy steels are relied on to perform a great number of services in the metalworking industries. Some of the nonferrous metals and alloys are also capable of being hardened above their normal condition either through heat treatment or cold working and find countless uses in product design and manufacturing.

9.1 DEFINITIONS OF HEAT-TREATING TERMS

Age hardening A process of aging that increases hardness and strength. Age hardening usually follows rapid cooling or cold working.

Annealing A process involving heat and cooling, usually applied to induce softening.

Austenite A term designating a metallurgical phase of steels, e.g., austenitic stainless steel.

Carburizing To introduce carbon while the steel is molten or while it is in the solid state by heating it in contact with carbonaceous material below its melting point.

Case hardening A process of hardening a ferrous alloy so that the "case," or surface layer, is much harder than the interior of the part. The typical case-hardening processes are carburizing and quenching, cyaniding, carbonitriding, nitriding, induction hardening, and flame hardening. Cases of Rockwell C 55 to 60 are readily obtained in medium- to high-carbon steels.

Flame hardening A process of hardening a ferrous alloy by heating it above the transformation range by means of a flame and then by cooling as required.

Hardenability The property of a ferrous alloy that determines the depth and distribution of hardness induced by quenching.

Induction hardening A process of hardening a ferrous alloy by heating it above the transformation range by means of electrical induction and then cooling as required.

Killed steel Steel which is deoxidized with silicon or aluminum in order to reduce the oxygen content so that no reaction occurs between carbon and oxygen during solidification.

Martensite An unstable constituent in quenched steel formed without diffusion only during cooling below a certain temperature. Martensite is the hardest of the transformation products of austenite.

Nitriding A process of case hardening in which a ferrous alloy, usually of special composition, is heated in an atmosphere of ammonia or in direct contact with a nitrogenous material to produce surface hardening by absorbing nitrogen, without quenching.

Normalizing Heating a steel part of heavy section to a temperature 100°F above the critical range and then cooling in still air.

Pickeling Chemical or electrochemical removal of surface scale and oxides.

Quench hardening Heating a steel within or above the transformation range and cooling at a rate faster than the critical rate to increase the hardness substantially. Usually involves the formation of martensite.

Solution heat treatment A process in which an alloy is heated to a suitable temperature, held at this temperature long enough for certain constituents to enter into solid solution, and then cooled rapidly to hold the constituent in solution. The metal is left in a supersaturated state which is unstable and may subsequently exhibit age hardening.

Spheroidizing Any process of heating and cooling that produces a round or globular form of carbide in steels.

Strain hardening An increase in hardness and strength caused by plastic deformation at temperatures lower than the recrystallization range.

Stress relieving A process of reducing residual stresses in a metal part by heating the part to a suitable temperature and holding this temperature for a sufficient time. This process is applied to relieve stresses induced by casting, quenching, normalizing, machining, cold working (i.e., springs), or welding.

Temper A condition produced in a metal or alloy by mechanical or thermal treatment and having characteristic structure and mechanical properties. In addition to the annealed temper, conditions produced by thermal treatment are the solution heat-treated temper and the heat-treated and artificially aged temper.

The critical temperature points for hardening steel are called the *decalescence point* and the *recalescence point.* These critical temperature points have a direct relation to the hardening of steel. Unless the hardening temperature passes the decalescence point, no hardening can take place, and unless the steel is cooled suddenly before its temperature reaches the recalescence point, no hardening can take place. These critical temperature points (decalescence and recalescence) vary for different types of steels and must be determined by tests. This variation in critical temperature points makes it necessary to heat different steels to different hardening temperatures together with the proper quench, or cooling rate.

The maximum temperature to which a steel is heated before quenching to harden is called the *hardening temperature.* The hardening temperatures for steels of various carbon contents may be summarized generally by the following:

Carbon content, %	Hardening temperature, °F
0.65–0.80	1450–1550
0.80–0.95	1410–1460
0.95–1.10	1390–1430
1.10 and over	1380–1420

Average Hardening and Tempering Temperatures of Steels. Generally, the average hardening temperature range for carbon and alloy steels is from 1375 to 1575°F. The average tempering range for steels is from 300 to 700°F.

9.2 HEAT TREATMENT OF CARBON STEELS AND ALLOY STEELS

Treatments for Heat-Treating Grades of Carbon Steels

Figure 9.1 lists data for heat treating the common-usage heat-treating grades of carbon steels. The tempering temperatures are not shown, and tempering is not mandatory on many applications. When tempering is required, the tempering range varies from 250 to 450°F.

Heat Treatments for Directly Hardenable Grades of Alloy Steels

Figure 9.2 lists data for heat treating the common-usage directly hardenable grades of alloy steels. Tempering temperatures range from 250 to 350°F.

Quenching Baths and Temperatures

The purpose of the quenching bath is to cool the heated steel at a rate that is faster than the critical cooling rate. To obtain different cooling rates, different quenching baths are used, including

SAE Steel	Normalizing Temperature (°F)	Annealing Temperature (°F)	Hardening Temperature (°F)	Quenching Medium
1025 1030	.....	.....	1575-1650	Water or brine
1035	.....	.....	1525-1575	oil or water
1036	1600-1700	.....	1525-1575	oil or water
1038 1039* 1040	1600-1700	.....	1525-1575	oil or water
1041	1600-1700 and/or	1400-1500	1474-1550	oil
1042 1043* 1045* 1046* 1050*	1600-1700	.....	1475-1550	oil or water.
1052 1055 1060 1065 1070 1074	1550-1650 and/or	1400-1500	1475-1550	oil
1078	.....	1400-1500	1450-1500	water or brine
1080 1090	1550-1650 and/or	1400-1500■	1450-1500	oil ♦
1095	....	1400-1500■ 1400-1500■	1450-1500 1500-1600	oil, water or brine oil
1137	1600-1700 and/or	1400-1500	1525-1575	oil or water
1140	1600-1700	.....	1500-1550	oil or water
1141 1144	1600-1700	1400-1500	1475-1550	oil
1146	1600-1700	.....	1475-1550	oil or water

NOTES: * = Commonly used on parts where induction hardening is employed. However, all steels from SAE 1030 up may have induction hardening applications. ♦ = May be water or brine quenched by special techniques such as partial immersion, or time-quenched; otherwise, they are subject to quench cracking. ■ = Spheroidal structures are often required on these high-carbon steels for machining purposes and should be cooled very slowly or be isothermally transformed to produce the desired structure.

FIGURE 9.1 Table of treatments for heat-treating grades of carbon steels.

SAE Steel	Normalizing Temperature (°F)	Annealing Temperature (°F)	Hardening Temperature (°F)	Quenching Medium
1330	1600-1700 and/or	1500-1600	1525-1575	water or oil
1335 1340	1600-1700 and/or	1500-1600	1500-1550	oil
4037	.....	1525-1575	1500-1575	oil
4047	.....	1450-1550	1500-1575	oil
4130	1600-1700 and/or	1450-1550	1600-1650	water or oil
4137 4140 4145 4150	1600-1700 and/or	1450-1550	1550-1600	oil
4340	1600-1700 & temper	1100-1225	1475-1525	oil
5130 5132 5140 5150	1650-1750 and/or	1450-1550	1500-1550	water or caustic oil oil
50100 51100 52100	....	1350-1450	1425-1475 1500-1600	water oil
6150	1650-1750 and/or	1550-1650	1600-1650	oil
9260	.....	.....	1500-1650	oil
8630	1600-1700	1450-1550	1550-1650	water or oil
8637 8640	1600-1700 and/or	1450-1550	1525-1575	oil
8645 8650	1600-1700 and/or	1450-1550	1500-1550	oil

NOTE: Except as noted, the steel is to be tempered to the required hardness. The exceptions are gears made from steels 4037 and 5150; temper at 350 to 400°F.

FIGURE 9.2 Table of heat treatments for directly hardenable grades of alloy steels.

- Water bath (70 to 100°F bath temperature)
- Water spray (ambient temperature)
- Brine solution (9% by weight or 0.75 lb sodium chloride per gallon of water)
- Caustic soda solution [5% solution by weight potassium hydroxide (lye)]
- Oil (ambient or 90 to 140°F)
- Oil over water
- Molten salt bath (nitrate salts) to 700°F
- Forced air

Tempering and Case Hardening Steels

Tempering. The purpose of *tempering* or *drawing* is to reduce the brittleness of hardened steel and to remove the internal stresses caused by the sudden cooling or quenching done in the hardening operation. In the tempering operation, the hardened steel is heated to a certain temperature and then quench cooled. Steel in the fully hardened condition consists mainly of martensite. When reheated to a temperature of approximately 300 to 750°F, a softer and tougher structure called *troostite* is formed. When the hardened steel is reheated to a temperature from 750 to 1290°F, a structure called *sorbite* is formed, which has less strength than troostite but which is much more ductile.

Tempering Temperatures for Steels

The color of the oxide coating on the reheated steel was used in the past as a means of determining the tempering temperature. Since this color may be affected by the composition of the steel, it is not considered as the most reliable method of determining the tempering temperature, although it served satisfactorily for many years. Today, the optical (infrared) pyrometer is used to determine the temperature to very accurate limits. Figure 9.3 shows the temperature of plain carbon steels by the surface color indicated.

Figure 9.4 shows the temperature of metals by the color of the radiation or glow of the metal. Metals below 900°F do not produce a radiated color, but only the color of the oxide coating, if present. This color scale may be used in an emergency situation, where a fair indication of hardening temperature may be made for hardening plain carbon steels, and Fig. 9.3 surface color equivalent temperatures may be used for tempering.

Degrees Fahrenheit	Degrees Celsius	Color of Steel
430	221.1	Very pale yellow
440	226.7	Light yellow
450	232.2	Pale straw-yellow
460	237.8	Straw-yellow
470	243.3	Deep straw-yellow
480	248.9	Dark-yellow
490	254.4	Yellow-brown
500	260.0	Brown-yellow
510	265.6	Spotted red-brown
520	271.1	Brown-purple
530	276.7	Light purple
540	282.2	Full purple
550	287.8	Dark purple
560	293.3	Full blue
570	298.9	Dark blue
640	337.8	Light blue

FIGURE 9.3 Tempering temperatures by color indication for plain carbon steels.

Temperature (°F)	Temperature (°C)	Color
932 - 1,022	500 - 550	Dull red
1,202 - 1,382	650 - 750	Dark red
1,562 - 1,742	850 - 950	Bright red
1,922 - 2,102	1,050 - 1,150	Yellowish red
2,282 - 2,462	1,250 - 1,350	Dull white
2,642 - 2,822	1,450 - 1,550	Bright white

FIGURE 9.4 Color scale for metal temperatures. Radiated color (glow).

Case Hardening Carbon and Alloy Steels

The case-hardening operation is performed on steels when a tough outer layer is required on the part and through-hardening of the steel is not required. This process has applications on parts that must be hardened to prevent wear on the outer surfaces only, while the core remains at its normal condition. Case hardening of low-carbon steels is a two-step operation. First, the outer layer of the part is *carburized* by introducing carbon into the surface, and second, the outer carburized layer is heat treated to form a hard "case." Surface or case hardening processes include

- Carburization
- Cyanide hardening
- Nitriding
- Carbonitriding
- Induction hardening
- Flame hardening

Low-carbon steels containing 0.10 to 0.20 percent carbon are suitable for carburized case-hardening operations.

Case-Hardening Depths

Light case	0.003 to 0.015 in
Medium case	0.015 to 0.040 in
Heavy case	0.040 to 0.250 in

Case Depth Measurements. A production method for determining the depth of the case-hardened layer consists of air cooling a test pin from the carburizing temperature, reheating to 1475°F, and quenching in oil. This treatment refines the effective case and leaves the core material coarse grained. Etching the fractured surface of the test pin in an aqueous solution of 7 percent concentrated nitric acid produces good contrast between the case-hardened layer and the core material. The test pin is made of the same material as the case-hardened parts being checked and consists of a cylindrical pin 0.500 in in diameter, 3.5 in long, with a 0.468-in-diameter groove (0.060 in wide), approximately 0.30 in in from the end of the pin. Test rings are usually 1.500 in outside diameter with a 0.75-in hole through the center of the ring and a thickness of 0.375 in.

Other methods of determining the case depth include:

- Reading of case depth with a Brinell microscope
- Step grinding and reading the Rockwell hardness at different depths
- Bluing of the fractured surface
- Carbon-cut analysis
- The martensite-start method

Cyanide Case-Hardening Precautions. Cyanide compounds are deadly poisons. Care should be taken so that even minute portions are not consumed accidentally. In storage, keep cyanide-containing salts separate from acids, because mixing these chemicals will produce deadly hydrocyanic gas. Do not add nitrate or nitrite salts to those containing cyanide because of the danger of explosion.

Surface Hardness versus Case Depth. Figure 9.5 shows several different common steels and the case depth to be expected from the Brinell hardness reading taken on the surface of the case-hardened part.

Heat Treatments for Carburizing Grades of Carbon Steels

Figure 9.6 shows the heat-treating process for carburizing grades of some of the common carbon steels.

Heat Treatments for Carburizing Grades of Alloy Steels

Figure 9.7 shows the heat-treating process for some of the common carburizing grades of alloy steels.

9.3 HEAT TREATMENT OF TOOL STEELS

Table 9.1 shows the data required for the heat treatment of all of the standard tool steel types. Table 9.2 provides notes and explanations for Table 9.1.

Steel	SAE 1020		SAE 6118		SAE 1020		SAE 4617		SAE 8620		SAE 4820	
Quench Temp (°F)	1675	1500	1675	1500	1675	1500	1675	1500	1675	1500	1675	1500
Quench Medium	Oil	Oil	Oil	Oil	Water	Water	Oil	Oil	Oil	Oil	Oil	Oil
Brinell												
156	.....	0.000										
167	.....	0.010										
170	.....	0.014										
179	0.000	0.019										
187	0.007	0.024										
192	0.009	0.026	.....	.....	.....	0.000						
197	0.011	0.029	.....	.....	.....	0.003						
207	0.016	0.035	.....	.....	.....	0.005						
217	0.023	0.041	.....	.....	.....	0.009	.....	0.000				
229	0.030	0.049	.....	.....	0.000	0.013	.....	0.004				
241	0.035	0.069	.....	0.000	0.005	0.018	.....	0.009				
255	0.040	0.077	.....	0.003	0.008	0.023	.....	0.015				
269	0.046	.....	.....	0.007	0.015	0.027	0.000	0.021				
285	0.053	.....	.....	0.013	0.022	0.034	0.007	0.024	.....	0.000		
302	0.060	.....	.....	0.018	0.027	0.039	0.010	0.028	.....	0.005		
321	0.074	.....	0.000	0.023	0.034	0.046	0.013	0.032	0.000	0.009		
341	.....	.....	0.004	0.029	0.041	0.054	0.017	0.035	0.003	0.012	.....	0.000
363	.....	.....	0.009	0.036	0.048	0.062	0.021	0.039	0.005	0.016	0.000	0.004
387	.....	.....	0.014	0.043	0.056	.....	0.024	0.041	0.010	0.022	0.008	0.010
412	.....	.....	0.019	0.048	0.062	.....	0.030	0.045	0.016	0.027	0.012	0.015
444	.....	.....	0.026	0.060	.....	.....	0.033	0.048	0.022	0.035	0.024	0.022
460	.....	.....	0.030	.....	.....	.....	0.035	0.051	0.026	0.039	0.028	0.026
477	.....	.....	0.034	.....	.....	.....	0.037	0.054	0.030	0.043	0.033	0.030
495	.....	.....	0.038	.....	.....	.....	0.039	0.056	0.034	0.047	0.038	0.035
512	.....	.....	0.042	.....	.....	.....	0.043	0.058	0.037	0.050	0.043	0.039
532	.....	.....	0.046	.....	.....	.....	0.051	0.060	0.042	0.055	0.049	0.043
555	.....	.....	0.051	.....	.....	.....	0.064	.....	0.047	0.061	0.055	0.049
578	.....	.....	0.057	.....	.....	.....	.....	.....	0.051	0.066	0.063	0.054
600	.....	.....	0.062	.....	.....	.....	.....	.....	0.055	.....	.....	0.060
627	.....	.....	.....	.....	.....	.....	.....	.....	0.061	.....	.....	.....

FIGURE 9.5 Case-depth table as determined by Brinell hardness; tabulated values are in inches and indicate case depths.

SAE Steel	Normalizing Temperature (°F)	Carburizing Temperature (°F)	Cooling Method	Reheat Temp. (°F)	Cooling Medium	Second Reheat (°F)	Cooling medium	Temper (°F)
1010	.....	1650-1700	water/brine	.....	.....	.....	.....	250-400
1015	.....	1650-1700	oil or water	1400-1450	water/brine	.....	.....	250-400
1018	.....	1650-1700	cool slowly	1650-1700	oil/water	1400-1450	water/brine	250-400
1020	.....	1500-1650	air/oil	.....	.....	.....	.....	optional
1024	1650-1750	1650-1700	oil	.....	.....	.....	.....	250-400
1025	.....	1650-1700	water/brine	.....	.....	.....	.....	250-400
1026	.....	1500-1650	oil/water	.....	.....	.....	.....	optional
1027	.....	1350-1575	air/oil	.....	.....	.....	.....	optional
1030	.....	1500-1650	oil/water	.....	.....	.....	.....	optional
1117	.....	1650-1700	cool slowly	1650-1700	oil/water	1400-1450	water/brine	250-400
1118	.....	1500-1650	oil/water	.....	.....	.....	.....	optional

NOTE: Even when the tempering temperatures are shown, the tempering operation is not mandatory on many applications. Tempering is generally employed for a partial stress relief and improves resistance to cracking from grinding operations. Higher temperatures than those shown may be used where the hardness specification on the finished parts permits. For more extensive data on hardening these classes of steels, see Section 5 - Materials and Their Uses, (steel section).

FIGURE 9.6 Heat-treatment table for carburizing grades of common carbon steels.

SAE Steel	Alternate Pretreatments Normalize■	+ Temper♦	Cycle Anneal	Carburizing Temperature (°F)	Cooling Method	Reheat Temp (°F)	Cooling Medium	Tempering Temp (°F)●
4023 4027 4028	Yes	.....	Yes	1650-1700	oil quench	.....	.....	250-350
4118	Yes	.....	.....	1650-1700	oil quench	.....	.....	250-350
4320	Yes	.....	Yes	1650-1700	oil quench	1425-1475	oil	250-350
4620				1650-1700	cool slowly	1475-1525	oil	250-350
4820	.....	.....	.....	1650-1700	oil quench	.....	.....	250-350
5120	Yes	.....	.....	1650-1700	cool slowly	1500-1550	oil	250-350
8615	Yes	.....	Yes	1650-1700	oil quench	1475-1525	oil	250-350
8617						1525-1575		
8620					cool slowly	1475-1525		
8622					cool slowly	1525-1575		
8720				1500-1650	oil quench	.....	.....	

NOTE: ■ Normalizing temperatures should be not less than 50°F higher than the carburizing temperature, followed by air cooling. ♦ After normalizing, reheat to temperature of 1000 to 1250°F and hold for approximately 4 hours. ● The tempering treatment is optional. For more extensive data on hardening these classes of steels, see Section 5 - Materials and Their Uses, (steel section).

FIGURE 9.7 Heat-treatment table for carburizing grades of common alloy steels.

TABLE 9.1 **Typical Heat Treatments for Tool Steels**

Tool steel type	Forging		Annealing			Hardening		
	Start, °F	Do not forge below °F	Temp., °F	Cooling rate, °F Max/h	Brinell hardness	Heating rate	Preheat temp., °F	Hardening temp., °F
High Speed: Tungsten Types								
T1	1950–2150	1750	1600–1650	40	217–255	Rapid from preheat	1500–1600	2300–2375(a)
T2	1950–2150	1750	1600–1650	40	223–255	Rapid from preheat	1500–1600	2300–2375(a)
T4	1950–2150	1750	1600–1650	40	229–269	Rapid from preheat	1500–1600	2300–2375(a)
T5	1950–2150	1800	1600–1650	40	235–285	Rapid from preheat	1500–1600	2325–2375(a)
T6	1950–2150	1800	1600–1650	40	248–302	Rapid from preheat	1500–1600	2325–2375(a)
T8	1950–2150	1800	1600–1650	40	229–255	Rapid from preheat	1500–1600	2300–2375(a)
T15	1950–2150	1800	1600–1650	40	241–277	Rapid from preheat	1500–1600	2200–2300(a)
High Speed: Molybdenum Types								
M1	1900–2100	1700	1500–1600	40	207–235	Rapid from preheat	1350–1550	2150–2225(a)
M2	1900–2100	1700	1600–1650	40	212–241	Rapid from preheat	1350–1550	2175–2250(a)
M3 (1; 2)	1900–2100	1700	1600–1650	40	223–255	Rapid from preheat	1350–1550	2200–2250(a)
M4	1900–2100	1700	1600–1650	40	223–255	Rapid from preheat	1350–1550	2200–2250(a)
M6	1900–2100	1700	1600	40	248–277	Rapid from preheat	1450	2150–2200(a)
M7	1900–2100	1700	1500–1600	40	217–255	Rapid from preheat	1350–1550	2150–2225(a)
M10	1900–2100	1700	1500–1600	40	207–255	Rapid from preheat	1350–1550	2150–2225(a)
M30	1900–2100	1700	1600–1650	40	235–269	Rapid from preheat	1350–1550	2200–2250(a)
M33	1900–2100	1700	1600–1650	40	235–269	Rapid from preheat	1350–1550	2200–2250(a)
M34	1900–2100	1700	1600–1650	40	235–269	Rapid from preheat	1350–1500	2200–2250(a)
M36	1900–2100	1700	1600–1650	40	235–269	Rapid from preheat	1350–1550	2225–2275(a)
M41	1900–2100	1700	1600–1650	40	235–269	Rapid from preheat	1350–1550	2175–2220(a)
M42	1900–2100	1700	1600–1650	40	235–269	Rapid from preheat	1350–1550	2125–2175(a)
M43	1900–2100	1700	1600–1650	40	248–269	Rapid from preheat	1350–1550	2100–2150(a)
M44	1900–2100	1700	1600–1650	40	248–285	Rapid from preheat	1350–1550	2190–2240(a)
M46	1900–2100	1700	1600–1650	40	235–269	Rapid from preheat	1350–1550	2175–2225(a)
M47	1900–2100	1700	1600–1650	40	235–269	Rapid from preheat	1350–1550	2150–2200(a)
Hot Work: Chromium Types								
H10	1950–2100	1650	1550–1650	40	192–229	Moderate from preheat	1500	1850–1900
H11	1950–2100	1650	1550–1650	40	192–235	Moderate from preheat	1500	1825–1875

TABLE 9.1 Typical Heat Treatments for Tool Steels (*Continued*)

Hardening		Tempering					
Minutes at temp.	Quench medium (b)	Temp., °F	Tempered Rc hardness	Depth of hardening	Heat-treat distortion	Hardening safety	Resist. to decarb.
High Speed: Tungsten Types							
2–5	O, A, or S	1000–1100(c)	65–60	Deep	A or S = low; O = medium	High	High
2–5	O, A, or S	1000–1100(c)	66–61	Deep	A or S = low; O = medium	High	High
2–5	O, A, or S	1000–1100(c)	66–62	Deep	A or S = low; O = medium	Medium	Medium to high
2–5	O, A, or S	1000–1100(c)	65–60	Deep	A or S = low; O = medium	Medium	Low
2–5	O, A, or S	1000–1100(c)	65–60	Deep	A or S = low; O = medium	Medium	Low
2–5	O, A, or S	1000–1100 (c)	65–60	Deep	A or S = low; O = medium	Medium	Medium
2–5	O, A, or S	1000–1200(c)	68–63	Deep	A or S = low; O = medium	Medium	Medium
High Speed: Molybdenum Types							
2–5	O, A, or S	1000–1100(c)	65–60	Deep	A or S = low; O = medium	Medium	Low
2–5	O, A, or S	1000–1100(c)	65–60	Deep	A or S = low; O = medium	Medium	Medium
2–5	O, A, or S	1000–1100(c)	66–61	Deep	A or S = low; O = medium	Medium	Medium
2–5	O, A, or S	1000–1100(c)	66–61	Deep	A or S = low; O = medium	Medium	Medium
2–5	O, A, or S	1000–1100(c)	66–61	Deep	A or S = low; O = medium	Medium	Low
2–5	O, A, or S	1000–1100(c)	66–61	Deep	A or S = low; O = medium	Medium	Low
2–5	O, A, or S	1000–1100(c)	65–60	Deep	A or S = low; O = medium	Medium	Low
2–5	O, A, or S	1000–1100(c)	65–60	Deep	A or S = low; O = medium	Medium	Low
2–5	O, A, or S	1000–1100(c)	65–60	Deep	A or S = low; O = medium	Medium	Low
2–5	O, A, or S	1000–1100(c)	65–60	Deep	A or S = low; O = medium	Medium	Low
2–5	O, A, or S	1000–1100(c)	65–60	Deep	A or S = low; O = medium	Medium	Low
2–5	O, A, or S	1000–1100(d)	70–65	Deep	A or S = low; O = medium	Medium	Low
2–5	O, A, or S	950–1100(d)	70–65	Deep	A or S = low; O = medium	Medium	Low
2–5	O, A, or S	950–1100(d)	70–65	Deep	A or S = low; O = medium	Medium	Low
2–5	O, A, or S	1000–1160(d)	70–62	Deep	A or S = low; O = medium	Medium	Low
2–5	O, A, or S	975–1050(d)	69–67	Deep	A or S = low; O = medium	Medium	Low
2–5	O, A, or S	975–1100(d)	70–65	Deep	A or S = low; O = medium	Medium	Low
Hot Work: Chromium Types							
15–40(e)	A	1000–1200(c)	56–39	Deep	Very low	Highest	Medium
15–40(e)	A	1000–1200(c)	54–38	Deep	Very low	Highest	Medium

TABLE 9.1 Typical Heat Treatments for Tool Steels (*Continued*)

Tool steel type	Forging: Start, °F	Forging: Do not forge below °F	Annealing: Temp., °F	Annealing: Cooling rate, °F Max/h	Annealing: Brinell hardness	Hardening: Heating rate	Hardening: Preheat temp., °F	Hardening: Hardening temp., °F
H12	1950–2100	1650	1550–1650	40	192–235	Moderate from preheat	1500	1825–1875
H13	1950–2100	1650	1550–1650	40	192–229	Moderate from preheat	1500	1825–1900
H14	1950–2150	1700	1600–1650	40	207–235	Moderate from preheat	1500	1850–1950
H19	1900–2100	1650	1600–1650	40	207–241	Rapid from preheat	1500	2000–2200(a)
Hot Work: Tungsten Types								
H21	1950–2150	1650	1600–1650	40	207–235	Rapid from preheat	1500	2000–2200(a)
H22	1950–2150	1650	1600–1650	40	207–235	Rapid from preheat	1500	2000–2200(a)
H23	1950–2150	1800	1600–1650	40	212–255	Rapid from preheat	1550	2200–2300(a)
H24	1950–2150	1750	1600–1650	40	217–241	Rapid from preheat	1500	2000–2250(a)
H25	1950–2150	1700	1600–1650	40	207–235	Rapid from preheat	1500	2100–2300(a)
H26	1950–2150	1750	1600–1650	40	217–241	Rapid from preheat	1600	2150–2300(a)
Hot Work: Molybdenum Type								
H42	1900–2050	1700	1550–1650	40	207–235	Rapid from preheat	1350–1550	2050–2225(a)
Cold Work: High-Carbon High-Chromium Types								
D2	1850–2000	1700	1600–1650	40	217–255	Very slowly	1500	1800–1875
D3	1850–2000	1700	1600–1650	40	217–255	Very slowly	1500	1700–1800
D4	1850–2000	1700	1600–1650	40	217–255	Very slowly	1500	1775–1850
D5	1850–2000	1700	1600–1650	40	223–255	Very slowly	1500	1800–1875
D7	2050–2125	1800	1600–1650	40	235–262	Very slowly	1500	1850–1950
Cold Work: Medium-Alloy Air-Hardening Types								
A2	1850–2000	1650	1550–1600	40	201–235	Slowly	1450	1700–1800
A3	1850–1950	1650	1550–1600	40	207–229	Slowly	1450	1750–1850
A4	1850–2000	1650	1360–1400	25	200–241	Slowly	1250	1500–1600
A6	1900–2050	1600	1350–1375	25	217–248	Slowly	1200	1525–1600
A7	1925–2100	1800	1600–1650	25	235–269	Very slowly	1500	1750–1800
A8	1950–2100	1700	1550–1600	40	192–241	Slowly	1450	1800–1850
A9	1950–2100	1700	1550–1600	25	212–248	Slowly	1450	1800–1875
A10	1800–1925	1600	1410–1460	15	235–269	Slowly	1200	1450–1500
Cold Work: Oil-Hardening Types								
O1	1800–1950	1550	1400–1450	40	183–212	Slowly	1200	1450–1500
O2	1800–1925	1550	1375–1425	40	183–217	Slowly	1200	1400–1475
O6	1800–1950	1500	1410–1450	20	183–217	Slowly	—	1450–1500
O7	1800–2000	1600	1450–1500	40	192–217	Slowly	1200	W=1450–1525(b) O=1500–1625(b)
Shock-Resisting Types								
S1	1850–2050	1600	1450–1500	40	183–229	Slowly	1200	1650–1750
S2	1850–2050	1600	1400–1450	40	192–217	Slowly	1200	1550–1650
S5	1850–2050	1600	1425–1475	25	192–229	Slowly	1400	1600–1700
S6	1850–2050	1600	1475–1525	25	192–229	Slowly	1400	1650–1750
S7	1950–2050	1700	1500–1550	25	187–223	Slowly	1200–1300	1700–1750

TABLE 9.1 Typical Heat Treatments for Tool Steels (*Continued*)

Hardening		Tempering					
Minutes at temp.	Quench medium (b)	Temp., °F	Tempered Rc hardness	Depth of hardening	Heat-treat distortion	Hardening safety	Resist. to decarb.
15–40(e)	A	1000–1200(c)	55–38	Deep	Very low	Highest	Medium
15–40(e)	A	1000–1200(c)	53–38	Deep	Very low	Highest	Medium
15–40(e)	A	1100–1200(c)	47–40	Deep	Low	Highest	Medium
2–5	A or O	1000–1300(c)	57–40	Deep	A = low; O = medium	High	Medium
			Hot Work: Tungsten Types				
2–5	A or O	1100–1250	54–36	Deep	A = low O = medium	High	Medium
2–5	A or O	1100–1250	52–39	Deep	A = low; O = medium	High	Medium
2–5	O	1200–1350(c)	47–34	Deep	Medium	High	Medium
2–5	O	1050–1200(c)	55–45	Deep	A = low; O = medium	High	Medium
2–5	A or O	1050–1250(c)	44–35	Deep	A = low; O = medium	High	Medium
2–5	O, A, or S	1050–1250(c)	58–43	Deep	S, A = low; O = medium	High	Medium
			Hot Work: Molybdenum Type				
15–45(e)	O, A, or S	1050–1200(c)	60–50	Deep	S, A = low; O = medium	Medium	Medium
			Cold Work: High-Carbon High-Chromium Types				
15–45(e)	A	400–1000	61–54	Deep	Lowest	Highest	Medium
15–45(e)	O	400–1000	61–54	Deep	Very low	High	Medium
15–45(e)	A	400–1000	61–54	Deep	Lowest	Highest	Medium
15–45(e)	A	400–1000	61–54	Deep	Lowest	Highest	Medium
30–60(e)	A	300–1000	65–58	Deep	Lowest	Highest	Medium
			Cold Work: Medium-Alloy Air-Hardening Types				
20–45(e)	A	350–1000	62–57	Deep	Lowest	Highest	Medium
25–60(e)	A	350–1000	65–57	Deep	Lowest	Highest	Medium
15–90(e)	A	350–800	62–54	Deep	Lowest	Highest	Medium to high
20–45(e)	A	300–800	60–54	Deep	Lowest	Highest	Medium to high
30–60(e)	A	300–1000	67–57	Deep	Lowest	Highest	Medium
20–45(e)	A	350–1100	60–50	Deep	Lowest	Highest	Medium
20–45(e)	A	950–1150	56–35	Deep	Lowest	Highest	Medium
30–60(e)	A	350–80	62–55	Deep	Lowest	Highest	Medium to high
			Cold Work: Oil-Hardening Types				
10–30	O	350–500	62–57	Medium	Very low	Very high	High
5–20	O	350–500	62–57	Medium	Very low	Very high	High
10–30	O	350–600	63–58	Medium	Very low	Very high	High
10–30	O or W	350–550	64–58	Medium	W = high O = very low	W = low O = very high	High
			Shock-Resisting Types				
15–45	O	400–1200	58–40	Medium	Medium	High	Medium
5–20	B or W	350–800	60–50	Medium	High	Low	Low
5–20	O	350–800	60–50	Medium	Medium	High	Low
10–30	O	400–600	56–54	Medium	Medium	High	Low
15–45(e)	A or O	400–1150	57–45	Deep	A = lowest; O = low	A = highest; O = high	Medium

TABLE 9.1 Typical Heat Treatments for Tool Steels (*Continued*)

	Forging		Annealing			Hardening		
Tool steel type	Start, °F	Do not forge below °F	Temp., °F	Cooling rate, °F Max/h	Brinell hardness	Heating rate	Preheat temp., °F	Hardening temp., °F
Special Purpose: Low-Alloy Types								
L2	1800–2000	1550	1400–1450	40	163–197	Slowly	—	W=1450–1550 O=1550–1700
L6	1800–2000	1550	1400–1450	40	183–255	Slowly	—	1450–1550
Water-Hardening Types								
W1	1800–1950(f)	1500	1360–1450(f)	40	156–201	Slowly	(g)	1400–1550(h)
W2	1800–1950(f)	1500	1360–1450(f)	40	156–201	Slowly	(g)	1400–1550(h)
W5	1800–1950(f)	1500	1360–1450(f)	40	163–201	Slowly	(g)	1400–1525(h)
						Hardening		
						Carburizing temp., °F		Hardening temp., °F
Mold Types								
P2	1850–2050	1550	1350–1500	40	103–123	1650–1700		1525–1550(i)
P3	1850–2050		1350–1500	40	109–137	1650–1700		1475–1525(i)
P4	1850–2050	1600	1600–1650	25	116–128	1775–1825		1775–1825(i)
P5	1850–2050	1550	1550–1600	40	105–131	1650–1700		1550–1600(i)
P6	1950–2150	1700	1550	15	183–217	1650–1700		1450–1500(i)
P20	1850–2050	1600	1400–1450	40	149–212	1600–1650		1500–1600
						Solution Treating		
						Heating rate	Preheat temp.	Solution temp., °F
P21	2000–2100	1750		Do not anneal		Slowly	Do not preheat	1300–1350

TABLE 9.1 Typical Heat Treatments for Tool Steels (*Continued*)

Hardening		Tempering					
Minutes at temp.	Quench medium (b)	Temp., °F	Tempered Rc hardness	Depth of hardening	Heat-treat distortion	Hardening safety	Resist. to decarb.
Special Purpose: Low-Alloy Types							
10–30	O or W	350–1000	63–45	O = medium	O = medium W = high	O = medium W = low	High
10–30	O	350–1000	62–45	Medium	Low	High	High
Water-Hardening Types							
10–30	B or W	350–650	64–50	Shallow	High	Low	Highest
10–30	B or W	350–650	64–50	Shallow	High	Low	Highest
10–30	B or W	350–650	64–50	Shallow	High	Low	Highest
Mold Types							
15	O	350–500	64–58(k)	Medium	Low	High	High
15	O	350–500	64–58(k)	Medium	Low	High	High
15	A	350–900	64–58(k)	Medium	Very low	High	High
15	O or W	350–500	64–58(k)		W = high O = low	High	High
15	A or O	350–450	61–58(k)		A = very low, O = low	High	High
15	O	900–1100	37–28	Medium	Low	High	High
		Aging					
		Temp., °F	Hardness Rc				
60–180	A or O	950–1025	40–30	Deep	Lowest	Highest	High

TABLE 9.2 Notes and Explanations for Table 9.1

(a) When the high temperature heating is carried out in a salt bath, the range of temperatures should be about 25°F lower than that shown.

(b) A = air quench
B = brine quench
O = oil quench
S = salt bath quench
W = water quench

(c) Double tempering recommended for not less than 1 h at temperature each temper.

(d) Triple tempering recommended for not less than 1 h at temperature each temper.

(e) Times shown apply to open furnace heat treatment. For pack hardening, a common rule is to heat for ½ h per inch of cross section of the pack.

(f) Forging, normalizing, and annealing temperatures of carbon tool steels are given as ranges because they vary with carbon content. The following temperatures are recommended:

Forging
0.50 to 1.25% C: the range shown
1.25 to 1.40% C: low side of range

Normalizing
0.60 to 0.75% C: 1500°F
0.75 to 0.90% C: 1450°F
0.90 to 1.10% C: 1600°F
1.10 to 1.40% C: 1600 to 1700°F

Annealing
0.60 to 0.90% C: 1360 to 1400°F
0.90 to 1.40% C: 1400 to 1450°F

(g) For large tools and tools having intricate sections, preheating at 1050–1200°F is recommended.

(h) Varies with carbon content as follows:
0.60–0.80% C: 1450–1550°F
0.85–1.05% C: 1425–1550°F
1.10–1.40% C: 1400–1525°F

(i) After carburizing

(k) Carburized case hardness

SOURCE: "AISI Steel Products Manual—Tool Steels," April 1976.

9.4 HEAT TREATMENT OF ALUMINUM ALLOYS

The heat treatment of aluminum alloys is a precision operation. The temperature/time cycles are critical and must be carried out by using the proper equipment and controls. The general types of heat treatments applied to aluminum and its alloys are

1. Preheating or homogenizing to reduce chemical segregation of cast structures and to improve their workability
2. Annealing to soften strain-hardened (work-hardened) and heat-treated alloy structures, to relieve stresses, and to stabilize properties and dimensions
3. Solution heat treatments to effect solid solution of alloying constituents and improve mechanical properties
4. Precipitation heat treatments to provide hardening by precipitation of constituents from solid solution

Solution Heat Treatment of Aluminum Alloys

The solution heat treatment of aluminum alloys improves mechanical properties by developing the maximum practical concentration of the hardening constituents in solid solution. This requires heating the aluminum-alloy part to a temperature close to the eutectic temperature, holding it there long enough to effect the desired solution, and then quenching fast enough to retain the desired solid solution.

The usual quenching medium for aluminum alloys is water. In quenching some products, water below 100°F provides the required quench rate for optimal properties of the alloy being heat treated. In other alloys or products, the water may be heated to temperatures above 100°F or even to the boiling point to control distortion and residual internal stresses. Hot oil is also used for some applications because it provides a quenching rate that improves resistance to stress corrosion. The water quench is either by total immersion or by spray.

The recommended conditions of temperature and time for heat treating some of the common aluminum alloys, produced by various methods, are given in Figs. 9.8 through 9.13.

Precipitation Heat Treatment of Aluminum Alloys

The rate of precipitation from the supersaturated solid solution existing immediately after quenching increases as the temperature of the metal is raised above room temperature. Precipitation occurs naturally at room temperature, providing useful degrees of precipitation hardening for some alloys. Precipitation heat treatment generally denotes treatment at an elevated temperature and is often called *artificial aging.* See Figs. 9.8 through 9.13 for the precipitation-hardening temperature/times for those alloys which may be precipitation hardened (artificially aged).

9.5 HEAT TREATING BERYLLIUM-COPPER ALLOYS

Beryllium-copper alloys are important metallic products which over recent years have found many applications in various industries, including the electrical, electronics, and mechanical industries. Beryllium-copper alloys have many desirable characteristics, including corrosion resistance, good electrical conductivity, good spring properties, good heat conductivity, and excellent stress fatigue resistance. Beryllium is also alloyed with nickel, called

Alloy Designation	Temper	Solution Heat Treatment Metal Temperature (°F)	Precipitation Heat Treatment Metal Temp. (°F)	Time @ Temp. Hours
2014	T4, T6, T651	935	320	18
2024	T3, T4, T6	920	375	16
6061	T4, T6	985	320	18
7075	T6, T651	900	250	24
7178	T6, T651	875	250	24

FIGURE 9.8 Heat-treatment table for aluminum-alloy sheet and plate.

Alloy Designation	Temper	Solution Heat Treatment Metal Temperature (°F)	Precipitation Heat Treatment Metal Temp. (°F)	Time @ Temp. Hours
2014	T4, T6, T651	935	320	18
2024	T351	920	.....	.....
6061	T4, T6, T651	980	350	8
7075	T6, T651	870	250	24
7178	T6, T651	870	250	24

FIGURE 9.9 Heat-treatment table for extruded aluminum-alloy rod, bar, shapes, and tube.

beryllium-nickel alloys, whose tensile strengths are quite high when the alloys are cold worked. Many products that formerly used spring-tempered phosphor-bronzes are now using the beryllium-copper alloys.

The element beryllium is toxic, as are its alloys. The OSHA safety regulations prescribe the conditions

Alloy Designation	Temper	Solution Heat Treatment Metal Temperature (°F)	Precipitation Heat Treatment Metal Temp. (°F)	Time @ Temp. Hours
2011	T3	975	.....	.....
2014	T4, T6, T651	935	320	18
2024	T3, T4,	920	.....	.....
	T6	920	375	16
6061	T4,	985	.....	.....
	T6, T651	985	320	18
7075	T6, T651	915	250	24
7178	T6, T651	900	250	24

FIGURE 9.10 Heat-treatment table for rolled or drawn aluminum-alloy wire, rod, bar, shapes, and tube.

Alloy Designation	Temper	Solution Heat Treatment Metal Temperature (°F)	(Quench water temp.)	Precipitation Heat Treatment Metal Temp. (°F)	Time @ Temp. Hours
2014	T4	935	(140-160)	.....	.....
	T6	935	(140-160)	340	10
2024	T4	920	(Room)	.....	.....
	T6	920	(Room)	375	18
6061	T6	985	(Room)	350	8
7075	T6	880	(140-160)	250	24
7178	T6	870	(Room)	250	24

FIGURE 9.11 Heat-treatment table for aluminum-alloy forgings.

Alloy Designation	Temper	Solution Heat Treatment Metal Temp. (°F)	Solution Heat Treatment Time Hours	Precipitation Heat Treatment Metal Temp. (°F)	Precipitation Heat Treatment Time Hours
142	T61	960	6	450	2
220	T4	810	16	.....	.....
319	T6	940	10	310	4
355	T6	980	10	310	4
356	T6	1000	10	310	4
357	T6	1000	10	350	6
363	T6	940	10	310	4
A750	T5	.....	.....	430	8

NOTE: Many of these alloys are covered by ASTM B26. Solution heat treatment is followed by quenching in water at 150 to 212°F. A boiling water quench is recommended when minimum quenching stresses and distortion are required. Maintain steady furnace temperatures for the solution heat treatment process.

FIGURE 9.12 Heat-treatment table for sand- and plaster-cast aluminum alloys.

Alloy Designation	Temper	Solution Heat Treatment Metal Temp. (°F)	Solution Heat Treatment Time Hours	Precipitation Heat Treatment Metal Temp. (°F)	Precipitation Heat Treatment Time Hours
142	T61	960	6	400	4
319	T6	940	6	310	4
354	T61	980	8	310	10
355	T6	980	6	310	4
356	T6	1000	8	310	4
357	T6	1000	8	350	6
359	T61	1000	10	310	10
750/A750	T5	.....	.....	430	8

NOTE: Many of these alloys are covered by ASTM B108. The solution heat treatment is followed by quenching in water at 150 to 212°F. Boiling water is recommended when minimum quenching stresses and distortion are required. Maintain steady furnace temperatures for the solution heat treatment process.

FIGURE 9.13 Heat-treatment table for permanent-mold cast aluminum alloys.

under which these alloys should be processed. The percentage of the population to which beryllium causes toxicity is small, but safeguards must be taken. The toxic effects of beryllium are seen as chronic pulmonary (lung) problems.

In the heat treatment of beryllium-copper alloys, the product is furnace heated to a prescribed temperature and held at this temperature for a number of hours, followed by still-air cooling to room temperature. Figure 9.14 shows the popular and common beryllium-copper alloys and their heat-treatment procedures. Figure 9.15 defines the ASTM temper designations used in Fig. 9.14.

Cold Work Hardening of Metals and Alloys

Many metals and alloys obtain their physical properties through the process of work hardening. The work-hardening processes include drawing, rolling, stretching, and hammering. Materials such as phosphor-bronze acquire their spring characteristics through the action of cold rolling or drawing. The percentage of reduction of the cross-sectional area of the material determines the degree of "temper" in the material. Figure 9.15 listed the percentage of reduction that determines the temper designation for beryllium-copper alloys. The phosphor-bronze alloys also receive their temper designations from the percentage of reduction that takes place in the cold-rolling operation (quarter hard, half hard, full hard, spring, and extra spring).

Stainless steels such as AISI types 303, 304, and others may acquire very high tensile strengths and spring properties through the cold-rolling process. The 300 series stainless steels are termed *austenitic* and

Alloy Designation (UNS system)	Temper (*)	Heat Treatment Time/Temperature	Tensile Strength Kpsi
25	(TF00)	3 hrs @ 600 °F	140-175
(C17200)	(TH01)	2 hrs @ 600 °F	150-185
	(TH02)	2 hrs @ 600 °F	185-215
	(TH04)	2 hrs @ 600 °F	190-220
190	(TM00)	Mill	100-110
(C17200)	(TM01)	Mill	110-120
	(TM02)	Mill	120-135
	(TM04)	Mill	135-150
	(TM05)	Mill	150-160
	(TM06)	Mill	155-175
	(TM08)	Mill	175-190
290	(TM00)	Mill	100 minimum
(C17200)	(TM02)	Mill	120 minimum
	(TM04)	Mill	140 minimum
	(TM06)	Mill	155 minimum
	(TM08)	Mill	175 minimum
165	(TF00)	3 hrs @ 600 °F	150-180
(C17000)	(TH01)	2 hrs @ 600 °F	160-190
	(TH02)	2 hrs @ 600 °F	170-200
	(Th04)	2 hrs @ 600 °F	180-210
	(TM00)	Mill	100-110
	(TM01)	Mill	110-120
	(TM02)	Mill	120-135
	(TM04)	Mill	135-150
	(TM05)	Mill	150-160
	(TM06)	Mill	155-175
3 (C17510)	(TF00)	2-3 hrs @ 900 °F	100-130
and	(TH04)	2-3 hrs @ 900 °F	110-135
10 (C17500)			
174	(TH04)	Mill	110-130
(C17410)			

NOTE: (*) = ASTM alphanumeric code for product tempers. Mill = product normally heat treated at the mill. For heat treatment, the part must be fully annealed, then accurately held at the indicated temperatures for the times shown. Uniform heating of the parts within the furnace MUST be maintained for a successful heat treatment. Annealed strip which is "dead-soft" offers maximum deep drawing and die-forming capabilities.

FIGURE 9.14 Heat-treatment table for beryllium-copper alloys.

ASTM Designation	Description	Cold Rolled Thickness Reduction in Percent
TB00	Solution annealed	0
TD01	Quarter hard	11
TD02	Half-hard	21
TD03	Three-quarters hard	29
TD04	Hard	37
TF00	The suffix "T" added to temper designations indicates that the material has been age-hardened by the standard heat treatment.	
TH01		
TH02		
TH03		
TH04		
TM00	Mill hardened to specific property ranges, no further heat treatment required.	
TM01		
TM02		
TM04		
TM05		
TM06		
TM08		

NOTE: Temper designations are defined in the specification ASTM B601, "Standard practice for temper designations for copper and copper alloys".

FIGURE 9.15 ASTM temper designations (for Fig. 9.14).

cannot be heat treated by using conventional heat-treating processes. The spring properties of these and many other metals and alloys are obtainable only through cold-working processes. The *martensitic* stainless steels (400 series) may be heat treated using conventional heat-treating methods. The 440 series stainless steels are popular martensitic stainless steels which may be hardened to a high degree by heating and quenching at specified temperatures and cooling rates.

CHAPTER 10

ELECTROPLATING AND OTHER FINISHES FOR METALS

Commercial products and equipment in all classifications require finishes of one type or other. These finishes range from basic oxide coatings to the various paints and plastics to electrodeposited metals such as copper, chromium, nickel, etc.

The finish or plating used on any particular part should contribute to the engineering qualities of the final product and not merely to its cosmetic appeal. The desired finish could include weather protection, resistance to corrosive chemicals, heat resistance, electrical conductivity, wear resistance, and improved lubrication qualities.

It is the design engineer's responsibility to specify the finish characteristics and specifications on a part or assembly. Designers should be aware of the types of finishes and plating processes that are commercially available and how to specify them on the design and detail part drawings. Arbitrary selection of a finish or plating and its thickness range can lead to many design problems relating to corrosion, cost, and dimensional interference.

This chapter will familiarize the designer, engineer, and other personnel in the metalworking industries with the common finishing processes, procedures for specifying thicknesses of platings, and the appropriate industrial standard specifications that control these finishes.

10.1 ELECTROPLATING DATA AND SPECIFICATIONS

Finishes

The finishing processes and methods in common use for engineering applications are as follows:

Mechanical finishing

- Sanding or grinding
- Brushing (scratch and satin)
- Sandblasting
- Shot peening and tumbling (metal or ceramic ball)
- Burnishing
- Mechanical powder plating
- Polishing

Chemical finishing

- Etching
- Bonderite
- Alodine
- Iridite
- Phosphating
- Passivating (stainless steels)

- Black oxide
- Blueing
- Teflon coating
- Other plastic coatings

Electrolytic oxides

- Sulfuric acid anodize (Alumilite, Alcoa process)
- Chromic acid anodize (aerospace and other applications)
- Martin hardcoat anodize (Martin-Marietta process)
- Electropolishing

Note: Anodizing may be performed on aluminum, magnesium, titanium, and zinc.

Electroplating

- Copper plate
- Cadmium plate
- Chromium plate (bright, hard, and black)
- Nickel plate (bright, black, and chromium-nickel)
- Gold plate
- Silver plate
- Tin plate (tin-cadmium, tin-lead, and tin-nickel)
- Indium plate
- Rhodium plate
- Palladium plate
- Zinc plate (nickel-zinc, zinc with chromate)
- Lead plate

The base metals and their alloys that are commonly electroplated are iron, steel, stainless steel (more commonly passivated), aluminum, copper, brass, bronze, titanium, and magnesium.

Hot-Dip Plating

- Hot-dip tin
- Hot-dip zinc
- Hot-dip aluminum
- Hot-dip lead

Galvanic/Electromotive Series. The electromotive series for common engineering materials is given in Fig. 10.1.

Electrochemical Equivalents for Various Metals. The electrochemical equivalents and other related data for the various plating metals and alloys are shown in Fig. 10.2. If you study the data for chromium, you will see

Engineering Material	Potential
Magnesium & magnesium alloys	-2.37v
Aluminum (1100 series)	-1.66
Zinc	-0.76
Chromium	-0.74
Type 304 stainless steel (active)	------
Type 316 stainless steel (active)	------
Steel, iron, cast iron	-0.41
Aluminum (2024)	------
Cadmium	-0.40
Nickel	-0.23
Tin	-0.14
Lead	-0.13
Copper	+0.34
Silver solder	------
Nickel (passive)	------
Chromium-iron (passive)	------
Type 304 & 316 Stainless steel (passive)	------
Silver	+0.80
Titanium	------
Platinum	+1.20
Gold	+1.50

FIGURE 10.1 Galvanic/electromotive series for common engineering materials.

Metal	mg/coulomb	gm/amp hr	oz/amp hr	Specific Gravity	oz/sq ft for 0.001 in	amp hr to deposit 0.001"/sq ft ♦	Symbol
Cadmium	0.58	2.10	0.074	8.64	0.72	9.73	Cd
Chromium	0.09	0.32	0.011	7.10	0.59	51.8	Cr
Copper	0.33	1.19	0.042	8.92	0.74	17.7	Cu
Gold	0.68	2.45	0.087	19.3	1.61	18.6	Au
Iron	0.29	1.04	0.037	7.90	0.66	17.9	Fe
Lead	1.07	3.87	0.136	11.3	0.94	6.91	Pb
Nickel	0.30	1.10	0.039	8.90	0.74	19.0	Ni
Palladium	0.28	0.10	0.035	12.0	0.10	28.6	Pd
Platinum	0.51	1.82	0.065	21.4	1.78	27.6	Pt
Rhodium	0.27	0.10	0.034	12.5	1.04	30.8	Rh
Silver	1.12	4.02	0.142	10.5	0.88	6.20	Ag
Tin	0.31	1.11	0.039	7.30	0.61	15.6	Sn
Zinc	0.34	1.22	0.43	7.10	0.59	13.7	Zn

NOTE: ♦ Equals specific gravity x 0.08323

FIGURE 10.2 Electrochemical equivalents and related data, based on 100% current efficiency.

why this metal is a difficult metal to plate, since it takes much electrical energy for thin coatings. Cadmium, on the other hand, is an easy metal to plate, but it is a toxic element, as are its salts.

Electroplating practice. The table in Fig. 10.3 lists the common plating metals and alloys and their possible baths, uses, and plating requirements. Reference data describing the different plating characteristics will be shown in later sections. Many of the plating processes and procedures are proprietary, their procedures being closely guarded trade secrets. Platings such as bright chromium, black chromium, gold, nickel-chromium, and others use proprietary procedures which are not published in any book. Plating has been called a "black art," and you will understand why when you try to obtain information on the exact processes and procedures used in some of the industrial plating practices. Plating procedures are difficult and must be closely controlled to exacting standards if the finished article is to be of high quality.

Electroplating Data and Specifications

Electroplating and Oxide Layer Thickness Ranges. Many plating thickness specifications call for plating thicknesses in micrometers (μm), but others may be specified in mils (0.001 in = 1 mil)

$$1\ \mu\text{m} = 0.00004\ \text{in} \quad \text{or} \quad \frac{4}{100}\ \text{mils (0.04 mil)}$$

$$25.4\ \mu\text{m} = 0.001\ \text{in} = 1\ \text{mil}$$

$$5\ \mu m = 0.0002\ in = 0.2\ mil = \frac{2}{10}\ mil$$

Both measurement units are given in the following listings, which summarize in outline form thickness specifications for various platings and coatings on a number of base-metal types.

Anodic Coating Thickness (Anodized Parts)

Aluminum

ASTM B 580-79

Type A hard coat	2.0 mil (50 μm)
B architectural	0.7 mil (18 μm)
D auto exterior	0.3 mil (8 μm)
E interior, a	0.2 mil (5 μm)
F interior, b	0.1 mil (3 μm)
G chromic acid	0.04 mil (1 μm)

Mil-A-8625C

Chromic acid	0.05–0.3 mil (1.3–8 μm)
Sulfuric acid	0.1–1.0 mil (3–30 μm)
Hard coat	0.5–4.5 mil (13–114 μm)

Magnesium

AMS 2478B

Acid, full coat	0.9–1.6 mil (23–41 μm)

Mil-M-45202C

Light HAE	0.1–0.3 mil (3–8 μm)
Heavy HAE	1.3–1.7 mil (33–43 μm)

Plating Metal	Typical Uses	Plating Solution Type	Plating Bath Temp. °F	Current Density Amps/ft^2	Volts DC	Throwing Power	Time to Deposit 0.001"
Cadmium	Protection	Cyanide	70-95	15-45	1-4	Good	20 min.
Chromium	Decorative, Engineering (Hard), Cylinder liners	Chromic acid	120	200-250	6-8	Poor	2 hours
Copper	Electroforming, Undercoat Stop-off for casehardening	Acid Cyanide Rochelle	75-120 75-100 140-160	15-40 5-15 20-60	1-2 1.5-3 2-3	Fair Good Good	35 min 90 min 45 min
Gold	Decorative, Electronics, PC boards	Cyanide Proprietary	120-160	5-15	2-6	Good	....
Indium	Bearing surfaces	Cyanide Sulfate Fluoborate	70-75 70-75 70-90	10-150 20 50-100	....	Good Poor Good	....
Iron	Electroforming, Repair	Chloride Sulfate	190 70-75	60 20	....	....	20 min 1 hour
Lead	Protection Bearing surfaces	Fluoborate	70-75	10-80	0.5	Good	40 min
Nickel	Protection, Decorative, Base coat for chromium	Sulfate-chloride Sulfamate Fluoborate	75-100	Varies	0.5-3	Fair	30 min
Rhodium	Decorative, Optical	Sulfate Phosphate	110-120	10-80	2.5-5	......	......
Silver	Protective, Decorative, Electrical contacts	Cyanide	80	5-15	1	Good	......
Tin	Protective,	Sulfate	70-75	40	1-3	Fair	15 min

	Food & dairy,	Fluoborate	75-100	50	......	Good	10 min
	Bearings,	Stannate	150-190	40	4-8	Excellent	30 min
Zinc	Protective	Sulfate	75-100	15-400	......	Fair	10 min
		Cyanide	100	10-50	......	Good	40 min
Alloys							
Brass	Rubber bonding, Decorative	Cyanide	75-100	3-10	2-3	Good	......
Bronze	Decorative, Base for chromium, Stop-off for steel	Cyanide-stannate	155	20-100	3-6	Excellent	30 min
Lead-tin	Bearings Solderability, Electrotyping	Fluoborate	70-75	60	1-2	Good	......
Tin-zinc	Solderability	Cyanide-stannate	150	10-75	4-5	Excellent	30 min
Tin-nickel	Printed circuits	Chloride-fluoride	150	25	1-2	Excellent	30 min

FIGURE 10.3 Electroplating practices table, average operating conditions.

Electroplating Thicknesses and Specifications

Cadmium plate

Range: 0.2 to 1.0 mil (5–25 μm)

Specification reference: ASTM A 165-80

Note: Cadmium metal is toxic, as are its salts. Use may be restricted, due to OSHA regulations.

Chromium Plate

Range: 0.01 to 6.0 mil (0.3–150 μm)

Decorative: 0.01 mil (0.3 μm)

Engineering: 2 mil (50 μm)

Specify hard chrome on the drawing for corrosion resistance.

Heavy chrome plating (4–6 mil) range will have porosity.

Specification reference: Fed QQ-C-320B

Copper Plate

Range: 0.1 to 5 mil (3–125 μm)

Specification reference: Mil-C-14550A

Used as a base for other platings.

Gold Plate

Range: 0.02 to 1.5 mil (0.5–38 μm) in eight thickness classes

Specification reference: Mil-G-45204B

Nickel Plate

Range: 0.2 to 7.9 mil (5–200 μm) engineering coatings

Specification reference: ASTM B 689-81 (may be soft or hard deposited within other specifications)

Nickel with Chromium Plate on Steel

Range: 0.4 to 1.6 mil nickel; 0.01 mil chromium

Kinds of coatings include bright, dull, and layered.

Specification reference: ASTM B 456-79

Also covers nickel with chromium on copper and copper alloys.

Rhodium Plate

Range: 0.008 to 0.25 mil (0.2–6.4 μm) engineering coatings

Specification reference: ASTM B 634-78

Palladium Plate

Range: 0.05 to 0.2 mil (1.3–5 μm)

Specification reference: ASTM B 679-80

Silver Plate

Range: 0.04 to 1.6 mil (1–40 μm)

Types include mat finish, bright finish, tarnish resistant (chromate-treated)

Specification reference: ASTM B 700-81

Note: Silver plating of electric current–carrying parts such as switch blades, breaker contacts, and other sliding parts is generally given as 0.2 mil (5 μm), except where industry specifications may call for 3 mil (76 μm). The 3-mil coating is more effective on surfaces subject to heavy friction, such as sliding electrical contacts.

Tin Plate

Range: 0.2 to 1.2 mil (5–30 μm)

Specification reference: ASTM B 545-72

Note: On electrical switch gear and other industrial equipment, 3-mil thickness is usually specified for service at paper mills and other environments where hydrogen sulfide in the atmosphere corrodes silver and other platings.

Zinc Plating on Steel

Range: 0.2 to 1 mil (5–25 μm) service conditions 1 through 4

The standard general thickness on most parts is 0.2 mil (0.0002 in). For severe service, the plating thickness should be 13 to 15 μm or 0.00057 in, or 6⁄10 mil (0.0006 in). Hot-dip coatings of 1 mil (0.001 in) are given to parts that are subject to extremely severe service.

Specification reference: ASTM B 633-78

Note: Bright zinc plating on steel parts (usually sheet metal and machined parts) is often given a yellow chromate conversion coating after the zinc plate. This process imparts a gold, iridescent appearance. Many commercial parts are given this treatment for corrosion resistance in lieu of painting. Conversion coatings are also available for the following metals: silver, copper and its alloys, tin, aluminum, magnesium, zinc-base die-casting alloys, and electroplated chromium.

During the conversion process, a complex chromium metal gel forms on the part surface, which contains hexavalent and trivalent chromium. When the soft gel coating dries, it becomes hard and somewhat abrasion resistant. This then forms what is the preferred plating and coating for many engineering applications.

10.2 PLATING METALS: CHARACTERISTICS AND PROPERTIES

Copper

Melting point: 1083°C (1981°F)

Specific gravity: 8.96 at 20°C

Reddish colored, malleable, and ductile metal. Excellent conductor of heat and electric current. Good corrosion resistance. Used as a plating base on irons and steels, and for other plating metals such as chromium, tin, and nickel.

Cadmium

Melting point: 320.9°C (609.6°F)

Specific gravity: 8.65 at 20°C

Soft, bluish white metal, similar to zinc in appearance. Cadmium is toxic, as are its compounds. Being replaced by zinc in many plating applications due to its toxic nature. Used to plate mechanical fasteners, etc., but should not be specified unless required by special application.

Chromium

Melting point: 1890°C (3434°F)

Specific gravity: 7.19 at 20°C

Steel-gray, lustrous, and very hard metal. On a scale of 1 to 10 it would be 9, with only diamond at 10. Hexavalent chromium compounds are toxic. Excellent corrosion resistance for many applications. Attacked by hydrochloric acid, in which it dissolves. Heavily plated deposits become porous on the outer layers. Used to plate hand tools, surgical instruments, decorative trim, cutting tools and drills, and wear-resistant surfaces.

Nickel

Melting point: 1453°C (2647°F)

Specific gravity: 8.90 at 20°C

Nickel is hard, malleable, ductile, slightly ferromagnetic, and a fair conductor of heat and electric current. Silvery-white–colored metal that will take a high polish. Excellent corrosion resistance. Used to plate handguns, hand tools, decorative trim, and corrosion-resistant containers.

Gold

Melting point: 1063°C (1945°F)

Specific gravity: 19.32 at 20°C

Yellow, soft metal which is extremely malleable and ductile. Excellent conductor of heat and electric current. Excellent corrosion resistance. The metal alloys are measured in karats, 24 karat being pure gold. 18 karat = $^{18}/_{24}$ = 0.75 or 75 percent pure gold; 14 karat = $^{14}/_{24}$ = 0.583 or 58.3 percent pure gold, etc. Used to plate electrical contacts, printed circuits, watches, and other jewelry.

Silver

Melting point: 960.8°C (1761.4°F)

Specific gravity: 10.50 at 20°C

Brilliant white, lustrous metal. Harder than gold but still very malleable and ductile. Highest thermal and electrical conductivity of all metals. Stable in normal air, but tarnishes when exposed to ozone, hydrogen sulfide, and air containing sulfur. Used to plate electrical contacts, conducting surfaces such as bus bars, eating utensils, and jewelry.

Tin

Melting point: 231.9°C (449.4°F)

Specific gravity: (white tin) 7.31, (gray tin) 5.75 at 20°C

Silvery-white metal which is malleable and slightly ductile. Resists seawater and tap water but is attacked by strong acids, alkalies, and acid salts. Used to plate electrical contacts, bus bars, and corrosion-resistant containers and tubing.

Indium

Melting point: 156.6°C (313.9°F)

Specific gravity: 7.31 at 20°C

Very soft, silvery-white metal. May be alloyed to produce very low melting point alloys. The metal is toxic.

Rhodium

Melting point: 1966°C (3570°F)

Specific gravity: 12.41 at 20°C

Silvery-white metal with low electrical resistance. Highly resistant to corrosion. Plated rhodium is extremely hard. The base metal is also ductile. Used to plate electrical contacts.

Palladium

Melting point: 1552°C (2825.6°F)

Specific gravity: 12.02 at 20°C

Steel-white metal which is soft and ductile when annealed. Very hard when cold worked. Does not tarnish in air. Uses include engineering plating and plating of electrical contacts.

Zinc

Melting point: 419.4°C (786.9°F)

Specific gravity: 7.13 at 20°C

Bluish white lustrous, brittle metal. Fair conductor of electricity. Will burn in air at a red heat, producing clouds of white zinc oxide. Good corrosion resistance and used widely to plate irons and steels.

Summary of plating and finishing

Figure 10.4 lists most of the presently available and widely used plating processes and protective finishes, excluding paints and plastic coatings. These finishes have a wide range of uses in aerospace, commercial aircraft, automobiles, consumer products, and other industrial applications. The name of the finish, its applications, and its final appearance are all listed in the figure. For further information concerning the latest military specifications and American standard specifications for these finishes, consult the plating companies or finishers.

The effective corrosion resistance of selective platings and finishes is shown in Fig. 10.5. The thicknesses or ranges are shown in inches. As can be seen from the data in Fig. 10.5, hard chromium is one of the best engineering finishes for metal products, although it is expensive compared with the other common platings.

10.3 ANODIZING

Anodizing is a process of oxidation produced in an electrolytic bath. Aluminum and its alloys are most commonly anodized, although the anodizing process may be performed on magnesium, zinc, and titanium. This section will cover the anodize coatings and processes for aluminum and its alloys only. The anodic coating, being aluminum oxide, is extremely hard and abrasive, especially aluminum *hardcoat* anodized surfaces. The classifications for aluminum and aluminum-alloy anodic coatings are as follows:

Finish	Application	Appearance
Brass	Used on high strength fasteners where a decorative finish is required	Gold iridescent
Cadmium	General purpose plating. High degree of corrosion resistance especially in salt atmospheres. Soft finish not for threaded applications requiring frequent tightening cycles. Cadmium is toxic. Zinc recommended as replacement.	Silver-gray
Cadmium and clear chromate	Suited for applications for hand contact. Corrosion resistant	Clear
Cadmium and black chromate	Recommended where black color is desired. Note: chromate can be applied in red, green, blue or other color as suits the application.	Black or color
Cadmium and bronze chromate	Normally used in military applications for identification of plating type.	Gold iridescent
Cadmium and yellow iridescent chromate	Similar to bronze	Yellow iridescent
Cadmium and olive drab chromate	Good for paint bonding applications.	Dull olive
Cadmium and phosphate	Similar to cadmium and olive drab chromate	Black
Copper	Used on Allen threaded products. Also for high heat applications.	Copper-matte
Chromium, decorative	Decorative and corrosion applications. Plating is usually very thin.	Bright or satin
Chromium, hard	A most durable plating which is extremely hard. Expensive. Engineering aplications. Cutting tools and drills.	Satin
Lead	For soldering applications.	Matte
Nickel	Decorative appearance, excellent wear and corrosion resistance.	Matte or bright
Silver	Electrical properties, utensiles, appearance.	Matte or bright

Tin	Anti-seize properties, corrosion resistance, non-toxic.	Grayish-white
Zinc	Popular plating for many applications; hardware, steel sheets, machined parts.	Bluish-white
Zinc and clear chromate	Similar to cadmium and clear chromate. Corrosion resistance, economical.	Clear
Zinc and black chromate	Similar to cadmium and black chromate. Where black color is desired.	Black
Zinc and bronze chromate	Similar to cadmium and bronze chromate. Color applications.	Gold iridescent
Zinc and yellow iridescent chromate	Similar to cadmium and yellow iridescent chromate. Color applications. Popular engineering finish. Economical and attractive for industrial applications.	
Zinc and olive drab chromate	Similar to cadmium and olive drab chromate. Good paint adhesion properties.	Dull olive
Zinc and phosphate	Similar to cadmium and phosphate. Black for color applications.	Black
Anodic	Finishes for aluminum, magnesium, and titanium. Many types and thicknesses available.	Many colors
Black oxide	For threaded fasteners such as set-screws. Only mildly corrosion and rust resistant. Normally applied to irons and steels. Often oiled.	Black
Iron phosphate	For hardware which is severely handled. Decorative gray finish.	Oily gray-black (oil finish)
Manganese phosphate	For thread lubrication applications. For frictional contact applications. Supplied oiled or waxed for additional protection	Gray-black (oil finish)
Black nitrate	For stainless steel fasteners which require uniform black finish.	Dull or luster black
Passivation	For stainless steel fasteners and parts to prevent corrosion. Acid treatment.	Bright or matte
Zinc phosphate	For exposed fasteners requiring good paint adhesion properties. Dry or lubricated.	Dull gray (oil coating)
Teflon coating	Corrosion resistant and high lubricity. For non-galling anti-seizing applications. Used for bearings. New processes incorporate anodizing with teflon impregnation.	Gray to black

FIGURE 10.4 Summary of plating and finishing types and applications.

Plating or Surface Treatment	Specification	Type	Class	Minimum Thickness	Salt Spray Test (minimum hours)	
					White Corrosion	Rust
Anodize	Mil-C-5541A	.....	.....	.....	168	.....
	Mil-A-8625B	I, II	.....	.....	240	.....
Cadmium and Chromate	AMS-2400L1	.....	.....	0.0002-0.0003	100	.....
	AMS-2400L2	.....	.....	0.0002-0.0004	100	.....
	AMS-2400L3	.....	.....	0.0003-0.0005	150	.....
Cadmium and Chromate	QQ-P-416a	II	1	0.0005	96	.....
	QQ-P-416a	II	2	0.0003	96	.....
	QQ-P-416a	II	3	0.0002	96	.....
	Mil-C-8837	II	1	0.0005	96	.....
	Mil-C-8837	II	2	0.0003	96	.....
	Mil-C-8837	II	3	0.0002	96	.....
Chromium, hard	Mil-C-11436	.....	.....	0.0012	.....	100
Lead	Mil-L-13808	I, II	3	0.00025	.....	24
	Mil-L-13808	I, II	2	0.0005	.....	48
Manganese Phosphate, dry	Mil-P-16232C	M	3	0.0002	.....	1.5
Manganese Phosphate, oil	Mil-P-16232C	M	2	0.0002	.....	24
Tin	Mil-T-10727A	I	.....	0.0002	.....	24
Zinc	Mil-Z-325a	I	2	0.0005	.....	96
	Mil-Z-325a	I	3	0.0002	.....	36
Zinc Phosphate, dry	Mil-P-16232C	Z	3	0.0002	.....	2
	Mil-P-16232C	Z	2	0.0002	.....	48
Zinc and Chromate	QQ-Z-325a	II	2	0.0005	96	.....
	AMS-2402E	.....	.....	0.0002-0.0003	100	.....
	QQ-Z-325a	II	3	0.0002	96	.....
	AMS-2402E	.....	.....	0.0003-0.0005	150	.....
Zinc and Phosphate	QQ-Z-325a	III	2	0.0005	.....	96
	QQ-Z-325a	III	3	0.0002	.....	36

FIGURE 10.5 Corrosion resistance of platings; minimum thicknesses are given in inches.

Classifications

Type I: Chromic acid anodizing, conventional coatings produced from chromic acid baths. Shall not be applied to alloys containing over 5 percent copper, over 7 percent silicon, or when alloying elements exceed 7.5 percent.

Type IB: Chromic acid anodizing, low-voltage process, 20 V. Heat-treatable alloys, such as T4, T6, etc., should be tempered prior to anodizing.

Type II: Sulfuric acid anodizing, conventional coatings produced from sulfuric acid baths. Heat-treatable alloys, such as T4, T6, etc., should be tempered prior to anodizing.

Type III: Hard anodic coatings (hardcoat). Shall not be applied to alloys containing over 5 percent copper or over 8 percent silicon unless agreed on by the supplier. Heat-treatable alloys, such as T4, T6, etc., should be tempered.

Classes

Class 1: Nondyed, natural, including dichromate sealing

Class 2: Dyed

Standard Specifications for Anodized Aluminum and Aluminum Alloys

ASTM B 244, Thickness of anodic coatings, measurement of

ANSI/ASTM B 137, Weight of coatings on anodized aluminum, measurement of

ASTM B 117, Method of salt spray (fog) testing

Sealing Anodized Aluminum and Aluminum Alloys. All types of anodizing must be sealed using any of the following methods after the electrolytic anodized coating is applied:

- Immersion in an aqueous solution of 5% sodium dichromate (15 min at 90 to 100°C)
- Immersion in deionized water (15 min at 100°C)
- Immersion in an aqueous solution of nickel or cobalt acetate (100°C for 15 min)
- Teflon impregnation processes (for sealing and lubricity)

Anodic Coating Design Data

Radii of curvature on anodized parts

Nominal coating thickness, in	Radius of curvature (outside/inside), in
0.001	approx. 0.032
0.002	approx. 0.062
0.003	approx. 0.093
0.004	approx. 0.125

Thickness Ranges of Anodic Coatings on Aluminum and Aluminum Alloys

Coating type	Thickness range, in
I and IB	0.00002 to 0.0003
II	0.00007 to 0.0010
III	0.0005 to 0.0045

Minimum Thickness (Typical) of Anodic Coatings on Aluminum and Alloys. See Fig. 10.6 for minimum typical anodic coating thicknesses on various aluminum alloys per type.

Alloy Designation	Thickness of Coating, inches	
	Type I and IB	Type II
1100	0.000029	0.000093
2024-T4	......	0.000125
2024-T6	0.000044	......
3003	0.000035	0.000103
5052	0.000033	0.000098
5056	0.000021	......
6061-T6	0.000034	0.000099
7075-T6	0.000040	......
Alclad 2014-T6	0.000045	......
Alclad 7075-T6	0.000041	......
295-T6	......	0.000107
356-T6	......	0.000102
514	......	0.000086

Note: Anodic coating types I, IB and II are normally applied as thin coats, while type III (Hardcoat) is normally applied as the thicker coatings. See table of anodic thicknesses following.

FIGURE 10.6 Minimum anodic coating thicknesses for various aluminum alloys.

Maximum Thicknesses of Anodic Coatings on Various Aluminum Alloys. See Fig. 10.7 for maximum attainable thicknesses of anodic coatings on selected aluminum alloys.

Alloy Designation	Anodic Coating Thickness - inches	Color
380	0.0006	Gray
360	0.001	Gray
319	0.0014	Gray
1100	0.0017	Bronze
2011	0.0021	Light gray
2014	0.0025	Gray
2019	0.0028	Bronze
6262	0.0031	Black
6061	0.0035	Black
2024	0.0040	Bronze
2017	0.0045	Bronze
6063	0.0050	Black
5052	0.0053	Black
2618	0.0055	Gray
2219	0.0058	Gray
218	0.0062	Gray
7079	0.0065	Dark gray
355	0.0075	Gray
7075	0.0082	Bronze
Almag 35	0.0100	Gray
356	0.0120	Light gray

Note: All above coatings may be dyed black. Coatings over 0.003 inches thick tend to chip and become milky in color and should be used only in the salvage of parts.
Source: Anodic, Inc., Stevenson, CT 06491.

FIGURE 10.7 Maximum anodic coating thicknesses on various aluminum alloys.

Design Notes for Anodized Parts

1. A 2-mil hardcoat anodic coating (0.002 in) will penetrate the part 0.001 in and protrude from the part 0.001 in. A 1.000-in-diameter part which is anodized 2 mils (0.002 in) will have a finished diameter of 1.002 in. Half the coating is inside the part and half is on the outside.
2. Avoid blind holes in parts.
3. Avoid hollow weldments (drill 0.250-in-diameter weep holes in the part).
4. Avoid steel inserts.
5. Avoid sharp corners (see radii chart preceding).
6. Avoid heavy to thin cross sections on the part.
7. Allow for the anodic coating in your design tolerances on the part.

Anodic Coating Specifications

Name	Hardcoat anodize	Chromic anodize	Sulfuric anodize
Army/Navy	Mil-A-8625, Ty-3	Mil-A-8625, Ty-1	Mil-A-8625, Ty-2
G.E.	AMS-2468D	AMS-2470H	AMS-2471D
Boeing	Code-302	Code-300	Code-301
IBM	41-207	41-204	41-203
Grumman	G-9031	9030B	G-9032

Hardcoat Anodize Processes

- Martin
- Alumilite
- Alpha

- Mae
- Sanford
- Boeing
- Scionic
- Hardas
- Imperv-X

Hardcoat/Teflon Processes

- Amphodize
- Hardtef
- Analon
- Tufram
- Polylube
- Lukon
- Nituf
- Kalon
- Ptfe
- Smoothcoat
- Sanfordize
- Hardlube

Hardcoat Anodize Physical Data

Hardness	65 to 70 Rockwell C scale (harder than hard chrome)

Color	Dark gray to black
Dielectric strength	800 V/mil
Machining	Grinding, lapping, polishing, and honing
Sealing	Dichromate, nickel, or cobalt acetate, hot water or Teflon impregnation
Resistivity	10^8 to 10^{12} Ω·cm

Note: Parts which have been anodized cannot be reanodized, nor can the anodic coating be made thicker.

10.4 COLORING PROCESSES FOR METALS, ALLOYS, AND PHOSPHATING STEELS

This section lists some of the coloring processes used to color metals and alloys. Coloring of metals may be for appearance as well as for additional protection against corrosion and rusting. These processes have evolved over a span of many years and are widely used in the metal finishing industries. Most of the processes involve the use of chemical action together with the application of heat.

It should be noted that the finished appearance of the metal being colored is affected to a great extent by how well the bare metal itself is finished and polished. For example, in the blueing process for handguns and rifles, the base metal is finished very carefully and is highly polished prior to blueing. If the metal parts are not finished and highly polished, the blueing process will not hide the defects in the base metal.

Coloring Metals (Ferrous and Nonferrous)

Black Oxide. Irons and steels may be given a black finish or black oxide finish by either of two processes:

1. Heat the ferrous metal part to approximately 500 to 700°F and plunge it into a container of good-quality machine oil. Reheat and plunge the part into the oil a number of times until the desired depth of black is obtained. Use precautions on hardened and tempered parts.
2. A thin black oxide coating also may be applied to iron and steel parts by immersing the parts in a boiling solution of sodium hydroxide and mixtures of nitrates and nitrites (sodium nitrate, potassium nitrate, etc.).

Blueing Irons and Steels. Many ferrous metal parts receive a blueing operation to enhance their appearance and also as a rust-preventive coating. Handguns, rifles, and shotguns normally receive a blueing operation unless chromium or satin nickel are specified. The following blueing process imparts a fine blue to blue-black finish: Clean the metal with a potassium bichromate–sulfuric acid solution, and then wash with ammonium hydroxide solution and wipe dry with a clean, lint-free cloth. Apply ammonium polysulfide until the desired depth of blueing is obtained. The finish may be made nearly black by repeating the process of applying the ammonium polysulfide. A light machine oil or silicone cloth wipe will then additionally provide the surface with excellent corrosion and rust resistance.

Phosphating Irons and Steels. Three types of phosphate coatings (conversion coatings) are given to ferrous metals: zinc phosphate, iron phosphate, and manganese phosphate. Zinc phosphate coatings vary in color from light to dark gray; iron phosphate, dark gray; and manganese phosphate, dark gray, becoming black with service. The phosphate coatings are used for paint bases, for an aid in cold working the metal, and for rust prevention.

The phosphatizing solution contains approximately 3 to 5 percent phosphoric acid by volume. The part to be phosphatized is first thoroughly chemically cleaned and dipped into a solution of 2 percent by volume hydrochloric acid for approximately 15 s and then water rinsed and dried. The part is then immersed in the phosphatizing bath for the time required to impart the desired coating. Zinc plates and zinc parts are also phosphatized in the same manner. Stainless steels and certain alloy steels cannot be phosphatized.

Passivation of Copper and Copper Alloys. A passivation coating may be applied to copper and its alloys, which imparts a blue-green color or patina. This coating is corrosion-resistant. The passivation solution may be made by using the following solution composition:

1. 6 lb ammonium sulfate
2. 3 oz copper sulfate
3. 1.4 fluid oz. technical ammonia (sp. gr. = 0.09)
4. 6.5 gal water

Apply the solution with a spray to the chemically cleaned copper or copper-alloy parts in six applications, drying between applications. The patina appears after approximately 6 h and continues as the part weathers.

Copper and Copper-Alloy Blackening (Alloys with More than 85 Percent Copper). To color copper and its alloys black, use the following procedure: Make a solution consisting of 4 oz arsenious oxide, 8 fluid oz hydrochloric acid (sp. gr. = 1.16), and 1 gal water. Then

1. Heat the solution to 175 to 200°F.
2. Immerse the copper or copper-alloy parts until a uniform black color is obtained.
3. Brush the parts while wet, dry the parts, and apply a protective clear finish such as polyurethane varnish.

Coloring Brasses. Brass may be given a green color with the following procedure: Mix a solution containing 1 oz ferric nitrate, 6 oz sodium thiosulfate, and 1 gal water. Then

1. Heat the solution to 160 to 180°F.
2. Immerse the brass parts in the solution until the desired color is obtained.
3. Dry and coat with a clear protective finish such as clear lacquer or polyurethane varnish.

10.5 PHOSPHATING, AND CLEANING SOLUTIONS AND FORMULAS

The following solutions and formulas for cleaning and finishing metals are for use by metal fabricating facilities for mechanisms and small, moving detail parts generated mainly for use in mechanisms and other ferrous metal products. These solutions, formulas, and finishing processes should be used on all ferrous metal parts where a zinc, zinc and yellow chromate, or other electroplated finish is neither desired or required. Zinc-plated and chromated parts and other parts that are electroplated usually increase in size, due to the thickness of the electroplated deposit. A study should be made, and the recommended parts that do not require a zinc, zinc and yellow chromate, or other electroplated finish should be phosphated in lieu of electroplating.

Phosphated metals (most ferrous metals and ferrous alloys) do not gain a significant plated thickness to interfere with mating parts on close assembly work. The simple phosphate coatings (that use only phosphoric acid and water as a 2 to 4 percent solution) are normally only 1.5 to 2.0 microinches thick. Many industrial parts and products may receive a metallic phosphate finish in lieu of a zinc and yellow chromate or other electroplated finish, which may be expensive and require the services of outside vendors (platers). Since the phosphated finishes are not electrodeposited metals and use nontoxic chemicals, most metal fabricators could possibly apply these coatings and finishes "in-house." The resultant cost savings could amount to thousands of dollars a year, and at the same time give the fabricator complete control over the scheduling and manufacturing of the small parts.

The chemicals, metals, residues, and solutions used in phosphating processes are relatively nontoxic and may easily be neutralized for disposal into sewage systems, or carried to disposal stations. Because they involve coatings and not electrodeposited processes (some of which are highly toxic and can cause environmental problems), the phosphating processes may not require a plating license.

Pickling Formulas

Pickling solutions are used to chemically remove surface scale and oxides from metals and alloys. Typical formulas include:

- One part sulfuric acid to 5 or 10 parts water, by volume. Use for cleaning steel and other ferrous metals.
- Sulfuric acid and nitric acid; 1 part of each acid with 5 to 10 parts water, by volume. Use for bright-cleaning brass alloys.
- Sulfuric acid and anhydrous ferric sulfate; 1 part acid to 5 or 10 parts water, by volume, plus small amount of ferric sulfate (1/10th part or less).
- Chromium phosphate and triethanolamine. Use for preparing aluminum for painting.
- Phosphoric acid (75 percent); 1 part acid to 25–35 parts water, by volume. Use for cleaning and phosphating steels in preparation for painting.

Note: Pickling baths must be maintained at a temperature of 145 to 185°F. Keep the pickling times short, because hydrogen embrittlement can occur if the metal remains in the pickling bath for an excessive amount of time. Pickling-bath times vary from 3 to 15 minutes, with the shorter times preferred. As a precaution to avoid hydrogen embrittlement, the pickled parts may be oven heated at 300°F for 30 minutes.

Phosphate Coatings (Ferrous Metals and Alloys)

The four major types of phosphate coatings for ferrous metals and alloys include:

1. Zinc phosphate
2. Iron phosphate
3. Manganese phosphate
4. Lead phosphate (a relatively new process)

Formulas for phosphating solutions and directions for their proper use are given in the following expositions and procedures.

Zinc Phosphating. Zinc phosphate coatings are used on many military and commercial articles, including small arms, knives, heavy weapons and field artillery, tools, general mechanism parts, and a host of other fabricated metal applications. Zinc phosphating is also called "Granodizing."

To prepare 16 fluid oz (480 ml) of zinc phosphate concentrate, obtain the following ingredients:

- 33 gm powdered zinc
- 436 ml phosphoric acid (75 to 85 percent solution)
- 110 ml distilled water
- 10 to 12 gm sodium nitrite (accelerator). (The amount of sodium nitrite may need to be adjusted to suit the application.) Potassium nitrite may be used in lieu of sodium nitrite.

For a smaller base quantity of concentrate, use the following measures:

- 16.5 gm powdered zinc
- 218 ml phosphoric acid (75 to 85 percent solution)
- 55 ml distilled water

- 5 to 6 gm sodium nitrite (accelerator). (The amount of sodium nitrite may need to be adjusted to suit the application.) Potassium nitrite may be used in lieu of sodium nitrite.

Procedure:

Dissolve the powdered zinc in the phosphoric acid, then add the distilled water and sodium nitrite. Mix thoroughly and filter through a stainless steel filter screen or paper filter. (Heat and outgassing occur during the acid/zinc mixing and dissolution.)

The finished zinc phosphate concentrate may then be mixed in the following phosphating solutions:

Solution I (normal):	16 fluid oz concentrate to 6.75 gal of water; or 1 fluid oz of concentrate to 54 oz of water.
Solution II (heavy):	16 fluid oz concentrate to 5 gal of water; or 1 fluid oz of concentrate to 40 oz of water.

Processing temperature is 170 to 180°F (recommended temperature is 175°F). Distilled water will yield the best results, but "soft" tap water is sufficient.

Processing time is 3 to 10 minutes. Longer processing times (10 minutes or more) deposit darker-colored and denser coatings and may be necessary for hardened steel parts such as firearm receivers and bolts. The phosphating process is usually self-terminating; chemical action will stop when the part is sufficiently coated.

Immediately rinse the phosphated parts in cold running water for 30 seconds. *Do not* allow the phosphating solution to dry on the phosphated part; otherwise, difficult-to-remove white deposits of zinc salts will form on the parts.

The parts may then be dipped into a mild solution of chromic acid and water (1 oz chromic acid to 1 gal of water).

Rinse the parts again in cold running water.

Thoroughly dry the parts with compressed or blown hot air. Bake critical parts as outlined herein.

Coat the parts with a light oil, an oil-kerosene mix, or WD40 lubricant.

Remove or drain excess oil from the parts. The parts may now be assembled or stored for future use.

Note: As a safety precaution, hardened and highly stressed parts that have been phosphate-treated for 10 minutes or longer should be baked at 300°F for 30 minutes to prevent hydrogen embrittlement.

Iron Phosphating (Reference Only). Iron phosphating, also known as "Coslettizing," uses powdered iron and phosphoric acid in the solution. Accelerator chemicals such as sodium nitrite, potassium nitrate, or potassium nitrite are not normally required for iron phosphating processes.

Manganese Phosphating (Reference Only). Manganese phosphating, also called "Parkerizing," uses powdered iron, manganese phosphate, and phosphoric acid in the solution.

Bonderizing. Bonderizing is performed on ferrous metals and alloys in a solution consisting of phosphoric acid, a catalyst, and water. This process is used to impart a rough, tough, frosted surface providing excellent paint adhesion. The phosphated coating is rust-resistant.

Note: Phosphate coatings are on the general order of 0.0002 in (²⁄₁₀ mil) thick. All phosphating solutions are used or processed at temperatures of 145 to 200°F. Immediate washing in cold water is recommended after the phosphate coating is applied, to prevent white streaks or blemishes from appearing on the surface of the metal or alloy.

Phosphate coatings are most applicable for low-, medium-, and high carbon steels in both the annealed and hardened conditions. Ferrous metal parts that have been highly hardened and some of the higher-alloy steels are more difficult to phosphate-coat. On these steels the phosphate coating will be light gray in color, unless the part is immersion-coated for longer periods (10 minutes or longer). Stainless steels and other high-nickel-chromium steels cannot be phosphated.

The more iron a ferrous material contains, the thicker and darker will be the phosphate coating. Heavy zinc phosphate coatings are the easiest to apply, with some of the processes requiring only 3 to 5 minutes to complete. Ferrous materials that are case-hardened cannot normally be phosphated. If high-alloy ferrous parts need a heavier or darker coating, they may be processed in phosphating solutions up to double the time normally required, i.e., up to 10 to 15 minutes for the solutions used in the zinc phosphating process detailed in this disclosure.

10.6 ETCHING METALS

The solutions used to etch various metals are often called *mordants,* and some of the solutions used in the etching processes are as follows.

Etching Irons, Steels, and Zinc Plate

Nitric Acid Solution. A popular solution for etching irons, steels, and zinc plates consists of a solution of 2 oz 50% nitric acid mixed into 15 oz water. This is a 1/16 solution. Always pour the acid into the water; never pour water into the acid. Pouring water into acid may cause a mixing reaction, which could cause the acid to splash from the mixing container. The solution is relatively slow acting unless used with a splash-type etching machine. For hand etching, a stronger solution can be used, but the etching cut becomes rough at the edges of the etching action. Solutions as strong as 1/8 can be used: 1 part concentrated nitric acid to 7 parts water by volume.

Etching Copper and Copper Alloys

Ferric chloride is often used to etch copper and its alloys. A 40°Bé solution of ferric chloride used at 75 to 80°F etches copper cleanly and not too rapidly so as to produce a rough etched edge. A 40°Bé solution is made by mixing 20 oz ferric chloride (anhydrous) with water to make a final volume of 1 liter (1000 ml). The specific gravity of this solution can be from 1.37 to 1.38. The designation Bé (Baumé) is pronounced "bow-may" and is defined with respect to specific gravity by the following equation:

$$\text{Specific gravity} = \frac{145}{145 - {}^\circ\text{Bé}}$$

With this equation, you may determine the degrees Baumé if you know what specific gravity you want or have or you may determine the specific gravity required of the solution if you know the degrees Baumé you wish to produce.

Note: The reaction of ferric chloride produces a great amount of heat when it is dissolved in water. The water temperature should be between 50 and 75°F prior to mixing the ferric chloride solution.

Solutions of ferric chloride as low as 30°Bé are used for fast etching of copper and its alloys, while the 40 to 42°Bé solutions are used for etching intaglio printing plates or photogravure work. These solutions also have been used to etch stainless steels. The ferric chloride bath should be contained in a glass, plastic, or wooden tank, since the solution is highly corrosive. Do not dispose of these solutions directly into standard drainage systems. The solutions should be diluted in water and neutralized with a base chemical such as sodium bicarbonate. Do not allow the solutions to come into contact with the skin; rubber gloves should be worn at all times during use of these solutions.

Printed circuit boards for electronic applications have been produced using ferric chloride solutions for many years. Using the lower degree Bé solutions produces a fast and accurate etching action, or cut, and the ferric chloride is economical and long lasting. Perchlorate chemicals are also used for etching printed circuit boards.

Frank Short's Etching Solution. The Frank Short etching solution has been used by printing plate makers and other metalworkers for years and is a slow-acting but very accurate etchant (mordant). The solution may be made as follows:

- 88 parts by volume of distilled water
- 2 parts by volume potassium chloride
- 10 parts by volume of hydrochloric acid (concentrated)

This is a two-part mixture and must be made by mixing two solutions and then pouring the two solutions together.

Solution 1: Mix 1 oz potassium chloride into 10 oz water at 190 to 200°F

Solution 2: Mix 5 oz hydrochloric acid into 34 oz distilled water

Then pour solution 1 into solution 2 in a well-ventilated area. *Caution:* Chlorine gas is evolved when the two solutions are mixed. Allow the solution to set for 30 min to 1 h before using. Store in a glass bottle with a rubber or glass stopper.

The part to be etched is lowered into a shallow glass pan or tank so that the solution covers the part to be etched by approximately ½ in. The etching action may be observed directly, since the etching solution is clear and transparent.

Copper and its alloys, irons and steels, and chromium-plated parts may be etched with this solution. Use adequate ventilation because chlorine gas in small amounts is evolved during the etching action. This solution is often used to check the etching mask for holes prior to fast etching in ferric chloride solutions. A copper part will begin to show a frosting effect in a few seconds after immersion in the solution. Holes in the etching mask may then be seen and repaired prior to the final etching.

A final note: Titanium heaters are used in etching baths for ferric chloride solutions.

Etching aluminum and aluminum alloys

Most aluminum alloys can be etched using sodium hydroxide solutions of varying strengths. Sodium hydroxide is commonly called *caustic soda* or *lye* and is a low-cost chemical. Sodium hydroxide is very corrosive or caustic and is poisonous. When aluminums are etched using sodium hydroxide solutions, one of the reaction products is hydrogen gas, which is highly explosive when mixed with normal air. Proper venting must be employed when using this process for etching aluminum alloys.

The etching action is relatively rapid, especially if the sodium hydroxide solution is above 1/16 (1 part sodium hydroxide to 15 parts water). Strong solutions produce a violent foaming reaction and substantial amounts of hydrogen gas.

CHAPTER 11

SOCIETIES, ASSOCIATIONS, INSTITUTES, AND SPECIFICATION AUTHORITIES

Following is a listing of recognized specification authorities, societies, and institutes from which the design and manufacturing engineer or metalworker can obtain specifications and standards covering many areas of engineering design and manufacturing.

Many of the standards are revised periodically by these authorities to keep pace with changing technology. It is therefore suggested that copies of the American standards you require to perform your job function be obtained directly from the standards organizations that generate them. In this way you are assured of using the most recent standard and its revisions. Updates to the standards listings are distributed periodically to those on the organization mailing lists, or annual standards listing manuals may be ordered directly from the standards organizations.

Many of the standards organizations publish handbooks and manuals related to their particular fields that are technically excellent. Indexes to the standards are also available for pinpointing your area of interest in the form of annual standards catalogs.

11.1 STANDARDS ORGANIZATIONS AND THEIR ACRONYMS

AAR
Association of American Railroads
59 East Van Buren Street
Chicago, IL 60605

ACA
American Chain Association
1000 Vermont Avenue NW
Washington, DC 20005

AEMS
American Engineering Model Society
Box 2066
Aiken, SC 29801

AFBMA
Anti-Friction Bearing Manufacturers Association
2341 Jefferson Davis Highway, Suite 1015
Arlington, VA 22202

AGMA
American Gear Manufacturers Association
1901 North Fort Meyer Drive, Suite 1000
Arlington, VA 22209

AISC
American Institute of Steel Construction
400 North Michigan Avenue
Chicago, IL 60611

AISI
American Iron and Steel Institute
1000 16th Street NW
Washington, DC 20006

The Aluminum Association
818 Connecticut Avenue NW
Washington, DC 20006

AMS (Aeronautical Materials Specifications)
Society of Automotive Engineers, Inc.
400 Commonwealth Drive
Warrendale, PA 15096

ANSI
American National Standards Institute, Inc.
11 West 42nd Street
New York, NY 10036

API
American Petroleum Institute
300 Corrigan Tower Building
Dallas, TX 75201

ASLE
American Society of Lubrication Engineers
838 Busse Highway
Park Ridge, IL 60068

ASM
American Society for Metals
Metals Park, OH 44073

ASME
American Society of Mechanical Engineers
345 East 47th Street
New York, NY 10017

ASMMA
American Supply and Machinery Manufacturers Association
1230 Keith Building
Cleveland, OH 44115

ASTM
American Society for Testing and Materials
1916 Race Street
Philadelphia, PA 19103

AWS
American Welding Society
550 NW Lejeune Road
P.O. Box 351040
Miami, FL 33135

BSA
Bearing Specialists Association
221 North LaSalle Street, Suite 2026
Chicago, IL 60601

CDA
Copper Development Association, Inc.
405 Lexington Avenue
New York, NY 10017

EEI
Edison Electric Institute
750 Third Avenue
New York, NY 10017

Federal Specifications
U.S. Navy Supply Depot
5801 Tabor Avenue
Philadelphia, PA 19120

FEMA
Farm Equipment Manufacturers Association
230 South Bemiston, Suite 809
St. Louis, MO 63105

IEEE
Institute of Electrical and Electronics Engineers
345 East 47th Street
New York, NY 10017

IFI
Industrial Fasteners Institute
1505 East Ohio Building
1717 East 9th Street
Cleveland, OH 44114

IIE
Institute of Industrial Engineers
25 Technology Park
Atlanta, Norcross, GA 30092

Ingot Number
Brass and Bronze Ingot Institute
33 North LaSalle Street, Room 3500
Chicago, IL 60602

ISA
Instrument Society of America
P.O. Box 12277
Research Triangle Park, NC 27709

Military Specifications
U.S. Naval Supply Depot
5801 Tabor Avenue
Philadelphia, PA 19120

MPIF
Metal Powder Industries Federation
105 College Road East
Princeton, NJ 08540

MPTA
Mechanical Power Transmission Association
3451 West Church Street
Evanston, IL 60203

NASC
National Aerospace Standards Committee
1321 Fourteenth Street
Washington, DC 20005

NEMA
National Electrical Manufacturers Association
2101 L Street NW, Suite 300
Washington, DC 20037

NFPA (National Electrical Code)
National Fire Protection Association
Batterymarch Park
Quincy, MA 02269

NIBA
National Industrial Belting Association
1900 Arch Street
Philadelphia, PA 19103

NIST
National Institute of Standards and Technology
Gaithersburg, MD 20899

NLGI
National Lubricating Grease Institute
4635 Wyandotte Street
Kansas City, MO 64112

NSC
National Safety Council
444 North Michigan Avenue
Chicago, IL 60611

RIA
Robot Institute of America
P.O. Box 1366
Dearborn, MI 48121

RMA
Rubber Manufacturers Association
1901 Pennsylvania Avenue NW
Washington, DC 20006

SAE
Society of Automotive Engineers
400 Commonwealth Drive
Warrendale, PA 15096

SME
Society of Manufacturing Engineers
One SME Drive
P.O. Box 930
Dearborn, MI 48128

SMI
Spring Manufacturers Institute
380 West Palatine Road
Wheeling, IL 60090

TWI
The Welding Institute
P.O. Box 5268
Hilton Head Island, SC 29928

UL
Underwriters Laboratories, Inc.
333T Pfingsten Road
Northbrook, IL 60062

Standards and Approval Agencies and Acronyms

AGA	American Gas Association
AGMA	American Gear Manufacturers Association
AHAM	Association of Home Appliance Manufacturers
AMCA	Air Movement and Control Association
ANSI	American National Standards Institute
ARA	American Refrigeration Association
ARL	Applied Research Laboratories
ASHRAE	American Society of Heating, Refrigeration, and Air-Conditioning Engineers
ASME	American Society of Mechanical Engineers
ASSE	American Society of Sanitary Engineering
CEC	California Energy Commission
CGA	Canadian Gas Association

CSA	Canadian Standards Association
DIN	Deutschland Ingineering Normalization (German Engineering/Industrial Standard)
DOT	Department of Transportation
ETL	ETL Testing Laboratories
FSEC	Florida Solar Energy Center
FM	Factory Mutual
GAMA	Gas Appliance Manufacturers Association
HVI	Home Ventilating Institute
IAPMO	International Association of Plumbing and Mechanical Officials
NEC	National Electrical Code
NEMA	National Electrical Manufacturers Association
NFPA	National Fire Protection Association
NSF	National Sanitation Foundation
OPEI	Outdoor Power Equipment Institute
OSHA	Occupational Safety and Health Administration
SRCC	Solar Rating and Certification Corporation
UL	Underwriters' Laboratories
USDA	United States Department of Agriculture
AEC	Atomic Energy Commission
AN	Army-Navy Standard

ASA	American Standards Association (Now ANSI)
IEC	International Electrotechnical Commission
ISO	International Standards Organization
JAN	Joint Army-Navy Standard
MS	Military Standard
NASA	National Aeronautics and Space Administration
USAS	USA Standard (Now ANSI)

11.2 APPROVAL ASSOCIATIONS AND THEIR TRADEMARKS

See Figure 11.1.

11.3 QUALITY CONTROL SYSTEMS AND ISO 9000 SERIES

Various standards for controlling the quality of industrial and consumer products have been implemented both nationally and internationally. These standards are important to manufacturers, product design engineers, and engineering departments. In the United States, a system known as *total quality management* (TQM) has been in effect for some years. Other quality control systems have been implemented nationally and on a company-level basis. These systems help the consumer as well as the manufacturer.

Internationally, the quality control or quality assurance system of most importance to American manufacturers is the ISO 9000 system. On January 1, 1993, the European Community (EC), consisting of 12 countries, officially formed a unified market. This began the adoption of the ISO 9000 system. All United States companies wishing to sell products to the European market will use the ISO 9000 system as their quality standard.

ISO 9000 establishes a *minimum* quality standard that may be expanded on, based on future requirements. The U.S. Department of Defense may replace standards MIL-I-45208A and MIL-Q-9858A with the ISO 9000 system. All U.S. companies are today required to be ISO 9000 compliant and certified in order to continue to do business with the EC nations. Note that the American TQM system will exceed the stated ISO 9000 guidelines. Sixty countries have adopted ISO 9000 as their national quality standard.

The EC Consumer Council Ministers have adopted a general product safety directive. This directive imposes general safety requirements on *all* consumer products sold in the EC market. U.S. exporters to the EC must familiarize themselves with the EC product liability regulations. Use of the ISO 9000 standard may be mandatory in the near future in order for U.S. companies to sell their products to the EC and other nations that have adopted the ISO 9000 standard.

ISO 9000 system

The ISO 9000 system is divided into a series of standards, including

AGA	American Gas Association
AGMA	American Gear Manufacturer's Assn.
AHAM	Association of Home Appliance Manufacturers
AMCA	Air Movement and Control Association
ANSI	American National Standards Institute
ARI	Air-Conditioning & Refrigeration Institute
ARL	Applied Research Laboratories
ASHRAE	American Society of Heating, Refrigerating, and Air-conditioning Engineers
ASME	American Society of Mechanical Engineers
ASSE	American Society of Sanitary Engineering
CGA	Canadian Gas Association
CSA	Canadian Standards Association
CSA	Canadian Standards Association, National Recognized Testing Laboratory
CE	Conformité Européenne
DIN	Deutschland Ingineering Normalization
DOT	Department of Transportation
ETL	ETL Testing Laboratories
FM	Factory Mutual
GAMA	Gas Appliance Manufacturers' Association

FIGURE 11.1 Trademarks of approval associations.

HVI	Home Ventilating Institute
IAPMO	International Association of Plumbing and Mechanical Officials
ISO	International Organization for Standardization
NEC	National Electrical Code
NEMA	National Electrical Manufacturers Association
NFPA	National Fire Protection Association
NSF	National Sanitation Foundation Internat
OPEI	Outdoor Power Equipment Institute
OSHA	Occupational Safety and Health Administration
SAE	Society of Automotive Engineers
SRCC	Solar Rating and Certification Corporation
SSPMA	Sump and Sewage Pump Manufacturers Association
UL	Underwriters Laboratories (Recognized Component Mark)
UL	Underwriters Laboratories (Recognized Component Mark for Canada)
UL	Underwriters Laboratories
UL	Underwriters Laboratories (Listing Mark Canada)
UL	Underwriters Laboratories (Listing Mark Classified products)
USDA	United States Department of Agriculture

- ISO 9001
- ISO 9002
- ISO 9003
- ISO 9004

ISO 9001, 9002, and 9003 are audited quality control standards, while ISO 9004 contains the guidelines to quality management elements and is not audited. The ISO standard your company may require depends on your manufacturing and product setup. The breakdown of the ISO 9000 system is as follows:

ISO 9001	Model for quality assurance in design and development, production, installation, and servicing
ISO 9002	Model for quality assurance in production and installation
ISO 9003	Model for quality assurance in final inspection and test
ISO 9004	Guidelines to quality management elements

Additional information on the entire ISO 9000 system may be obtained from the following organizations:

National Technical Information Service (NTIS)
5285 Port Royal Road
Springfield, VA 22161
Tel. (703) 487-4650
Fax (703) 321-8547

Registration Accreditation Board (RAB)
611 East Wisconsin Avenue
Milwaukee, WI 53202
Tel. (414) 272-8575

U.S. Government Printing Office (GPO)
Superintendent of Documents
Washington, DC 20402
Tel. (202) 783-3238
Fax (202) 275-2529

11.4 LISTING OF ANSI STANDARDS BY CATEGORY

1. Thread systems
2. Fastening and joining devices
3. Machining practices
4. Tools and tooling
5. Mechanical components
6. Welding
7. Heat treatment
8. Tolerances and fits
9. Drawing symbols and formats
10. Gaging and inspection

Category 1: Thread Systems

ANSI B1.9-1973, Buttress inch screw threads

ANSI B1.10-1958 (R1988), Unified miniature screw threads

ANSI B1.11-1958 (R1989), Microscope objective thread

ANSI B1.18M-1982 (R1987), Metric screw threads for commercial mechanical fasteners—Boundary profile defined

ANSI B1.20.3-1976 (R1982), Dryseal pipe threads (inch)

ANSI B1.20.4-1976 (R1982), Dryseal pipe threads (metric translation of B1.20.3-1976)

ANSI/ASME B1.1-1989, Unified inch screw threads (UN and UNR thread form)

ANSI/ASME B1.5-1988, Acme screw threads

ANSI/ASME B1.7M-1984, Screw threads, definitions and letter symbols for

ANSI/ASME B1.8-1988, Stub Acme screw threads

ANSI/ASME B1.12-1987, Screw threads—Class 5 interference fit thread

ANSI/ASME B1.13M-1983 (R1989), Metric screw threads—M profile

ANSI/ASME B1.20.1-1983, Pipe threads (general purpose), inch

ANSI/ASME B1.20.7-1966 (R1983), Hose coupling screw threads

Category 2: Fastening and Joining Devices

ANSI B18.1.1-1972 (R1989), Small solid rivets (0.4375 in diameter and under)

ANSI B18.1.2-1972 (R1989), Large rivets (0.500 in diameter and over)

ANSI B18.2.1-1981, Square and hex bolts and screws, inch series

ANSI B18.2.3.1M-1979 (R1989), Screws, metric hex cap

ANSI B18.2.3.2M-1979 (R1989), Screws, metric formed hex

ANSI B18.2.3.3M-1979 (R1989), Screws, metric heavy hex

ANSI B18.2.3.5M-1979 (R1989), Bolts, metric hex

ANSI B18.2.3.6M-1979 (R1989), Bolts, metric heavy hex

ANSI B18.2.3.7M-1979 (R1989), Bolts, metric heavy hex structural

ANSI B18.2.3.8M-1981, Screws, metric hex lag

ANSI B18.2.4.1M-1979 (R1989), Hex nuts, style 1, metric

ANSI B18.2.4.2M-1979 (R1989), Hex nuts, style 2, metric

ANSI B18.2.4.3M-1979 (R1989), Hex nuts, slotted, metric

ANSI B18.2.4.4M-1982, Nuts, metric hex flange

ANSI B18.2.4.5M-1979 (R1990), Hex jam nuts, metric

ANSI B18.2.4.6M-1979 (R1990), Hex nuts, heavy, metric

ANSI B18.5.2.1M-1981, Bolts, metric round head, short square neck

ANSI B18.6.1-1981, Wood screws, inch series

ANSI B18.6.2-1972 (R1983), Slotted head cap screws, square head set screws, and slotted headless set screws

ANSI B18.6.3-1972 (R1983), Slotted and recessed head machine screws and machine screw nuts

ANSI B18.6.4-1981, Screws, tapping and metallic drive, inch series, thread forming and cutting

ANSI B18.7-1972 (R1980), Semitubular rivets, full tubular rivets, split rivets, and rivet caps, general purpose

ANSI B18.8.1-1972 (R1983), Clevis pins and cotter pins

ANSI B18.8.2-1978 (R1989), Pins—Taper pins, dowel pins, straight pins, grooved pins, and spring pins, inch series

ANSI B18.9-1958 (R1989), Plow bolts

ANSI B18.11-1961 (R1983), Miniature screws

ANSI B18.17-1968 (R1983), Wing nuts, thumb screws, and wing screws

ANSI B18.22M-1981, Washers, metric plain

ANSI B18.22.1-1965 (R1981), Plain washers

ANSI/ASME B18.1.3M-1983 (R1989), Metric small solid rivets

ANSI/ASME B18.2.2-1987, Square and hex nuts, inch series

ANSI/ASME B18.2.3.4M-1984, Screws, metric hex flange

ANSI/ASME B18.2.3.9M-1984, Metric heavy hex flange screws

ANSI/ASME B18.3-1986, Socket cap, shoulder and set screws, inch series

ANSI/ASME B18.3.1M-1986, Screws, socket head cap, metric series

ANSI/ASME B18.3.3M-1986, Hexagon socket head shoulder screws, metric series

ANSI/ASME B18.3.4M-1986, Screws, hexagon socket button head cap, metric

ANSI/ASME B18.3.5M-1986, Hexagon socket flat countersunk head cap screws, metric series

ANSI/ASME B18.3.6M-1986, Screws, hexagon socket set, metric series

ANSI/ASME B18.5.2.2M-1982, Bolts, metric round head square neck

ANSI/ASME B18.6.5M-1986, Metric thread forming and thread cutting tapping screws

ANSI/ASME B18.6.7M-1985, Metric machine screws

ANSI/ASME B18.13-1987, Screw and washer assemblies—SEMS, inch series

ANSI/ASME B18.15-1985, Forged eyebolts

ANSI/ASME B18.21.1-1990, Lock washers

ANSI/ASME B18.21.2M-1990, Lock washers, metric series

Category 3: Machining

ANSI B5.8-1972 (R1988), Chucks and chuck jaws

ANSI B5.10-1981 (R1987), Machine tapers

ANSI B5.16-1952 (R1986), Accuracy of engine and tool room lathes

ANSI B17.1-1967 (R1989), Keys and keyseats

ANSI B17.2-1967 (R1978), Woodruff keys and keyseats

ANSI B74.2-1982, Shapes and sizes of grinding wheels and shapes, sizes, and identification of mounted wheels.

ANSI B74.13-1990, Markings for identifying grinding wheels and other bonded abrasives

ANSI B94.2-1983 (R1988), Reamers

ANSI B94.3-1965 (R1984), Straight cutoff blades for lathes and screw machines

ANSI B94.7-1980 (R1987), Hobs

ANSI B94.8-1967 (R1987), Inserted blade milling cutter bodies

ANSI B94.11M-1979 (R1987), Twist drills, straight shank and taper shank combined drills and countersinks

ANSI B94.21-1986 (R1987), Gear shaper cutters

ANSI B94.49-1975 (R1986), Spade drill blades and spade drill holders

ANSI/ASME B5.1M-1985, T-slots—Their bolts, nuts, and tongues

ANSI/ASME B94.6-1984, Knurling

ANSI/ASME B94.9-1987, Taps, cut and ground threads

ANSI/ASME B94.19-1985, Milling cutters and end mills

Category 4: Tools and Tooling Accessories

ANSI B5.25-1978 (R1986), Punch and die sets, inch

ANSI B5.25M-1980 (R1986), Punch and die sets, metric

ANSI B94.14-1968 (R1987), Punches—Basic head type

ANSI B94.14.1-1977 (R1984), Punches—Basic head type, metric

ANSI B94.33-1974 (R1986), Jig bushings

ANSI B107.6-1978 (R1987), Box, open end, combination and flare nut wrenches, inch series

ANSI B107.9-1978 (R1987), Box, open end, combination and flare nut wrenches, metric series

ANSI B107.10M-1982 (R1988), Socket wrenches, handles and attachments for hand, inch and metric series

ANSI/ASME B107.5M-1987, Socket wrenches, hand, metric

ANSI/ASME B107.8M-1984, Adjustable wrenches

ANSI/ASME B107.19-1987, Pliers, retaining ring

Category 5: Mechanical Components

ANSI B29.2M-1982 (R1987), Inverted tooth (silent) chains and sprockets

ANSI B29.6M-1983 (R1988), Steel detachable link chains and sprockets

ANSI B29.10M-1981 (R1987), Heavy duty offset sidebar transmission roller chains and sprocket teeth

ANSI B29.15-1973 (R1987), Heavy duty roller type conveyor chains and sprocket teeth

ANSI B29.19-1976 (R1987), A and CA550 and 620 roller chains, attachments and sprockets

ANSI B92.1-1970 (R1982), Involute splines and inspection, inch version

ANSI B92.2M-1981 (R1989), Involute splines, metric module

ANSI/ASME B29.1M-1986, Precision power transmission roller chains, attachments and sprockets

Category 6: Welding

ANSI/AWS D9.1-90, Sheet metal welding code

ANSI/AWS D10.12-89, Recommended practices and procedures for welding low-carbon steels

ANSI/AWS D11.2-89, Guide for welding iron castings

ANSI/AWS D14.2-86, Machine tool weldments, specification for metal cutting

Category 7: Heat Treatment

ANSI/SAE AMS 2728, Heat treatment of wrought copper beryllium alloy parts

ANSI/SAE AMS 2756, Gas nitriding of steel parts

ANSI/SAE AMS 2757, Gaseous nitrocarburizing

ANSI/SAE AMS 2759, Heat treatment of steel parts, general requirements

ANSI/SAE AMS 2759/3, Heat treatment of precipitation corrosion resisting and maraging steel parts

ANSI/SAE AMS 2759/4, Heat treatment of austenitic corrosion resistant steel parts

ANSI/SAE AMS 2759/5, Heat treatment of martensitic corrosion resistant steel parts

ANSI/SAE AMS 2759/6, Heat treatment and gas nitriding of low-alloy steel parts

ANSI/SAE AMS 2760A, Heat treatment—Carbon, low-alloy, and specialty steels

ANSI/SAE AMS 2770D, Heat treatment of aluminum alloy parts

ANSI/SAE AMS 2775A, Case hardening of titanium and titanium alloys

Category 8: Tolerances and Fits

ANSI B4.1-1967 (R1987), Preferred limits and fits for cylindrical parts

ANSI B4.2-1978 (R1984), Preferred metric limits and fits

ANSI B4.3-1978 (R1984), General tolerances for metric dimensioned products

ANSI B89.3.1-1972 (R1988), Out-of-roundness, measurement of

ANSI B89.6.2-1973 (R1988), Temperature and humidity environment for dimensional measurement

ANSI Y14.5M-1982 (R1988), Dimensioning and tolerancing

ANSI/ASME B1.22M-1985, Gages and gaging practice for MJ series metric screw threads

ANSI/ASME B107.17M-1985, Gages, wrench openings, reference

Category 9: Drawing Symbols and Formats

ANSI Y10.20-1975 (R1988), Mathematic signs and symbols for use in physical sciences and technology

ANSI Y14.1-1980 (R1987), Drawing sheet size and format

ANSI Y14.7.1-1971 (R1988), Gear drawing standards—Part 1, For spur, helical, double helical, and rack

ANSI Y14.7.2-1978 (R1989), Gear and spline drawing standards—Part 2, Bevel and hypoid gears

ANSI Y14.17-1966 (R1987), Fluid power diagrams

ANSI Y14.36-1978 (R1987), Surface texture symbols

ANSI Y32.10-1967 (R1987), Graphic symbols for fluid power diagrams

Category 10: Gaging and Inspection

ANSI B4.4M-1981 (R1987), Inspection of workpieces

ANSI B89.3.1-1972 (R1988), Out-of-roundness, measurement of

ANSI/ASME B1.2-1983, Gages and gaging for unified screw threads

ANSI/ASME B1.3M-1986, Gaging systems for dimensional acceptability, inch and metric screw threads (UN, UNR, UNJ, M, and MJ)

ANSI/ASME B1.16M-1984, Gages and gaging for metric M screw threads

ANSI/ASME B1.19M-1984, Gages for metric screw

threads for commercial mechanical fasteners—Boundary profile defined

ANSI/ASME B107.17M-1985, Gages, wrench openings, reference

and environmental problems, as can the indiscriminate handling and disposal of toxic waste products.

In order to help protect the environment, the Environmental Protection Agency (EPA) was established. This agency generates and administers the rules and regulations designed to help protect the environment from poisonous and otherwise harmful chemicals and materials used throughout industry. It should be noted that many of the heavy metals and their compounds are toxic. Many organic, carbon-based materials and chemicals are also toxic, some to an extreme extent. Many materials and compounds are also nonbiodegradable, which means that they will remain in the environment indefinitely as waste products.

Bearing the preceding in mind, we can readily see why these regulating agencies and administrations are necessary in the modern industrial world. OSHA standards for use in any of the categories currently under regulation are available to employees. The standard you are interested in checking may be obtained directly from OSHA or your company risk manager. All manufacturing installations are required to keep the OSHA and EPA regulating standards on their premises and make them available to any employee interested in obtaining information from these regulatory standards.

If you have doubts pertaining to the possible hazards of the materials, substances, and processes you use or come into contact with during your job duties, check the applicable OSHA or EPA standards. Modern American industrial employees are no longer required to forfeit their health or lives because of their job requirements, which may be in violation of the OSHA and EPA regulatory standards.

12.1 TYPICAL INDUSTRIAL HAZARDS

The following list of safety and health hazards is only a minute example of the conditions, materials, and substances that a metalworking employee can be subjected to.

- Welding over cadmium-plated parts (highly toxic fumes from the cadmium)
- Case-hardening steel parts with cyanide compounds without proper ventilation and equipment
- Fumes from the electroplating processes
- Grinding and machining beryllium alloys (beryllium-copper, beryllium-nickel)
- Machining and grinding magnesium
- Machining plastics with fiberglass fillers without ventilation and evacuation equipment
- Skin contact with plating solutions
- Fumes from the metal etching processes
- Foot and hand injuries when handling large sheet metal parts
- Injuries on machine tools not equipped with guards and shields
- Injuries on stamping dies due to inadequate or missing shields
- Ultraviolet radiation burns and eye injuries from welding processes, and radiation from laser metal-cutting machines.
- Paint fumes in nondesignated painting areas

12.2 SAFETY PRACTICES WHEN OPERATING MACHINE TOOLS

Many grave injuries occur in the operation of machine tools, especially during manual operations. The following list of precautions will help to prevent injury to machine tool operators, machinists, and tool and die makers.

- Wear the proper eye protection.
- Know exactly what to expect your machine to do *before* you turn on the power switch.
- Do not wear neckties or any loose clothing while operating a machine tool; the sleeves of long-sleeve shirts also must be rolled up past the elbow.
- Do not wear rings, watches, neck chains, or wrist bracelets while operating machine tools.
- Do not wear gloves while operating machine tools.
- Know instinctively where the power switch is located on your machine tool so that it may be quickly switched off in an emergency.
- Keep your work area clear of obstructions.
- Other machining precautions are described in Chap. 6, "Machining: Machine Tools and Practices."

Index

About the Author

Ronald A. Walsh is one of McGraw-Hill's most successful writers. An electromechanical design engineer for more than 40 years, he wrote *Machining and Metalworking Handbook,* Second Edition, *Electromechanical Design Handbook,* Third Edition, and is the co-author of *Engineering Mathematics Handbook,* Fourth Edition, all published by McGraw-Hill. Former director of research and development at Powercon Corporation, Severn, Maryland, he now consults widely.